RETURN WA⁻
TO

WATER RESOURCES
CENTER ARCHIVES

JUN   1991

UNIVERSITY OF CALIFORNIA
BERKELEY

WATER RESOURCES CENTER ARCHIVES
UNIVERSITY OF CALIFORNIA,

Salinity in Irrigation and Water Resources

WATER RESOURCES CENTER ARCHIVES
UNIVERSITY OF CALIFORNIA,

CIVIL ENGINEERING

A Series of Textbooks and Reference Books

Editors

ALFRED C. INGERSOLL
*Associate Dean, Continuing Education
University of California, Los Angeles
Los Angeles, California*

CONRAD P. HEINS, Jr.
*Department of Civil Engineering
and Institute for Physical Science
and Technology
University of Maryland
College Park, Maryland*

KENNETH N. DERUCHER
*Department of Civil Engineering
Stevens Institute of Technology
Castle Point Station
Hoboken, New Jersey*

1. Bridge and Pier Protective Systems and Devices, *Kenneth N. Derucher and Conrad P. Heins, Jr.*
2. Structural Analysis and Design, *Conrad P. Heins, Jr., and Kenneth N. Derucher*
3. Bridge Maintenance Inspection and Evaluation, *Kenneth R. White, John Minor, Kenneth N. Derucher, and Conrad P. Heins, Jr.*
4. Salinity and Irrigation in Water Resources, *Dan Yaron*

Salinity in Irrigation and Water Resources

Edited by **Dan Yaron**
The Hebrew University of Jerusalem

MARCEL DEKKER, INC. New York and Basel

Library of Congress Cataloging in Publication Data
Main entry under title:

Salinity in irrigation and water resources.

(Civil engineering ; 4)
Includes index.
1. Saline irrigation. 2. Saline waters. 3. Soils,
Salts in. I. Yaron, Dan. II. Series.
TA1.C4523 vol. 4 [S619.S24] 624s [333.91'3] 80-29430
ISBN 0-8247-6741-1

COPYRIGHT © 1981 by MARCEL DEKKER, INC. ALL RIGHTS RESERVED

Neither this book nor any part may be reproduced or transmitted in any form or by any means, electronic or mechanical, including photocopying, microfilming, and recording, or by any information storage and retrieval system, without permission in writing from the publisher.

MARCEL DEKKER, INC.
270 Madison Avenue, New York, New York 10016

Current printing (last digit):
10 9 8 7 6 5 4 3 2 1

PRINTED IN THE UNITED STATES OF AMERICA

Contributors

Jay C. Andersen Department of Economics, Utah State University, Logan, Utah

Nathan Arad* Engineering Division, Mekorot Water Co., Ltd., Tel-Aviv, Israel

Leon Bernstein (Retired), U.S. Salinity Laboratory, Riverside, California

Eshel Bresler Soil Physics Division, Institute of Soils and Water, Agricultural Research Organization, The Volcani Center, Bet Dagan, Israel, and Faculty of Agriculture, Hebrew University of Jerusalem, Rehovot, Israel

Gideon Fishelson Department of Economics, University of Tel-Aviv, Tel-Aviv, Israel

Pinhas Glueckstern Desalting Division, Mekorot Water Co., Ltd., Tel-Aviv, Israel

R. J. Hanks Department of Soil Science and Biometeorology, Utah State University, Logan, Utah

Earl O. Heady Economics Department and Center for Agricultural and Rural Development, Iowa State University, Ames, Iowa

Charles W. Howe Department of Economics, University of Colorado, Boulder, Colorado

Ralph W. Johnson School of Law, University of Washington, Seattle, Washington

J. J. Jurinak Department of Soil Science and Biometeorology, Utah State University, Logan, Utah

Allen V. Kneese Department of Economics, University of New Mexico, Albuquerque, New Mexico and Resources for the Future, Washington, D.C.

Kenneth L. Leathers† Department of Economics, Colorado State University, Fort Collins, Colorado

---

*Present address: Maot Ltd., Consulting Engineers, Tel-Aviv, Israel

†Currently on assignment with the Colorado State University field party in Lesotho, South Africa

Contributors

Brian L. McNeal Department of Agronomy and Soils, Washington State University, Pullman, Washington

Charles V. Moore Economic Statistical Cooperative Service, U.S. Department of Agriculture, Department of Agricultural Economics, University of California, Davis, California

Kenneth J. Nicol Department of Economics, Iowa State University, Ames, Iowa

Gaylord V. Skogerboe Department of Agricultural and Chemical Engineering, Colorado State University, Fort Collins, Colorado

George S. Tolley Department of Economics, The University of Chicago, Chicago, Illinois

James C. Wade* Center for Agricultural and Rural Development, Iowa State University, Ames, Iowa

Robert J. Wagenet Department of Soil Science and Biometeorology, Utah State University, Logan, Utah

Wynn R. Walker Department of Agricultural and Chemical Engineering, Colorado State University, Fort Collins, Colorado

Aaron Wiener Tahal Consulting Engineers, Ltd., Tel-Aviv, Israel

Dan Yaron Department of Agricultural Economics and Management, The Hebrew University of Jerusalem, Rehovot, Israel

Jeffrey T. Young† Department of Economics, University of Colorado, Boulder, Colorado

Robert A. Young Department of Economics, Colorado State University, Fort Collins, Colorado

Jennifer E. Zamora Department of Economics, University of New Mexico, Albuquerque, New Mexico

---

*Present address: Department of Agricultural Economics, University of Arizona, Tucson, Arizona
†Present address: Department of Economics, Marshall University, Huntington, West Virginia

Preface

This book deals with the problem of salinity in irrigation and water resources which affects about one-third of the world's irrigated lands in humid as well as in arid and semi-arid regions.

The scope of the book stems from the conviction that complex real life problems cannot be solved by a single discipline and that an interdisciplinary approach is needed. By interdisciplinary we do not mean an intermingling among the different disciplines, but rather a system which decomposes the problems into distinct elements in such a way, that each belongs to a well-defined discipline, and the links between them are clearly understood. Each element or subject is analyzed and discussed using the techniques and the terminology of its own discipline; at the same time the highlights of the discussion are well understood by workers in other disciplines.

To workers in the field of water resources an interdisciplinary approach is not unfamiliar. There exists a long-standing record of cooperation between specialists in biological and agricultural sciences, engineering and social sciences, in the evaluation, development and management of diversified water projects all over the world.

This book follows the same framework, and the subject of salinity in irrigation and water resources is dicussed in a logical sequence by outstanding experts from the above disciplines. The contents of the book are reviewed in detail in the first chapter.

I would like to express my thanks, first of all, to the contributing authors. Compiling and editing a book such as this is a tedious task, one which could not have been completed without their full cooperation.

Secondly, as an agricultural economist, I would like to thank my colleagues from other disciplines with whom I have cooperated in the analyses of various irrigation and water resources problems. Special mention is due to soil scientists Professor E. Bresler, Dr. J. Shalhevet and Dr. H. Bielorai.

Finally, I would like to thank my wife, Giza, for her encouragement during the lengthy process of compiling and editing this volume.

Dan Yaron
The Hebrew University of Jerusalem

Contents

Contributors		iii
Preface		v
1	The Salinity Problem in Irrigation—An Introductory Review *Dan Yaron*	1
2	Evaluation and Classification of Water Quality for Irrigation *Brian L. McNeal*	21
3	Effects of Salinity and Soil Water Regime on Crop Yields *Leon Bernstein*	47
4	Irrigation and Soil Salinity *Eshel Bresler*	65
5	Fertilization and Salinity *J. J. Jurinak and Robert J. Wagenet*	103
6	Impact of Irrigation on the Quality of Groundwater and River Flows *Gaylord V. Skogerboe and Wynn R. Walker*	121
7	Economic Evaluation of Irrigation with Saline Water Within the Framework of Farm, Methodology and Empirical Findings: A Case Study of Imperial Valley, California *Charles V. Moore*	159
8	Physical and Economic Evaluation of Irrigation Return Flow and Salinity on a Farm *R. J. Hanks and Jay C. Andersen*	173
9	Economic Impacts of Regional Saline Irrigation Return Flow Management Programs *Robert A. Young and Kenneth L. Leathers*	201
10	The Measurement of Regional Economic Effects of Changes in Irrigation Water Salinity Within a River Basin Framework: The Case of the Colorado River *Charles W. Howe and Jeffrey T. Young*	215

11	Economic Cost and Trade-Offs in Improving Water Quality and Nonpoint Pollution Through Agriculture: An Interregional Approach *Earl O. Heady, Kenneth J. Nicol, and James C. Wade*	245
12	Cost Sharing and Pricing for Water Quality *Gideon Fishelson and George S. Tolley*	271
13	Legal and Institutional Approaches to Salinity Management *Ralph W. Johnson*	305
14	Desalination, A Review of Technology and Cost Estimates *Nathan Arad and Pinhas Glueckstern*	325
15	The Strategy of Water Resource Management and Development in the Arid Zone *Aaron Wiener*	363
16	The Future of Arid Lands *Allen V. Kneese and Jennifer E. Zamora*	379
Index		407

Salinity in Irrigation and Water Resources

1

The Salinity Problem in Irrigation
An Introductory Review

Dan Yaron

Hebrew University of Jerusalem, Rehovot, Israel

Introduction

Soil salinity problems and irrigation with saline water are widespread – it is estimated that they affect about one-third of all irrigated land (Yaron *et al.*, 1969). The areas involved include regions where the climate is humid (e.g., Holland, Sweden, Hungary, and the U.S.S.R.), as well as arid or semiarid regions (e.g., southwestern United States, Australia, India, and the Middle East). According to Evans (1974), 100,000 acres of land per year are no longer productive because of salinization. Most of the irrigated land in Iran and more than 50% of the irrigated land in Syria are affected by salinity. In India, 15 million acres are lost to agriculture because of salinity. Acute soil salinity problems, to the extent of complete elimination of agricultural areas, have been reported also from South Africa, and from Uzbekistan in the Soviet Union. In the United States about 28% of irrigated land suffers from depressed yields due to salinity.

The essence of the salinity problem stems from the fact that practically all irrigation waters contain some amount of dissolved salts. During irrigation, these salts tend to concentrate in the soil, unless they are leached by rainfall and/or irrigation water (in excess of evapotranspiration). Saline irrigation water (as defined in Chaps. 2 and 3), low soil permeability, inadequate drainage conditions, low rainfall, and poor irrigation management all contribute to the tendency of salt to accumulate in soils, which, in turn, adversely affects crop growth conditions and yield levels. As mentioned above, in the extreme, land can be totally lost with regard to agricultural production.

The salinity problem seems to be more widespread and acute in arid and semiarid regions than in humid ones. This is due to the need in the former regions for extensive irrigation, low rainfall, their relative scarcity of good-quality water, the subsequent necessity to utilize water resources of low quality, and the resulting deterioration in water quality of natural flows and aquifers. Note also that the salinity problem is often exacerbated in areas with drainage problems. Under such conditions, good management of the regional (watershed) water system on the one hand, and of irrigation and water use on the farm on the other, is essential for sustained agriculture.

Water Resource and Irrigation Systems – An Overall View

Problems involved in the development and management of any water resource system are quite complex and extend across several professional disciplines. The complexity is augmented significantly by consideration of salinity. Instead of one major dimension – water quantity – its sources, flows, and the effects of its relative availability on agricultural, industrial, domestic, and recreational uses, there are now two major dimensions – the *quantity* and the *salinity* of water.

The complexity of proper exploitation of water resources and management of irrigation systems calls for an interdisciplinary approach among the biological, agricultural, and soil sciences, hydrology, engineering, economics, sociology, and public administration, among others. Although there exist a few outstanding examples of interdisciplinary work in the field (e.g., Mass et al., 1966), the integration of the numerous disciplines into one well-coordinated entity is still a major problem. In order to maximize the benefits of analytical efforts toward solutions of water utilization problems, a well-defined framework of the system and its major elements is necessary. It seems that an approach based on systems analysis, with (1) clearly delineated subsystems and well-defined links between them, and (2) a clear definition of the major components of each subsystem and their interdependence upon each other, is necessary as a proper framework for any useful analysis.

A typical water resource irrigation system may be viewed as comprising the following major subsystems: (1) *goals and objectives;* (2) the *core* of the system, or the *input-output mechanism;* (3) *decision making,* and *controllable inputs* to the core; (4) *exogenous factors* or *uncontrolled inputs* (e.g., weather, the state of world economy, price levels); (5) *outputs* or *outcomes;* and (6) *evaluation of outputs* or a *feedback-and-control* unit. A block diagram representation of a water resource-irrigation system based on the above lines is shown in Fig. 1. The links between the subsystems (blocks) represent flows of (1) information; (2) orders and instructions; (3) water and salt; (4) other materials; and (5) funds.

The internal partitions and structure of the various subsystems are not shown in Fig. 1. In a more detailed representation each subsystem would be

Salinity Problem in Irrigation

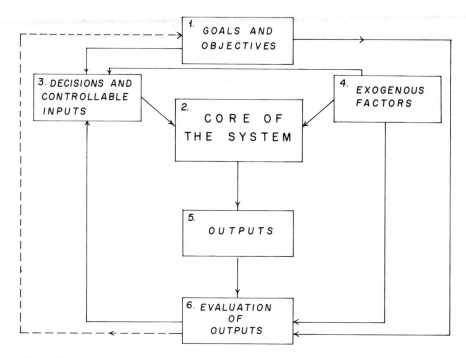

Fig. 1 Block-diagram presentation of a water resource irrigation system.

divided into its major components, and those divided again according to their constituent factors and so on. For example, the core could be divided into demand and supply components. The demand component could be further itemized as agricultural, industrial, domestic, and recreational demand. The supply component could be divided into sources (of water and of salinity), the conveying system, and the disposal system of excess water and salts.

This system description should be augmented by at least two additional dimensions. The first is the dimension of time. The representation of Fig. 1 is static and describes the system at a given point of time, including the flows between the major subsystems. However, a static presentation of a dynamic system is proper only under stationary conditions, and this is an exception rather than a general case. The flows representing the interactions among the subsystems are time-consuming and their effects may encounter lags of different time lengths. Moreover, the internal mechanism of almost all major elements in the system (not detailed in Fig. 1) has a time dimension.

Another dimension that should be incorporated is the level of the various socioeconomic entities in the hierarchical continuum of the system. For example, the lowest entity in the hierarchy is the farm. Higher entities are the irrigation district or the region, a multiregional entity (or a river basin), and the nation. Each of these entities is in itself an integral and well-defined system, divisible into subsystems of goals and objectives, decision-making mechanism, input-output mechanism, and so on. The lower an entity is in the hierarchy, the easier the definition and delineation of its subsystems. This is particularly so with regard to the "softer" parts of the system, such as goals and objectives or the decision-making subsystems. They can be evaluated with relative ease at the farm level, but their definition becomes increasingly complex and cumbersome at the higher levels of the hierarchy, such as a watershed or the nation.

A comprehensive presentation of the detailed structure of the elements of a water resource-irrigation system is outside the scope of this chapter, and we conclude with these introductory remarks. Most of the subsystems are discussed in detail in subsequent chapters, especially subsystems 1-3, 5, and 6. Subsystem 4 (*exogenous factors*) is considered only as a side issue.

The remaining sections of this introduction review the major issues with regard to salinity problems in irrigation systems in an attempt to integrate into a meaningful entity the specific topics that are discussed singly in detail later in this volume.

Basic Physical Relationships in Salinity in Irrigation

Some concentration of inorganic salts in water and in the soil is necessary as a source of nutrients for plants. It is the excess of salts in the soil solution that has a negative effect on crops. This distinction between saline and nonsaline water for irrigation purposes is somewhat arbitrary. According to Bernstein (see Chap. 3), "saline soil waters by definition contain 80 meq or more of salts per liter, although sensitive crops may be affected at half this salt concentration and tolerant ones at only twice this concentration." Furthermore, proper management of irrigation may significantly modify the effects of salinity on crop yields.

An evaluation and classification of water quality for irrigation is presented by McNeal in Chap. 2. A common characterization of the degree of salinity of water is by reference to the total dissolved solids (TDS) in terms of milligrams per liter. Using this criterion, the range of salt contents in irrigation waters is from 75 to 100 TDS mg/liter up to 4,000 TDS mg/liter. As noted in Chap. 2, it has been estimated that "approximately 80% of the irrigation waters used throughout the western United States have TDS levels less than 1,000 mg/liter, with nearly half containing less than 500 mg/liter." However, in some areas in the southwestern United States, the salinity content of irrigation water is considerably higher and in some situations approaches 4,000 mg TDS/liter (Pecos River, near Oila, Texas). In Israel, in the Beissan Valley, water containing 3,000 mg TDS/liter is used for irrigation of tolerant crops; and at N'eot Ha'kikar, close to the Dead Sea, the salinity level of irrigation water approaches 5,000 mg TDS/liter. The groundwater used for irrigation normally contains higher TDS levels than do river waters of the same region; the salt content of the soil solution is higher than that of the irrigation water.

Not only is the total salt content of importance in determining the suitability of water for irrigation, but also the composition of the salts. Factors to be taken into account include the existence of particular ions that may have toxic effects on crops (e.g., chloride, boron), or the balance among some ions which may harmfully affect the properties of soils (e.g., the balance between sodium and calcium plus magnesium). These problems are reviewed and discussed by McNeal in Chap. 2. However, McNeal notes that water quality is not an inherent property of the water itself, but rather a property intimately related to irrigation management practices.

A prerequisite for the understanding of irrigation management problems with saline water is a knowledge of the processes of salt accumulation and leaching in the soil during irrigation and of the response of plants to soil salinity and moisture. The overall complex of the physical and biological relationships involved in irrigation with saline water is considered as composed of two phases: (1) one in which the relationships among the *decision variables* (such as timing, quantity, and quality of irrigation water) and the *soil state variables* (such as soil salinity, soil moisture, etc.) are analyzed, and (2) a phase in which the soil state variables are related to the *target variables* such as quantity and quality of yield (Fig. 2).

The relationships among the soil state variables and the target variables are discussed in detail by Bernstein in Chap. 3. Crop response is affected by both soil salinity and moisture, and the relationship between these two factors is quite intricate (e.g., for a given soil salinity, salt concentration varies inversely with water content). The difficulties of simultaneously controlling soil moisture and salinity concentrations are discussed and the rationale underlying the experimental methods commonly practiced at the U.S. Salinity Laboratory in Riverside, California, are presented. These methods involve a series of small (4.27 x 4.27 m) plots of permeable soil in which salinity can be controlled; moisture, fertilizers, and so on, are supplied at levels and in a schedule designed to give maximal yields. Chapter 3 presents estimates of the salt tolerance of numerous crops based on these experiments. For forage crops, field crops, and vegetable crops, they are presented in terms of the estimated yield reduction at varying levels of soil salinity. It must be emphasized here that understanding of the experimental methods and their implications is essential for correct interpretation of the data reported.

Significant differences in tolerance exist between the salt-sensitive crops (mostly vegetables) and the relatively tolerant species such as cotton, barley, and sugar beets.

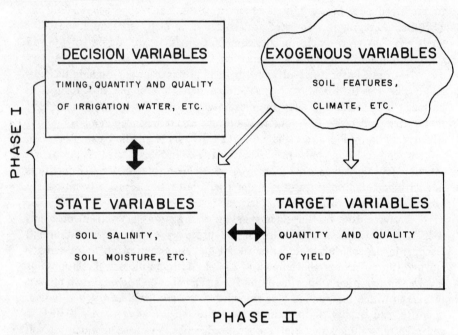

Fig. 2 Phases I and II in the irrigation process.

Relatively less information is available regarding salt tolerance of fruit crops (because of the compelxity of carrying out experiments with perennials and their reaction over time). Nevertheless, it is well established that fruit crops are generally affected by certain ions such as chloride and sodium, as well as by total salinity. Chapter 3 presents tolerance estimates of fruit crops to total salinity (expressed in terms of electrical conductivity) and to chloride. In some situations there is a high correlation between the total salinity and the cloride content of the soil solution and difficulties in interpretation of salt tolerance may exist.

Various refinements concerning salt tolerance during the various stages of plant growth, as well as the effects of interaction factors such as climate, soil fertility, and soil moisture, are discussed in this chapter.

Bernstein discusses in relative detail the interactive effects of irrigation frequency and soil salinity on crop yields and suggests that the effects of soil moisture (matric potential) and soil salinity (osmotic potential) are additive. Accordingly, it is possible to modify the salinity effect by frequent irrigation. Irrigation systems such as trickling, designed to maintain high moisture status in the soil, are advantageous from this point of view.

Bernstein points out that soil salinity is not generally uniform in the root zone, and he presents some experimental data suggesting that tolerance limits for soil salinity need to be specified separately for the upper root zone—generally with minimal salinity level—and the lower or (generally) the maximal salinity zone. Crop tolerance data accounting for this distinction are presented.

The understanding of soil chemistry and physics and the integration of crop tolerance with soil moisture and salinity provide the background for proper irrigation management practices in terms of irrigation methods, timing, quantity, and salinity of irrigation water. The application of the existing theories and the knowledge on salinity-irrigation relationships play a major role in this respect. These subjects are discussed by Bresler in Chap. 4 and by Jurinak and Wagenet in Chap. 5.

In the first part of Chap. 4, Bresler presents the theoretical background from which irrigation mangement decisions should be derived. The topics include theories regarding the behavior of salts in the soil solution, phenomena affected by the concentration and composition of the soil solution, such as swelling of soils and soil hydraulic conductivity, theories and models regarding rules of water flow and transports of solutes in soils.

The second part of Chap. 4 is management-oriented. It presents models designed for the estimation of leaching, accumulation, and distribution of salts under a variety of conditions. The major importance of these models is that they enable one to predict within reasonable limits of accuracy the variations of soil moisture and salinity over time and depth under various irrigation regimes. These models combined with crops' response functions to soil moisture and salinity (Chap. 3) constitute the basis for irrigation management decisions.

The application of the salt leaching models is discussed with respect to reclamation of saline soils by flooding, sprinkler, and trickle irrigation. It is shown that reclamation by flooding is generally inferior to the other methods. The major application of these models is related to the control of soil moisture and salinity during the crop-growing period. In this context Bresler presents several situations under which optimal combinations of water quantity, salinity, and timing are found with the aid of these models. For example, assuming that significant yield response to salinity occurs only about a certain critical threshold concentration, combinations of water quantity and salinity are found that maintain the soil salinity below that critical level during the growing season of a crop. Another example presents results of an analysis which suggests that in regions with Mediterranean climate (i.e., semiannual alternation of irrigation in dry summers and rainy winter seasons), it might be more efficient to control soil salinity by regulating the soil water regime rather than by leaching.

The interrelationship between fertilization and salinity is discussed by Jurinak and Wagenet in Chap. 5. Two factors underlying this interrelationship are the osmotic potential of the soil solution, which increases with additions of salts, and the influence of salinity on the uptake of both micro and macro nutrients by the plant's roots.

The effect of fertilization on the osmotic potential of the soil is discussed in relative detail. For example, soil salinity can be significantly increased by an application of nitrogen fertilizers, which are generally quite soluble and have little effect on the solubility of the other salts already present in the soil solution. On the other hand, phosphorus fertilization may decrease the total dissolved salts and the osmotic potential of the soil solution as a result of creation of insoluble calcium and magnesium phosphates. Another factor to be considered is the method of fertilizer application. For row crops the most efficient fertilizer use is achieved when the zone of placement is close to the row. However, when the fertilizer is applied by a method that localizes it in high concentrations in a small volume of soil, adverse osmotic effects can develop.

A measure of the relative effects of different fertilizers on the osmotic potential of the soil solution under identical conditions is the salt index discussed in this chapter. This index has obvious management-oriented implications with respect to the selection of fertilizers to be applied. For example, regarding nitrogen carriers, ammonia has salt index of 47.1 whereas that of ammonium nitrate is 104.7 (with sodium nitrate being the 100.0 reference base).

Other subjects reviewed by Jurinak and Wagenet in the first part of their chapter are fertility-salinity effects of irrigation water quality, salinity effects on fertilizer transformations, and salinity effects on nitrogen fixation.

Whereas the basic mechanisms involving fertilization and salinity, such as the solubility relationships of the applied fertilizer, osmotic effects of salts on water availability and microbiological activity, nutrient absorption by plants, and the influence of specific soil properties, are relatively well understood, difficulties in the understanding and interpretation of certain phenomena may arise when interactions between several mechanisms are encountered. Such interactions are discussed in the closing sections of Chap. 5, and the ultimate effects of fertilizing saline soils under field conditions are evaluated. The authors present various observations of a number of researchers and state that caution should be employed in extrapolation of the results presented, beyond the conditions from which they were developed. The understanding of the basic mechanisms regarding fertilization and salinity should be helpful in this respect.

In Chap. 6, by Skogerboe and Walker, irrigation having a wide interregional (watershed) scope is discussed. A distinction is made among three phases of irrigation projects: (1) the *delivery* phase; (2) the farm irrigation phase; and (3) the water-removal phase.

The discussion of the interrelationships among these phases, and, especially, of the balance of salts between the farm phase and the removal phase is of major interest. This is a typical case of "externalities" defined as a situation in which actions of one group of people have indirect effects (adverse or favorable) on others.

The essence of the externality problem is that return flows from irrigated farms drain to rivers or groundwaters which, in turn, constitute the sources for irrigation water of other farms. (Drainage water generally contains considerably higher concentrations of dissolved salts than the application water.)

As shown in Chap. 6, the quantities of salts transferred from one phase to another (i.e., from the delivery system to the farms and from the farms to the removal system) are subject to control. Sometimes, however, there is a conflict of interest among the various parties of the overall regional system. Where the course of water supply is a river, upstream farmers contribute to the pollution of downstream waters by the return flows from their fields. This tendency is exacerbated under conditions of availability of low-cost water, with no incentives for careful irrigation and avoidance of overirrigation. The tendency is even more acute when the source supplying the water is saline, and, from the point of view of the upstream farmers, excess irrigation, with a certain amount intended for leaching of salts is necessary. In such a case a clear conflict of interests exists—avoidance of leaching irrigation by upstream farmers involves them in a loss of income, whereas using leaching exacerbates the salination of downstream waters and correspondingly causes a loss of income to downstream farmers.

An analogous situation exists when the return flows are drained to groundwater or aquifers. Here, however, the process may be slowed down by the mixing that occurs with large volumes of water stored in the aquifers, and the salination may be gradual, extending over decades. Such a process of gradual salination of groundwater is observed in Israel; in such a case, the benefits from irrigation accrue to those currently farming, whereas the damage and income loss is borne by future generations.

As detailed in Chap. 6, reduction of the quantity of salt drained in return flows may be achieved by increasing irrigation efficiency on farms—using optimal irrigation schedules; improving irrigation systems by such means as lining canals and using pipes; and improving the interfarm water conveyance systems, thus decreasing seepage losses, and so on.

On the whole, some degree of return flow is necessary. Complete lack of soil leaching and drainage will result in a gradual buildup of salinity in the soil. In extreme situations, waterlogging—that is, a rise in groundwater levels—may occur. Thus, if the groundwater level reaches the soil surface and evaporates, the contained salts are left on the ground surface (this process is often exacerbated by capillary action).

If it were possible to drain the return flows and deposit the salts in the deeper soil layers, below the root zone of plants but above the groundwater level, an ideal solution would pertain, but unfortunately, nowadays this seems utopian. In some situations drainage of return flows to the sea may prove to be a useful solution.

Obviously, the gravity of the salt problem in return flows depends on the salt content of water at the delivery phase. Reduction of salinity at this phase can be achieved by diverting salty sources out of the system and introducing and mixing high-quality and desalinated waters.

What is the optimal quantity and salinity of water to be supplied to the farms in a certain region over a given span of time? What seasonal variation in quantity and salinity is to be recommended? What is the proper irrigation system for the farm? What is the proper quantity and quality of water and the irrigation schedules for the various crops on the various types of farms? What is the proper drainage system for a particular irrigation region? These are the major questions to be answered with respect to any irrigated area. The answers depend on numerous conditions such as types and sizes of farms, capital available, managerial capacity of the farms, prices of water outputs and inputs, and the regional infrastructure, among others.

There are two major difficulties in answering the above questions. The first is lack of sufficient information on the physical input-output relationships in irrigation with saline water. This difficulty can be overcome by further research.

The second difficulty is considerably harder; it is related to the conceptual and moral problems involved in the externalities discussed previously. The questions posed above are generally dealt with by engineers and economists used to the rigor of (1) problem definition, and (2) optimization of a given objective function subject to certain constraints that emerge from the subject matter of the problem. Since a conflict of interests is inherent in the system, the crucial issue arises—from whose point of view should optimization be sought? Group A of farms, Group B, etc., or optimization from the point of view of the society as a whole, at the cost of certain parts of the society?

Questions relating to this problem are discussed in Chap. 12, whereas in Chap. 13 the institutional and legal aspects involved in the implementation of any policy are reviewed. A prerequisite for the determination of a policy with respect to comprehensive management of a water system in which salinity is encountered is a proper identification of the problem, and an understanding and ability to estimate, within reasonable limits of accuracy, the benefits and losses that will accrue to the various parties involved, resulting from the alternative policies under consideration. These points are discussed in Chaps. 7 through 10 and 14.

Economic and Social Evaluation of Salinity in Irrigation

In the economic analysis and the social evaluation of systems of irrigation with saline water, a distinction is suggested among various types of models according to: (1) the level of the entity referred to, in the farm-region-nation hierarchy; (2) whether "externalities" exist and are taken into account; and (3) the range of time referred to.

The distinction between a farm and higher hierarchical levels is obvious. The problem of externalities has been reviewed in the previous section. From the point of view of time, a distinction is suggested between the following:

1. A "short-run" approach refers to relationships confined to a single irrigation season and does not take into consideration the effects of salt accumulation over time.
2. A "long-run" approach does take into account the effects of salt accumulation over time in soils, river flows, and aquifers. It comprises a succession of short-run processes, the initial conditions of which are affected by salt accumulation in previous periods; the decisions over any single season take into account the possible terminal conditions that may result, according to alternative decisions, and their effects on succeeding periods.

Although externalities may be incorporated into both the short- and long-run models, it is obvious that they are more meaningful in the long-run context.

Chapters 7 through 13 discuss the problems of salinity in irrigation from economic and social viewpoints; the scope of the analysis develops and broadens in sequence.

Chapter 7 by Moore presents an economic analysis of irrigation with saline water in the Imperial Valley of California, with reference to a farm unit and on the basis of a short-run (one-year) time horizon; externalities are not included in the analysis. Three types (sizes) of farms are considered, with varying quantities and salinity levels of water supplied to them. Utilizing information on the response of crops to soil moisture and salinity, alternative efficient combinations of water quantity and salinity for nine crops were formulated and incorporated into farm planning analyses performed with the aid of linear programming. The objective of the analysis for each type of farm was to maximize the farm's returns in a single year under the above conditions. Within this framework, the effect of different water salinities on the optimal crop composition, irrigation regime, and the farm's income is presented and discussed.

Chapter 8, by Hanks and Andersen, presents an analysis and empirical estimates of the substitution relationships, under a variety of conditions, between the quantity of salt in return flows and the income of the farm generating the return flows. The physical relationships involved are analyzed comprehensively, in reference to quantity and salinity of irrigation water, irrigation methods (sprinkling and flooding), irrigation frequency, uniformity of water distribution, transpiration and evapotranspiration, drainage, salt distribution in the soil profile, salt outflow to groundwater, and predicted yields of the relevant crops under selected situations. The physical model presented in this chapter can probably be classified as a very comprehensive one in its reference to almost all relevant variables as endogenous. The economic model is designed to maximize the farm income over a single year (a Vernal, Utah, farm provided the framework for the economic analysis), with alternative restriction levels on salt outflows resulting from irrigation. The real cost to the farm of restricting the salt outflow is estimated.

The data obtained from the economic analysis of each situation considered indicate optimal crop compositions and the optimal farm irrigation management policy. Although the economic analysis of this chapter is static in that it refers to a single year, the model can easily be adapted to a long-run analysis. Indeed, the analyses of the physical processes are extended over a series of years.

The authors deal explicitly with the substitution relationships between the farms' income and the amount of salt in the farms' return flows, and present the tools for quantitative analysis of the externality problem at the "exit gate" of the salt discharging farm.

Chapter 9, by Young and Leathers, presents a case study of economic impacts of management policies for controlling saline irrigation return flows within a regional framework. The study focuses on the Grand Valley, at the confluence of the Colorado and Gunnison Rivers, in west-central Colorado. The average annual salt pickup from the area's 56,000 acres of irrigated land is estimated at about 460,000 tons, or over 8 tons per irrigated acre. The relatively high level of salt contributions derives from seepage from the conveyance system and fields onto a marine-deposited shale formation that underlies most of the Valley's arable soils.

Policies for controlling salt pickup range from lining of water conveyance systems through improvement in irrigation efficiency to outright abandonment of farming on all or parts of the irrigated lands.

The analysis is based on linear programming models representing subcategories of farms in the study area. These models were assembled from data collected in a detailed field interview survey of the commercial farmers in the study area. Functions reflecting the direct costs of various nonstructural control alternatives for the study area are derived, and indirect impacts on related economic sectors in the region are traced.

An extension of the scope of the economic analysis to a case of interaction between two regions is provided in Chap. 10 by Howe and Young. The Colorado River Basin provides the terms of reference for empirical study, with the opportunities for reducing salinity present in the Upper Basin, and the major damages from increasing salinity occurring in the Lower Basin.

Considered are (1) additions to salt loading of the river by upper stream users and alternatives for reducing the quantities, for example by reducing irrigated areas or by changing cropping patterns; and (2) the damage caused to the downstream users, and alternatives for reducing that damage by adopting proper irrigation management practices. A supply schedule or reducing salt loading by the upper users is derived, that is, the cost accrued to the upper users as a function of reduced quantities of salts loaded into the river. The supply schedule is related to the resulting benefits to the downstream users.

Chapter 10 accounts for the fact that a reduction in irrigated agriculture in a region has both a direct effect on the region's economy and an indirect one, due to the reduction of the activity of other sectors linked to agriculture via the regional sectoral interdependence mechanism. Estimation of the ratios between the total (indirect plus direct) effects and direct effects only, called the multiplier, is discussed and empirical estimates are presented. Reference is made to an input-output model for the Lower Basin and the multiplier is computed making alternative assumptions with respect to the adaptive mechanism of the regional economic system. The direct and indirect effects for the Upper Basin are estimated using data from various sources. Whereas the empirical estimates presented fall within a somewhat wide range depending on the underlying assumptions, and additional information is called for, some clear-cut conclusions are derived. One is that improved irrigation management in the Upper Basin is socially worthwhile. However, since the benefits accrue to the farmers of the Lower Basin, the question of providing incentives to Upper Basin farmers is raised.

The comparative analysis of the cost of reducing salt loading versus the resulting benefits raises a serious dilemma in the case where the interacting parties belong to distinct generations, as may be when there is a continuous, but slow process of deterioration of groundwater quality due to return flows. The process can be slowed down and the real cost of the reduction in salt loading to the present generation can be estimated. However, the benefits derived from the slowdown will accrue only in decades ahead. In view of uncertainties regarding prices, possible developments in agricultural and irrigation technology, and especially the technology of desalination, any analysis of the present value of these far into the future is extremely difficult. Because of the difficulties involved in estimating future benefits from water resource as a whole, there is a tendency on behalf of some thinkers to promote the philosophy of conservation of natural resources as a dominating policy rule. Strict adherence to such a policy can lead to extremely conservative and costly policy decisions. But the crucial question still remains to what extent the present generation can debit future generations in terms of deteriorating water quality, and impose on them the development of revolutionary water-saving technologies and/or the development of costly projects such as water desalination.

Heady, Nicol, and Wade in Chap. 11 digress from the salinity problem per se, but in terms of methodology their contribution constitutes a continuation and an extension of the previous models to the national level, the highest in the farm-region-nation hierarchy. They describe a study aimed at the analysis of optimal land and water use and allocation in the United States as a whole, in relation to nonpoint pollution control (with the agricultural sector being its major source) and water environmental quality improvement. This is a large multiregional programming model with 223 producing areas, 51 water supply regions, 9 land classes in each production area, numerous cropping activities, and 25 commodity market regions.

The analysis, projected to the year 2000, is intended to provide a basis for the selection of an efficient interregional allocation of agricultural production such that (1) the cost of producing and transporting the various crop and livestock commodities will be minimized; (2) consumer demand (allowing for projected exports) will be satisfied; (3) agricultural technology, such as dry land or irrigated production, crop rotation, tillage methods, fertilizer levels, and so on, will be adapted to meet the above objectives. The pivotal parameter in the study is the permissible soil loss per acre for each soil resource group, which is related to reduction in nonpoint pollution. The permissible levels were varied parametrically from 3 to 5, 10 tons, and unlimited soil loss per acre.

The results of the study provide indications of changes in the allocation of agricultural production because of environmental soil loss restrictions, increased transfer of water from upstream to downstream regions, changes in the interregional distribution of farm income, higher prices of major agricultural commodities, and a reduction in consumer welfare resulting from the increased cost of living. The implications of these results are that intergroup and intergeneration income transfers are involved, and the real cost of these transfers to society versus the real benefit derived from lesser soil erosion should be weighed at the highest, national level. As mentioned previously, although the empirical subject matter of this chapter is not salinity, the methodology could easily be adapted to analysis of the salinity problem. In effect, in the closing sections of Chap. 11 the authors discuss the extension of their model needed to incorporate chemicals associated with salinity.

The issues discussed in Chaps. 8 through 11 indicate that when externalities exist, the market mechanism fails to provide satisfactory solutions from the point of view of the society. Numerous examples in support of this statement can be found in the field of water resources and natural resources in general. It is now generally recognized that in such situations society should intervene through its representatives, namely the government, at the national, regional, or municipal level as appropriate. The intervention comprises: (1) the selection of the favored solution; (2) the selection of the means designed to achieve it.

The selection of solutions is not easy, because most of them affect income distribution and lead to conflicts of interest. Any solution aims at the achievement of two major goals of the society—economic efficiency (i.e., increased

Salinity Problem in Irrigation

output of the society as a whole) and fair income distribution. The relative weight of these two goals is very hard to establish in most real-life situations. Conceptually their weighing involves moral value judgments. Politically they involve conflict among various pressure groups, each promoting its own interests.

Fishelson and Tolley (Chap. 12) discuss externality problems from the economic viewpoint. Within a rigorous framework, they show that a competitive market economy fails to provide the social optimum (considered as the maximal output achievable by the society), whereas it does not fail to do so in the absence of external effects. Several policies intended to impose socially optimal solutions on decision makers under various situations are presented and discussed. The problem of improving low-quality national water sources, when no interaction between polluters and those who bear the consequences exists, is considered. Cost-sharing arrangements among the various participants in water-quality improvement projects are discussed, so that pressures to deviate from the socially optimal scale of the project, its location, design, and operation may be eliminated.

The philosophy favored by the authors is that for social optimization, efficient solutions should be sought first (maximizing the total output of the society) and cost-sharing and pricing schemes consistent with the above solutions should be designed as a subsequent state. To achieve equity, policy measures that do not interfere with efficiency should be looked for. The authors suggest that most likely these equity measures will not be related to water. However, they do not specify the measures to be taken.

It should be noted that policy makers (and some economists) often claim that although the above philosophy is appealing, in many situations it is hard to identify and recommend policy means that will satisfy the above criteria. There is a tendency among farmers and other groups to object to direct income transfer payments, and many of the other equity-oriented policy means tend to affect efficiency.

Chapter 13 by Johnson discusses the salinity problem from the legal and institutional point of view. It emphasizes the complexity of the problem which is exacerbated by the fact that it usually arises at a relatively advanced stage of water allocation to various users, when water rights have been established and investments made. Furthermore, the problem results from the cumulative impact of many users.

The author is in favor of a regional, basinwide, or larger entity, with sufficient jurisdiction and power to (1) gather the needed information; (2) design the optimal plan; (3) implement that plan; and (4) decide upon cost allocation of the plan adopted.

The author examines the Colorado River Basin salinity management program as an example of the problem and of some possible solutions. In the Basin, many legal and practical constraints have affected the institutional structure developed for salinity control. The Colorado River Basin Salinity Control Forum, a voluntary organization composed of representatives from the Basin states, has proven quite effective in promoting the needed studies and developed plans for implementation of recommendations. It has worked in cooperation with the states involved, and with the federal government, which has the necessary power to carry out the overall plan.

Most of the salinity management activities in the Colorado River Basin at present are centered around "positive" actions such as improving farm irrigation management, and the control of water flows in canals, laterals, and drainage outlets, rather than taking "negative" policy means, in terms of restrictive regulations regarding water allotments and irrigation quotas. Although the second approach may become necessary in the future as salinity problems worsen, according to Johnson, "most water managers and other experts in the Colorado Basin believe the voluntary program will, in the long run be more effective." Johnson suggests that there is danger in too easy generalization about the applicability of U.S. water management practices to other countries. In the highly developed United States, with its strong economic system and educated populace, a substantially voluntary approach has achieved some success in the Colorado Basin. Whether this approach will continue to be effective for that Basin in the future, as the intensity of competition for water increases, is not at all clear. What does seem clear, in Johnson's view, is that in most places in the world such a voluntary program will not work, at least not as effectively.

Some experts suggest that economic incentives should be used to achieve water savings and increased irrigation efficiency. To provide incentives for efficiency, Howe and Orr (1974) propose a water rights purchase program, within which water rights could be purchased by farmers at a price determined at a level that would encourage water saving. A water pricing system based on similar principles, which provides incentives for savings and at the same time minimizes adverse effects, was established in Israel in 1972. The essentials of the system are as follows:

1. Presently existing water rights remain in effect.
2. Differential prices are charged for water as follows: (a) a relatively low price (price A) for 85% of the historically approved water quota; (b) for the remaining 15% of the water quota, a higher price (price B) is charged, set at a level to induce saving; (c) for any water use above 100% of the quota, a significantly high price (price C) is charged, intended to minimize the use of water above the approved quota.
3. Prices A, B, and C are determined on a regional basis in accordance with regional differences in water productivity.

It is important to note that by properly setting prices A and B, the goal of water saving can be achieved without affecting farmers' income. Incidentally, this pricing scheme is an example of a policy in which the conflict between the goals of efficiency and income distribution is minimized.

Future Prospects

In some situations, desalted seawater is the only reliable source of water supply. This is true for isolated islands, certain desert locations, and in other unique circumstances. In other situations desalination of seawater, of brackish water, or of water gradually salinated by the action of man, in the present is, and in the future may be, the best alternative for the solution of the salinity problem in irrigation under conditions of severe water scarcity or quality problems. One example is the planned Colorado River desalination plant at the United States-Mexico border. Another is a seawater desalination plant, considered as a contribution to the solution of the water problems of Israel toward the end of this century.

Desalination is discussed in Chap. 14 by Arad and Gluckstern. They show that in 1972 there were 812 land-based desalting plants of 25,000 gal/day (100 m^3/day) capacity or larger, in operation or under construction throughout the world, with a total desalting capacity of 348 MGD (1.3 x 10^6 m^3/day). Since then, a substantial increase in plant capacity has been reported, notable examples being a 48-MGD (180,000 m^3/day) plant in Hong Kong, and a 3-MGD (11,340 m^3/day) plant in Oklahoma. The above plants are intended primarily for urban water supply or for military installations.

In Chap. 14, the technology and the typical features of the major desalination processes, namely, distillation (multistage flash evaporation, multieffect distillation, and vapor compression) and membrane processes (electrodialysis and reverse osmosis), are reviewed, and the limitations of these processes discussed. The actual performance of 17 desalting systems at 11 different sites are described.

The future of desalted water as a source for irrigation on a wide scale depends on the cost per unit. In Chap. 14, the cost structure of the various processes is presented, including fuel consumption. It is shown that for small plants (1-MGD range) considering current technology, and making alternative assumptions with respect to the various cost components, the cost of desalination ranges from 1 to 1.7 \$/m^3 for seawater and from 0.35 to 0.50 \$/m^3 for brackish water in terms of 1975 dollars. For large, 100-MGD, nuclear dual-purpose (electrical power and desalting) plants, with the economic benefits of combined power and water production assigned to the latter, the estimated cost of desalination of seawater ranges between 0.26 \$/m^3 and 0.45 \$/m^3. Estimates of the economic value of irrigation water (marginal value product) under a variety of conditions suggest that increased efficiency and cost reduction in desalination are needed in order to justify desalted water as a potential source for irrigation, unless some specific and unique conditions prevail.

Chapters 15 by Wiener and 16 by Kneese and Zamora constitute the epilogue to the volume.

Weiner discusses the strategy of water resource planning and utilization under conditions of scarcity and deterioration of quality. His position is that in an era of rapid demographic and economic expansion, water resource management will be needed even in humid regions having an abundant supply of water. In arid regions, water resource management is indispensable if we wish to ensure an adequate supply for the most important uses and maintain the flow and quality of water resources for future generations.

In such regions, problems of biological and physicochemical pollution are liable to become acute at an early stage, and some of these deterioration processes may prove to be irreversible, or reversible only at high cost and after a long delay. To prevent deterioration of the resource, action will be required well ahead of the acute pollution stage.

The selection of a sufficiently comprehensive planning space is therefore necessary. Such a planning space should comprise, in addition to all the dimensions relevant to quantitative management, dimensions related to water quality, and it should extend over a sufficiently large geographic area. As scarcity becomes more acute, the planning space will have to be made more comprehensive and its area of coverage extended.

When scarcity reaches the point where it becomes nonfeasible to meet the potential demands of all use sectors, the resources base of the principal production processes into which water enters as an important raw material will have to be included in the planning space, and the demand will thereby become an endogenous variable in the water resource planning space. Institutional structure and legal administration will have to be adapted to the needs of water resource planning.

Arid regions will have to adopt systemic planning well before humid regions because the penalty for failing to do so will be much heavier in the former than in the latter areas.

Kneese and Zamora present a broad perspective of the situation of the world's arid lands, the trends affecting them, and possible future developments. They pose the hypothesis that aridity tends to affect adversely per capita income and verify it by means of statistical analysis of international data. They discuss the advantages that can be attributed to arid lands, for example, recreational sites, adaptability to special crops, military installations, residential areas, and footloose industry. However, the situation in most of the arid lands (the oil-rich arid areas are a notable exception) is dominated by negative factors. These factors are as a result of the distortion of the environmental equilibrium established over past centuries with relatively slight human activity. Accelerated human activity as a result of population growth, and the consequent growing demand for food, has distorted the historical low-activity-level equilibrium, resulting in the evolution of steppe areas into desert, deterioration of mountain environments, salt accumulation in rivers and groundwater aquifers, and declining wildlife population. The question whether a new equilibrium can be established on a higher human actiity level (taking into account technological progress) or whether population control and in some cases *reduction* of population is necessary, is answered by Kneese and Zamora in a pessimistic mood.

It is not unlikely that the philosophy of Kneese and Zamora may prove overly pessimistic in the future. One intuitively recalls the pessimism of the Malthusian era and the tremendous, unexpected technological progress that followed. The aim of this volume, however, is not to confront pessimism with optimism. It deals with one aspect of efficient utilization of natural resources by man—solutions of the problem of salinity in irrigation. It presents the essentials of the underlying physical relationships, describes the currently available technology, and discusses the economic and institutional problems involved in proper management of water resource with especial reference to the problem of salinity.

At the end of this review one is tempted to refer to the analogous situation of the ancient irrigation cultures in the Middle East, which flourished under adverse natural conditions that were overcome by the ingenuity of man. On the other hand, one must recall that these ancient cultures were destroyed by wars and the deterioration of social systems that became unable to maintain themselves at their former level of civilization. Which one is the proper historical analogue to our era?

Will the present efforts of farmers, technicians, scientists, and policy makers lead man to significant achievements in his struggle for well-being? Science, know-how, and rational activity on the part of individuals and nations will certainly contribute toward this end.

References

Evans, N. H. (1974), Finding Knowledge Gaps: The Key to Salinity Control Solutions. In J. E. Flack and C. W. Howe—(eds.), *Salinity in Water Resources*, Proceedings of the 15th Annual Western Resources Conference at the University of Colorado, Merriman Publishing Co., Boulder, Colo.

Howe, C. W., and Orr, D. V. (1974), Economic Incentives for Salinity Reduction and Water Conservation in the Colorado River Basin—In J.E. Flack and C.W. Howe (eds.), *Salinity in Water Resources*, Proceedings of the 15th

Annual Western Resources Conference at the University of Colorado, Merriman Publishing Co., Boulder, Colo.

Maass, A., Hufschmidt, M. M., Dorfman, R., Thomas, H. A., Jr., Marglin, S. A., and Fair, G. M. (1966), *Design of Water Resource Systems,* Harvard University Press, Cambridge, Mass.

Yaron, B., Danfors, E., and Vaadia, Y. (1969), *Irrigation in Arid Zones,* The Royal Institute of Technology, Department of Land Improvement and Drainage, Stockholm, Sweden, and the Volcani Institute of Agricultural Research, Department of Soils and Water, Beit-Dagan, Israel (draft ed.).

2

Evaluation and Classification of Water Quality for Irrigation

Brian L. McNeal

Washington State University, Pullman, Washington

Introduction

Before undertaking a mutlidisciplinary treatment of the impact of salinity on water resource use and development, it is necessary to describe and explain the usage of the more important water quality parameters associated with salinity. Initial workers in the area of salinity effects on plants used the term *water quality* in a straightforward manner to designate the total quantity of dissolved salts present in a given water. This is still a major usage of the term, but the concept has broadened as our knowledge of salinity effects has grown. In addition to such overall effects, there are also effects of specific ions on soil water transmissivity or plant growth. Furthermore, it is important to recognize the tendency for salts to precipitate or dissolve from the soil during irrigation, in order that the salt load and composition of irrigation and drainage waters can be predicted more accurately. Each of the above subjects is treated in this chapter, with the goal of providing the reader with a basic understanding of the relatively few concepts required to deal with most water quality problems related to salinity management. No attempt is made to provide a comprehensive review of the considerable research that has been conducted to date in this general area, although the papers cited should enable the reader to locate additional articles in areas of particular interest.

In addition to those aspects related to salinity, other facets of water quality have become increasingly important in recent years. Irrigation agriculture was conducted initially in sparsely settled areas, where high-quality waters were employed for surface irrigation using gravity-flow systems. Increased population pressures, however, have led to the common reuse of irrigation waters, with the resultant presence of organic wastes and bacteria, pests and plant pathogens, and suspended particles in the waters (Wilcox and Durum, 1967; Rhoades and Bernstein, 1971). Besides the public health hazard and loss in aesthetic value, the presence of organic wastes and dissolved nutrients in waters can lead to extensive growth of aquatic weeds and microflora, and hence can modify both the water requirements and water transmission capabilities of many irrigation projects. The loss of sediments from irrigated lands not only decreases the inherent productivity of the eroded land, but also renders the use of modern irrigation systems more difficult, with the sediments plugging drip irrigation lines and sprinkler nozzles, and accelerating the wear on sprinkler equipment. The potential corrosiveness of waters containing high concentrations of acids or salts also must be considered when using many modern water conveyence and application devices (Wilcox and Durum, 1967). Water reuse can lead as well to reduced water quality in situations where return flows dominated by surface runoff are of substantially higher temperature than mean river temperatures, with resultant effects on fish growth, recreation, or industrial cooling capabilities.

The reader should keep in mind such expanded definitions of "water quality" in many modern irrigation contexts. However, the term will be used here in its more traditional sense, although broadened slightly to include effects of some specific ions on potential irrigation usage.

Total Salinity Hazard

Total Salt Concentration

The water quality parameters of greatest overall importance to salinity management are the concentrations and total quantities of dissolved salts present. Historically, water chemists have determined the total dissolved solids (TDS) of water samples by evaporating filtered samples of water to dryness and weighing the quantity of salt remaining. Once a representative sample of surface water or groundwater had been collected (U.S. Salinity Laboratory Staff, 1954; Wilcox, 1955; Wilcox and Durum, 1967), the determination appeared to be a straightforward one, which could be performed by any reputable laboratory. However, several of the salts remaining after drying could retain variable amounts of water, depending on the drying temperature and the degree of rehydration between drying and weighing, so the quantity was never free from ambiguity. Values were reported traditionally as parts per million (ppm) TDS, although more precise modern usage has favored the use of milligrams per liter (mg/liter) as the reporting unit. The two sets of units essentially are identical for waters of importance to irrigation agriculture, although numerical values in milligrams per liter are somewhat larger than corresponding values in parts per million (because of solution densities appreciably greater than unity) for brines and some highly concentrated drainage waters and groundwaters.

All irrigation waters contain dissolved salts, with the total salt levels of most irrigation and drainage waters being in the range of a few hundred to a few thousand milligrams per liter. The lowest reported values for irrigation waters are levels of approximately 75 to 100 mg/liter, as typified in the western United States by the Columbia, Sacramento, and Feather Rivers (U.S. Salinity Laboratory Staff, 1954; Rhoades and Bernstein, 1971). The total quantity of dissolved solids normally increases with increasing distance from the river's headwaters, as a result of mineral weathering and accretion of atmospheric salts during water flow through the river system. Even for high-quality river waters, the levels of dissolved salts are commonly 10 to 20 times greater than those present in precipitation of the region (Rhoades and Bernstein, 1971).

More typical levels of dissolved salts for river waters in the western United States range from a value of approximately 200 mg/liter for the Snake River to 700 to 900 mg/liter for the Colorado, Arkansas, and Gila Rivers (U.S. Salinity Laboratory Staff, 1954). It has been estimated that approximately 80% of the irrigation waters throughout the western United States have TDS levels less than 1,000 mg/liter, with nearly half containing 175 to 500 mg/liter (Wilcox, 1959) and less than 10% containing more than 1,500 mg/liter (Wilcox and Durum, 1967). For comparative purposes, 1,000 mg/liter = 1.36 tons of salt per acre foot of water.

Waters containing the higher dissolved salt levels become increasingly more hazardous for irrigation of salt-sensitive crops or for application to fine-textured soils, through which water for salinity control may be passed only with difficulty. However, irrigation waters used in the Salt River Valley of central Arizona sometimes approach TDS levels of 2,000 to 3,000 mg/liter, and the Pecos River near Orla, Texas, has been used for irrigation even at TDS levels exceeding 4,000 mg/liter. The more saline waters can be used on an intermittent basis in humid regions where salts are flushed from the system during the rainy season (Wilcox and Durum, 1967), except for cases where absorption of salts by foliage during sprinkler irrigation would accentuate damage for crops that are particularly sensitive to salt injury (Bernstein, 1967).

Groundwaters used for irrigation normally contain higher TDS levels than do river waters from the same region. The lowest reported TDS values of well waters commonly used for irrigation are 200 to 300 mg/liter, with well waters having levels as high as 2,000 to 3,000 mg/liter rather commonly being used for irrigation in the western United States (Rhoades and Bernstein, 1971). The higher levels of dissolved salts in such waters are due both to selective withdrawal of water (while leaving behind the accompanying salts) by plants in the groundwater recharge area, and to dissolution of minerals in the soils and rocks through which water has flowed. Samples of soil solutions from irrigated areas commonly contain 2 to 10 times the TDS levels of corresponding irrigation waters, due to these same processes (Rhoades and Bernstein, 1971).

Electrical Conductivity

As indicated in the previous section, determination of the total quantity of dissolved salts present in a water is somewhat ambiguous, requires a substantial amount of time for the drying process, and necessitates relatively precise laboratory techniques to deal with the small quantities of salt commonly present in irrigation and drainage waters. As a consequence, a more popular general measure of the total quantity of dissolved salts in a given water is the electrical conductivity (EC) of that water. This determination involves placing a sample of water between two electrodes, imposing an electrical potential difference between the electrodes, and measuring the resistance of the solution. As the salt concentration of the solution increases, its ability to transmit electricity also increases. Results generally are converted to the reciprocal of the electrical resistance, in terms of electrical conductance, in order to produce a parameter that increases as the salt concentration increases. Because of the variations in results with variations in geometry of the measurement cell, a "cell constant" related to cell geometry is determined for each EC cell by measuring conductance values for standard solutions (usually KCl) of specified concentrations (U.S. Salinity Laboratory Staff, 1954; Bower and Wilcox, 1965). Results multiplied by this cell constant are reported as EC values, and are independent of the geometry of the measuring system. As such values are dependent on the temperature as well, results generally are converted to a reference temperature of 25°C, either by use of a standard conversion table (U.S. Salinity Laboratory Staff, 1954) or by an electrical compensator within the measuring unit itself.

The EC normally would be measured in mhos per centimeter (mho/cm), with the mho defined as a reciprocal ohm and with the unit of length resulting from the inclusion of the cell constant in the EC values. Because of the large amount of dissolved salt required to produce an EC of 1 mho/cm, most EC values for natural waters are reported in units of millimhos per centimeter (mmho/cm) or micromhos/cm (μmho/cm). The micromho per centimeter has been used traditionally for dilute salt solutions such as rainwaters and many irrigation waters, whereas the millimho per centimeter was proposed by Schofield in 1942 as the standard for soil solutions and drainage waters (Wilcox and Resch, 1963) and gradually is becoming the standard for most tabulations of EC values. The EC values of a large percentage of the suitable irrigation waters in the western United States can be related to the total quantity of dissolved salts by the following approximate relations (U.S. Salinity Laboratory Staff, 1954):

$$\text{TDS (mg/liter)} \approx 640 \times \text{EC (mmho/cm)} \tag{1}$$

and

$$\text{TDS (mg/liter)} \approx 0.64 \times \text{EC } (\mu\text{mho/cm}) \tag{2}$$

Hence, the EC values of many irrigation waters would occur in the range from 0.15 to 1.5 mmho/cm (150 to 1,500 μmho/cm, or 96 to 960 mg/liter), with the EC values of soil solutions and drainage waters correspondingly higher (Rhoades and Bernstein, 1971). Techniques have been developed in recent years to estimate EC values of the latter two types of solutions from in situ, nondestructive measurements (Oster and Willardson, 1971; Halvorson and Rhoades, 1974), using either buried electrodes in a matrix of porous ceramic or large sets of electrodes inserted into the soil surface just prior to the time of measurement. Unfortunately, both techniques are still in the category of research approaches, rather than serving as routine soil management tools.

Composition of Dissolved Salts

The actual EC of a given water depends not only on the total quantity of dissolved salts present, but also on the ionic composition of the solution. Techniques have been described recently for estimating EC values from ionic composition data (McNeal et al., 1970; Tanji and Biggar, 1972). Certain ions, such as chloride, function as more effective transmitters of electrical current than do other ions, such as sulfate (U.S. Salinity Laboratory Staff, 1954). However, the differences in EC that exist for single-salt solutions become of less importance for mixed-salt solutions, including most natural waters. Hence, a single relation between EC and total salt concentration (expressed in terms of chemical equivalents or milliequivalents per liter) generally is valid over a rather wide concentration range (U.S. Salinity Laboratory Staff, 1954).

Although earlier water composition data (as opposed to TDS or EC values) commonly were reported as parts per million or milligrams per liter of the respective ions, the current standard for major cations (calcium, magnesium, sodium, and potassium) and major anions (chloride, sulfate, bicarbonate, and carbonate) in salinity appraisals is the use of milliequivalents per liter (meq/liter). For a given ion, this unit is numerically equal to 1,000 times the normality of that ion in the solution. The weights of common ions and elements required to produce a concentration of 1 meq/liter for that substance are given in Table 1. Division of ion composition data in milligrams per liter by the appropriate equivalent weight values from the table will produce corresponding ion composition data in milliequivalents per liter. On the average, the effective "milli-

Table 1 Equivalent Weights of Solutes Commonly Reported in Salinity Appraisals

Solute	Symbol	Valence of ion or radical	Equivalent weight[a]
Calcium	Ca	+2	20.0
Magnesium	Mg	+2	12.2
Potassium	K	+1	39.1
Sodium	Na	+1	23.0
Ammonium	NH_4	+1	18.0
Ammonium-nitrogen	NH_4-N	+1	14.0
Bicarbonate	HCO_3	−1	61.0
Carbonate	CO_3	−2	30.0
Chloride	Cl	−1	35.5
Sulfate	SO_4	−2	48.0
Sulfate-sulfur	SO_4-S	−2	16.0
Nitrate	NO_3	−1	62.0
Nitrate-nitrogen	NO_3-N	−1	14.0

[a]Numerically equal to the milligrams per liter of solute required to produce a concentration of 1 meq/liter.

equivalent weight" of the cations plus the anions in typical irrigation and drainage waters ranges between 55 and 75 mg/liter, with the mean value for a large number of waters from the western United States found in one study to be 64 mg/liter (U.S. Salinity Laboratory Staff, 1954). It was this value, along with the approximate relation

$$\text{Total cations } or \text{ anions (meq/liter)} \approx \text{EC (mmho/cm)} \times 10 \qquad (3)$$

(which generally is accurate to within ±5% for irrigation waters and soil extracts in the EC range from 0.1 to 5.0 mmho/cm), which was used in the development of Eqs (1) and (2).

Methods for determining the composition of major solutes in irrigation and drainage waters have been compiled in several sources (U.S. Salinity Laboratory Staff, 1954; Bower and Wilcox, 1965; Wilcox and Durum, 1967), and can be performed by most modern analytical laboratories. In addition, rapid field tests for salinity have been proposed (Richards et al., 1956; Bower, 1972). Review of such analytical methods is beyond the scope of this volume. However, certain guidelines prove useful in assessing the reliability of specific data before incorporating them into salinity management models. Such guidelines (U.S.

Salinity Laboratory Staff, 1954; Bower and Wilcox, 1965) include the following:

1. The sum of all major cations (meq/liter) should approximately equal the sum of all major anions (meq/liter). Agreement should average within ±5% for high-quality laboratories, although continued exact agreement indicates that one major solute is being determined by difference (usually sodium for older data, and sulfate for more recent data). This latter practice may be used as a cost-saving feature even by some excellent laboratories, but serves as a warning that shortcuts are being taken to produce the reported data.
2. Equation (3) should hold true for most waters having EC values less than 5 mmho/cm.
3. Carbonate should not be present in greater than trace amounts unless the water pH exceeds 8.5.
4. Calcium plus magnesium should be low if carbonate is present in other than trace amounts for samples exposed to the atmosphere prior to analysis, although high values (several meq/liter) of both calcium plus magnesium and carbonate may coexist in waters exposed to high concentrations of gaseous carbon dioxide prior to analysis.

Analytical data adhering consistently to the above guidelines should be valid for most salinity appraisal estimates. Complete water composition data for all major cations and anions offer the greatest versatility and reliability for salinity appraisals. However, the expense of obtaining such data precludes their availability for many studies. Hence, overall estimates of water salinity, such as the EC, often are used for general salinity recommendations and water quality monitoring programs.

Osmotic Properties

Although the discussion to this point has centered around salt concentrations and estimates thereof, actual plant responses to salinity are related to the osmotic properties produced by the salts present. The osmotic pressure or osmotic properties produced by the salts present. The osmotic pressure (OP) or osmotic pressures produced by some common salts are listed in Table 2. Those salts that produce large numbers of ions per molecule, and that remain most completely dissociated into individual ionic components, are those that produce the greatest osmotic effects. However, the differences evident for single-salt solutions become less pronounced for mixed-salt solutions, so that large numbers of irrigation and drainage waters exhibit a common relationship between

Table 2 Typical Osmotic Pressures Produced by Some Common Salts[a]

Salt	meq/liter per atmosphere	mg/liter per atmosphere
NaCl	24	1,403
$CaCl_2$	32	1,776
Na_2SO_4	36	2,557

[a] From Bernstein (1961).

osmotic pressure (or osmotic potential), and electrical conductivity (U.S. Salinity Laboratory Staff, 1954). This relationship is given by the expression

$$\text{OP (atm)} = 0.36 \times \text{EC (mmho/cm)} \tag{4}$$

With a negative sign affixed, the OP values in atmospheres become nearly identical numerically to osmotic potential values in bars. Another way of viewing the relationships is that approximately 28 meq/liter of "typical" salinity is equivalent to 1 atm of osmotic pressure or 1 bar of osmotic potential.

A nearly 10-fold range in the salt tolerance of crop plants exists (Bernstein, 1967), with beans undergoing a 25% reduction in yield due to salinity at a soil solution EC value of 2 mmho/cm, whereas several of the grasses experience a similar yield reduction only at EC values of 16 to 18 mmho/cm. Plants also vary in their sensitivity to salinity with stage of growth and with climate (U.S. Salinity Laboratory Staff, 1954). However, plant growth inhibition from salinity is often well correlated with the osmotic potential of the soil solution. Divisions of the "Schofield scale" (delineating conditions of negligible, appreciable, substantial, severe, and very severe salinity effects on crop growth) were set at EC values for solutions from saturated soils of 2, 4, 8, and 16 mmho/cm (U.S. Salinity Laboratory Staff, 1954), which corresponds to osmotic potentials of approximately −0.7, −1.4, −2.9, and −5.8 bars, respectively. Hence, the generalization that growth reductions for many plants occur at osmotic potentials of −2 to −4 bars (Wilcox, 1959). Techniques have been developed to measure osmotic potentials directly in the field, or to infer them from EC values, as the two parameters are closely related [Eq (4)].

Salt Balance

Many early appraisals of salinity hazards centered around the concept of maintaining "salt balance" for all irrigation projects. Simply stated, this concept maintained that the quantity of salt removed in drainage waters from a project should be equal to or greater than the quantity of salt entering the project if the project were to maintain agricultural production indefinitely. The concept arose from the poor success of many historical irrigation projects, where saline drainage waters accumulated until the project became too saline to support economic crop production. Modern projects operating at "negative" salt balance (where more salt was entering than leaving the project on a sustained basis) were similarly regarded as doomed.

A classic long-term salt balance study for the Rio Grande River system was conducted by Wilcox and co-workers (Wilcox and Resch, 1963; Bower, 1974), following formalization of the salt balance concept by Schofield (U.S. Salinity Laboratory Staff, 1954). During the period of this study (1938-1962), operation of irrigation projects under conditions of favorable salt balance was common, because large quantities of water were passed through the projects with subsequent return of large amounts of soil salts to the associated river system. However, current trends appear to be toward the development of integrated water management approaches for river systems, with each project expected to minimize the amount of salinity contributed by its return flows (Van Schilfgaarde et al., 1974). As has been pointed out by Rhoades and Bernstein (1971) and by Bower (1974), minimization of the salinity in return flows is best accomplished by minimizing the amount of water leaving the project, even though this results in higher concentrations of salinity per unit of return flow. Under low return-flow conditions, many irrigation projects may operate indefinitely at an unfavorable salt balance, because much of the salt remaining in the project is precipitated as relatively insoluble carbonate or sulfate salts, which generally exert only minor effects on plant growth. Such precipitation of salts is treated in greater detail in a subsequent section of this chapter.

Another major problem with the concept of salt balance is the fact that uneven water distribution on individual irrigated fields leads to excessive leaching on some portions of a given field, and to accumulation of salts in other areas of the same field. In some cases, more than 30% of the applied water must pass through a project to overcome this effect and produce a positive salt balance for the entire project (Bower, 1974). However, one can approximate *average* changes in soil solution EC from data for salt balance (Wilcox and Resch, 1963). The salt balance approach also proves useful in assessing present patterns and long-term trends in the leaching of plant nutrients (Carter et al., 1971; Nelson and Weaver, 1971).

Sodium Hazard

Sodium Adsorption Ratio

It has been recognized from antiquity that some salt-affected soils develop a white surface crust of salts during fallow periods, whereas others develop instead a dark brown to black surface crust of salts mixed with dispersed organic matter. The latter soils, which are characterized by low water infiltration rates, have been termed "black alkali," "slick spot," "solonetz," or "sodic" soils. This condition results from an accumulation of adsorbed (exchangeable) sodium on exchange sites of soil minerals and organic matter, which causes the soil to disperse and become impermeable to water at low salt concentrations. Some salt-affected soils with adequate initial permeability also tend to seal or become relatively impermeable during leaching. Examination of such soils generally reveals an excess of exchangeable sodium, which causes the soil to disperse once excess salts have been leached from the soil profile.

Various means have been proposed for classifying the tendency of waters to produce sodic soils. Some early attempts used the soluble sodium percentage as a parameter, with the concentration of sodium (usually in milliequivalents per liter) being expressed as a percentage of the total concentration of cations in the water (Wilcox, 1955). However, at low salt concentrations, sodium may exceed 90% of the total cations present before any significant sodium hazard exists from a particular water.

The sodium adsorption ratio (SAR) has been proposed as a more useful measure of the sodium hazard of a water (U.S. Salinity Laboratory Staff, 1954). This parameter is defined by the relation:

$$SAR = \frac{(Na)}{[(Ca + Mg)/2]^{1/2}} \qquad (5)$$

where all concentrations are in milliequivalents per liter. As suggested in the previous paragraph, waters of low total salt concentration require substantially higher proportions of sodium than do waters of high total salt concentration to produce a given SAR value. If cation concentrations are given instead in milligrams per liter or parts per million, they must be converted to milliequivalents per liter by dividing by the appropriate equivalent weight values from Table 1. Because of the general relationship between EC and the sum of the cations [Eq. (3)], and because the concentration of potassium in most waters associated with irrigation agriculture is relatively small (with sodium, calcium, and magnesium making up the bulk of the remaining cations), the salinity and sodium hazards of a given water can be estimated from any two of the three parameters: EC, Na concentration, and (Ca + Mg) concentration (Richards et al., 1956).

Although sodic soils tend to have pH values of 8.5 or higher, particularly at low salt concentratons, soil pH in general is not a reliable indicator of soil sodium status (U.S. Salinity Laboratory Staff, 1954).

Numerous equations have been developed relating water composition to exchangeable-ion composition on soil mineral and organic surfaces (Bower, 1959). All of these equations employ a term similar to the SAR for exchange reactions involving Na and either Ca or Mg, although many deal in terms of ion activities ("effective concentrations") rather than in terms of ion concenrations per se. The use of ion activities is an unnecessary complication for many purposes related to water resource use and development, so the SAR has been employed for the purposes of this chapter. For more sophisticated calculations, the reader should refer to other sources (Tanji et al., 1967; Oster and McNeal, 1971; Dutt et al., 1972; Griffin and Jurinak, 1973).

The SAR is related to the exchangeable ion composition through the exchangeable sodium ratio (ESR), as defined by the relation

$$ESR = \frac{NaX}{CaX + MgX} = k_G \cdot SAR \tag{6}$$

where X refers to exchange sites and k_G is a Gapon-type constant having an average value of 0.015 for many soils from the irrigated portions of the western United States (U.S. Salinity Laboratory Staff, 1954). Despite a marked lack of theoretical basis, this useful relation has demonstrated validity for mixed-salt solutions containing both Mg and Ca, and an apparent insensitivity to total salt concentration over the range from 20 to 240 meq/liter (Bower, 1959).

The ESR is in turn related to the exchangeable sodium percentage (ESP) by the expression

$$ESP = 100 \times \frac{ESR}{1 + ESR} \tag{7}$$

For many soils from the western United States, the ESP of the soil and the SAR of the soil solution are approximately equal numerically up to ESP levels of 25 to 30. This is the ESP range of most interest to irrigation agriculture, for higher ESP levels generally are associated with soils so slowly permeable that major problems with crop production result. A nomogram relating water composition data to SAR and associated ESP values is available from several sources (U.S. Salinity Laboratory Staff, 1954; Wilcox, 1955; Wilcox and Durum, 1967). The ESP-SAR relationships have been studied for soils from other areas as well (Sinanuwong and El-Swaify, 1974). Because of potential errors in experimental ESP determinations, and because of the relative simplicity of the SAR determination, the SAR has gained increasing acceptance as the favored parameter for indicating the presence of sodic soil conditions.

The ESP of the surface foot of irrigated soil may approach that predicted from the SAR of the irrigation water (U.S. Salinity Laboratory Staff, 1954), but field ESP values generally are higher than predicted from irrigation water composition alone. This is a result of both the concentration of salts by evapotranspiration of water from the plant root zone and selective precipitation of calcium and magnesium salts during the evapotranspiration process. Each of these processes increases the SAR of the soil solution, and hence the equilibrium ESP of the associated soil. The SAR of a solution increases as the square root of the increase in total salt concentration of the solution during removal of water from the plant root zone, providing that no ion precipitation occurs during the concentration process. Thus, a fourfold concentration of an irrigation water during evapotranspiration will produce a doubling of the SAR. The ESP of soil equilibrated with that water will also double if the process is repeated often enough to permit approach to ion-exchange equilibrium with the more concentrated solution. Hence, the ESP in the lower root zone for field soils being irrigated with a given water may be as much as two to three times the SAR of the applied water.

In managing irrigation waters and drainage waters for salinity control, it is generally desirable to maintain the SAR of the irrigation water below a value of 10 (Wilcox, 1959; Wilcox and Durum, 1967). However, irrigation water SAR values as high as 15 to 20 may be approached for some coarse-textured soils without noticeable effects on soil permeability. The deleterious effects of SAR on soil permeability are reduced as the salt concentration is increased. In some cases, the salt concentration of the soil solution may be sufficient to offset the effects on soil permeability of irrigation water SAR values greater than 20. On the other hand, waters of low salt concentration (less than 50 mg/liter TDS) may disperse the soil even at low ESP levels (Wilcox and Durum, 1967). Factors associated with soil permeability decreases for sodic soils have been studied in considerable detail, and models for predicting such decreases have been developed (McNeal, 1974). Gypsum may be added to the soil or water periodically to correct tendencies toward soil dispersion. It should also be remembered during waste water disposal operations that irrigation water and municipal water requirements with respect to "hardness" (calcium and magnesium content) are often diametrically opposed (Rhoades and Bernstein, 1971), so that even partially softened water may require periodic applications of gypsum if it is to be used for irrigation agriculture (Wilcox and Durum, 1967).

Langelier Index (pH_c)

As indicated in the previous section, the sodium hazard of an irrigation water normally is expected to increase approximately as the square root of the increase in total salt concentration of the water during the irrigation and plant growth processes. However, this is true only if no slightly soluble salts are precipitated from the water. In practice, many irrigation waters contain sufficient quantities of sulfate and bicarbonate ions to produce precipitation of calcium sulfate or calcium carbonate in the plant root zone. Precipitation of such salts removes calcium from solution and hence may markedly increase the sodium hazard of the water.

The precipitation of calcium sulfate (gypsum) from soil solutions and irrigation waters has been studied in considerable detail. Gypsum is soluble in distilled water to the extent of approximately 30 meq/liter, or an EC of approximately 2 mmho/cm (because soluble, uncharged pairs of ions are present and yet do not conduct electricity in this particular case). However, this solubility is increased markedly in the presence of other salts. As an order-of-magnitude approximation, the maximum solubility of gypsum in irrigation waters and soil solutions can be inferred from the values of Table 3. Whenever the product of anticipated calcium concentration times anticipated sulfate concentration in a given solution would exceed the solubility product given in the table, it should be presumed that gypsum will precipitate. If more exact estimates are needed, work such as that of Tanji et al. (1967), Oster and McNeal (1971), or Dutt et al (1972) should be consulted.

The quantity of salt precipitated as gypsum may be important in agricultural areas such as the Imperial Valley of California, where large quantities of gypsum have precipitated in soils irrigated repeatedly with Colorado River

Table 3 Apparent Solubility Products[a] of Gypsum ($CaSO_4 \cdot 2H_2O$) in Some Natural Waters

Total salt concentration (meq/liter)	Apparent solubility product[b] (meq/liter)^2
32	1,000
60	1,500
120	2,500
240	3,300
480	4,000

[a]Estimated using an extended Debye-Hückel version of the procedure of Tanji (1969) to 120 meq/liter, with increasing reliance on the empirical apparent solubility product values of McNeal and Bower (unpublished data) at higher salt concentrations.
[b]Equal to (Ca concentration, meq/liter) × (SO_4 concentration, meq/liter), for a solution having all of its calcium and sulfate derived from gypsum, with remaining cations in the ratio 4 meq Na/1 meq Mg, and with all remaining anions assumed to be chloride.

water. However, soils containing substantial quantities of precipitated gypsum are not normally susceptible to sodium damage, because of the natural reclamation potential imparted to such soils by the gypsum. The primary importance of gypsum precipitation arises from the quantities of salts removed from solution during the precipitation process. Precipitation decreases the need to maintain a favorable salt balance when managing the salinity of the irrigation project in question. In areas exhibiting a tendency toward gypsum precipitation, an integrated long-term measure of the actual quantities of salts precipitated in the plant root zone under various management regimes can be obtained by monitoring drainage water compositions from tile drain effluents.

Of considerably more concern is the precipitation of calcium carbonate from irrigation waters containing high levels of bicarbonate. The reactions involved can be represented by the relation

$$Ca + 2HCO_3 \rightleftharpoons CaCO_3\downarrow + H_2O + CO_2\uparrow \tag{8}$$

The resultant $CaCO_3$ is so insoluble under most conditions (even when compared to gypsum) that the precipitated calcium exerts only relatively minor subsequent effects on the solution phase. As a result, soils containing precipitated $CaCO_3$ can still become highly sodic, and can be dispersed readily if soluble salts are then removed during any leaching process.

Several approaches have been used to predict $CaCO_3$ precipitation in soil water systems. Computer programs have been developed to consider all facets of the problem concurrently (Nakayama, 1968; Oster and Rhoades, 1975; Griffin and Jurinak, 1973), and several simplified parameters have been proposed as predictive aids (Wilcox et al., 1954; Bower et al., 1965; Wilcox and Durum, 1967; Rhoades, 1968). A reasonable compromise for predicting $CaCO_3$ precipitation is the use of the so-called Langelier saturation index. The approach was developed originally for predicting the precipitation of $CaCO_3$ scale in boilers. However, as shown by the work of Wilcox and Bower (Bower et al., 1965, 1968), the index also works well for soils. The Langelier index is defined by the relation

$$\text{Langelier index} = pH_a - pH_c = pH_a - [(pk'_2 - pk'_c)] + pCa + pAlk] \tag{9}$$

where pH_a can be set equal to the pH of the soil for most soil water systems (Bower et al., 1965); pH_c is the calculated pH that would exist if the water in question were equilibrated with solid-phase $CaCO_3$; $(pk'_2 - pk'_c)$ is a set of appropriate dissociation and solubility constants for carbonic acid and solid-phase $CaCO_3$, with an adjustment made for effects of total salt concentration on ion activities; pCa is the negative logarithm of the concentration (moles per liter) of soluble calcium; and $pAlk$ is the negative logarithm of the concentration of soluble bicarbonate plus carbonate (equivalents per liter). Positive values of the index indicate that $CaCO_3$ will precipitate from the water, whereas negative ones indicate that the water will dissolve $CaCO_3$. Information required to calculate pH_c values for irrigation waters and soil solutions is provided in Table 4. Further modifications of the index have included the substitution of pH 8.3 (the pH of a nonsodic calcareous soil equilibrated with the concentration of gaseous carbon dioxide normally present in the atmosphere) for pH_a, and the inclusion of soluble magnesium as well as soluble calcium in the pCa term of the expression (Bower et al., 1968). This modified pH_c value has been designated subsequently by the symbol pH_c^*. In using this approach, Bower et al., (1968) have shown that pH_c^* is linearly related to the fraction of bicarbonate precipitating from a water at any given leaching fraction (fraction of applied irrigation water actually passing through the plant root zone). The relevant figure from their work is reproduced here as Fig. 1. Bower et al., (1965) had shown previously that the fraction of bicarbonate precipitated from a water correlated well with the Langelier index, if the index was positive. Thus, the pH_c^* of a proposed irrigation water can be calculated using the information provided in Table 4, and the fraction of bicarbonate to be precipitated at any proposed leaching fraction can be estimated from the relations of Fig. 1. As each milliequivalent of bicarbonate precipitating must be accompanied by a milliequivalent of calcium

Table 4 Information Required to Estimate the pH_c of Natural Waters from Chemical Composition Data[a]

Concentration (meq/liter)	$(pk'_2 - pk'_c)$	pCa	pAlk
0.1	–	4.30	4.00
0.5	2.11	3.60	3.30
1	2.13	3.30	3.00
2	2.16	3.00	2.70
4	2.20	2.70	2.40
6	2.23	2.52	2.22
8	2.25	2.40	2.10
10	2.27	2.30	2.00
15	2.32	2.12	1.82
20	2.35	2.00	1.70
25	2.38	1.90	1.60
30	2.40	1.82	1.52
35	2.42	1.76	1.46
40	2.44	1.70	1.40
50	2.47	1.60	1.30

[a]From Bower et al., (1965). Relate $(pk'_2 - pk'_c)$ to total cation concentration, and pCa and pAlk to calcium and titratable base concentrations, respectively.

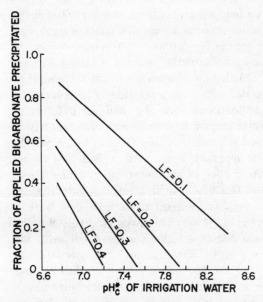

Fig. 1 Fraction of applied bicarbonate precipitated in lysimeter studies as related to pH_c^* of irrigation water and leaching fraction (LF). Crop = alfalfa. Soil = Pachappa sandy loam (Bower et al., 1968).

[Eq. (8)], a modified calcium concentration can be obtained for use in subsequent SAR calculations. The total salt concentration of the irrigation water also can be adjusted by subtracting the quantities of calcium and bicarbonate that are predicted to precipitate under the given set of conditions. The degree to which the irrigation project can operate at a negative salt balance without excess accumulation of salts in the plant root zone can then be predicted.

Recent studies (Rhoades et al., 1973; Oster and Rhoades, 1975) have examined the relative contributions of mineral weathering and $CaCO_3$ precipitation in modifying drainage water compositions of soils irrigated with various irrigation waters. The results in general substantiate the earlier findings, and demonstrate that the pH_c can be used as a semiquantitative index of the tendency toward $CaCO_3$ precipitation over a wide range of actual conditions. However, the studies have shown as well that the relations of Fig. 1 actually depend on the partial pressure of carbon dioxide present at the bottom of the root zone, with even larger percentages of bicarbonate precipitating as carbonate once the drainage waters leave the carbon dioxide-rich area at the bottom of the root zone and reequilibrate with the atmosphere via drain tile or drainage water sampling devices. It has also been shown that primary soil minerals can dissolve to the extent of 3 to 5 meq/liter under anticipated field conditions (Rhoades,

1968). An empirical equation incorporating a correction term for soil mineral weathering for the estimation of soil ESP values has been developed by Rhoades (1968), and Rhoades has also stressed the concept of leaching for exchangeable sodium control as well as for salinity control (Rhoades and Bernstein, 1971). Rhoades (1968) has pointed out that $CaCO_3$ precipitation, rather than mineral weathering, is the normal net effect for waters having pH_c values less than 8.3.

As no theoretical basis exists for including magnesium in pH_c^* estimates, such usage must be justified on an empirical basis only. Such predictive abilities are fortunate, for many early water analyses reported calcium and magnesium concentrations as a single joint sum, rather than as separate figures. Magnesium appears to be immobilized in many soils by a poorly defined mechanism (Bower, 1974), which may include precipitation as insoluble carbonates or silicates, or indirect "losses" from solution via ion exchange as the competing soluble calcium precipitates from the soil solution (Wilcox and Durum, 1967).

Residual Sodium Carbonate

Another empirical parameter that has been used to predict the potential for $CaCO_3$ precipitation from irrigation waters is the residual sodium carbonate (RSC). This approach was developed by Eaton (Wilcox, 1955) and received rather widespread usage for several years, including development of acceptable RSC ranges for irrigation waters (Wilcox et al., 1954; Wilcox and Durum, 1967). The main problem with this and similar concepts (such as Doneen's concept of "effective salinity," Wilcox and Durum, 1967) is an inherent assumption of complete precipitation of $CaCO_3$ from the irrigation water. As is evident from Fig. 1, the amount of $CaCO_3$ that precipitates depends on water characteristics and actual management conditions. Although the RSC concept may be useful qualitatively, it has limited quantitative value for salinity appraisals. It will not be considered further in this chapter.

Boron Hazard

In addition to the osmotic and soil structure effects discussed above, some dissolved ions can cause specific effects on certain high-value crops. Crops most susceptible to such injury are generally of woody type, including stone fruits, citrus, avocados, grapes, and both woody and nonwoody berry crops.

An element that often is treated separately because of its extremely hazardous nature is boron (B). This element is present in all irrigation waters (Wilcox, 1959), and is essential for plant growth processes at optimum levels of 0.2 to 0.5 mg/liter. However, it can become toxic to more sensitive crops at irrigation water boron concentrations of 0.3 to 1.0 mg/liter, and can produce toxicities for a wide variety of crops at irrigation water levels of only 1 to 4 mg/liter (Wilcox and Durum, 1967). Only a few surface waters in the western United States carry boron at levels toxic to plants, but many groundwaters contain toxic concentrations of this element (Wilcox, 1959). As with many other toxic elements, foliar damage includes necrotic spots on foliage, and tip and edge burn on leaves (Bernstein, 1961). The presence of such symptoms should prompt a reappraisal of water analyses to determine if toxic concentrations of some specific element may be present.

The assessment of boron hazard is quite straightforward. The boron concentration of irrigation waters normally is determined colorimetrically (Bower and Wilcox, 1965), and is reported in units of milligrams per liter. Essentially all of the boron tolerance data that have been developed similarly are reported in milligrams per liter, so one must only estimate the average amount by which the boron may be concentrated during plant growth processes in order to isolate instances where boron may constitute a potential plant growth hazard. In cases of high boron levels, the selection of boron-tolerant crops (U.S. Salinity Laboratory Staff, 1954; Bernstein, 1961; Wilcox and Durum, 1967) is a wise management practice.

A source of potential problems when relating irrigation water composition to soil water composition is the fact that boron is adsorbed by soils, although the boron adsorption capacities of most soils are low. Such adsorption is influenced by the presence of certain amorphous soil minerals and by soil surface area (Hatcher et al., 1967; Griffin and Burau, 1974). As a result of adsorption, the boron concentration of the soil solution will build up less slowly than predicted from the relative ratio of irrigation and drainage water volumes, but will also decrease more slowly than will the total salt concentration during a leaching operation. For example, it has been observed for soils from the Coachella Valley of California that it takes approximately three times as much water to lower the relative boron concentration to the same extent as the relative lowering of the total salt concentraton (Reeve et al., 1955). Thus, although 1 ft of water may remove 80% of the salts from the first foot of soil during ponded leaching of a typical salt-affected soil profile, 3 ft of water may be required to remove the same percentage of boron from the same soil layer.

Other Specific Ion Hazards

In addition to the boron hazard, many tree fruits, woody vines, and berry crops are susceptible to damage from specific ions such as sodium and chloride. As with boron, the toxicity of these elements tends to produce distinctive foliar symptoms, such as necrotic spots and marginal or tip burn (Bernstein, 1961).

Some species, such as citrus and many tree fruits, are particularly susceptible to foliar damage when irrigation waters containing toxic concentrations of Na or Cl are applied directly to foliage from sprinkler lines. Most species susceptible to specific ion damage can exhibit considerable variation in potential toxicity from one rootstock to another. Hence, the grafting of susceptible vegetation onto tolerant rootstocks may produce acceptable crop growth even on soils having moderately high Na and Cl levels.

As with boron hazard, Na and Cl hazards normally are tabulated directly in terms of Na or Cl concentrations of either the irrigation water or the soil solution (U.S. Salinity Laboratory Staff, 1954; Bernstein, 1961). Thus, it is only necessary to estimate the relative amount by which the irrigation water will be concentrated in the soil during crop growth. Because of adsorption of Na on soil exchange sites, either the Gapon equation [Eq. (6)] or a similar approach must be used to estimate dissolved and exchangeable levels for this ion, when such are needed (Tanji et al., 1967; Rhoades, 1968; Oster and McNeal, 1971; Dutt et al., 1972). Some tree fruits encounter Na toxicity at ESP levels as low as 5% (Pearson, 1960; Bernstein, 1967), although many field crops evidence no apparent Na toxicities even at ESP levels of 40 to 60%.

In addition to the above specific ion problems, other toxicities have been reported for specialty crops (and particularly for some woody species). The potential for specific ion toxicities should be considered whenever susceptible species are being grown in areas where high concentrations of toxic ions are known or suspected to be present. In particular, workers have recorded bicarbonate (Bernstein, 1967), lithium (Bradford, 1963), and selenium (Rhoades and Bernstein, 1971) toxicities in several portions of the western United States. Bicarbonate toxicity has been reported only in areas of exceptionally high bicarbonate levels, but the latter two elements may be toxic at levels as low as 0.1 to 0.2 mg/liter. Nearly one-fourth of the 400 California waters analyzed by Bradford (1963) contained levels of lithium potentially toxic to citrus. A final specific ion worthy of note is fluoride, which may be associated with highly saline groundwater in areas of geothermal activity (Wilcox, 1955). This species is of interest primarily because of its association with human dental problems. Like boron, it is weakly adsorbed by soil minerals, with the adsorption greatest for highly weathered soils.

Classification of Water Quality for Irrigation

Many early books on irrigation agriculture included a diagram summarizing the author's best estimates of the salinity and sodium hazards associated with various types of irrigation waters (U.S. Salinity Laboratory Staff, 1954; Wilcox, 1955; Wilcox and Durum, 1967). Waters then were placed in "pigeonholes," with a given "classification" being assigned to each water. However, this approach did not allow properly for various management decisions on the part of the grower. Good growers could carry out sustained economic crop production using relatively poor-quality irrigation waters, whereas poor managers might encounter salinity problems with relatively good-quality waters. As a result, in recent years the trend has been away from the assignment of a specific water "quality" to a given irrigation water, and toward consideration instead of water quality in the context of proposed management practices. In other words, as stated by Bernstein (1967), the question is no longer one of "How good is the water?" but rather of "What can be done with this water?"

The basic philosophy of Bernstein's (1967) approach centers around the use of the leaching requirement (LR), which can be defined by the relation

$$LR = \frac{EC_{iw}}{EC_{dw}} \left(or \frac{SAR_{iw}}{SAR_{dw}} \text{ or } \frac{C_{iw}}{C_{dw}} \right) = \frac{D_{dw}}{D_{iw}} \tag{10}$$

where iw refers to irrigation water, dw refers to drainage water, EC is the electrical conductivity, SAR is the sodium adsorption ratio, C is the concentration of a specific toxic element, and D refers to equivalent surface depth of water. The concept of leaching requirement thus can be used either for salinity control, for exchangeable-sodium control, or for specific ion control in the soil-plant-water system.

Bernstein's extension of the leaching requirement concept was to calculate a potential leaching fraction (LF), based on soil infiltration rate (I). This LF must be adequate to meet both crop evapotranspiration rate (E) needs and the limiting (maximum) leaching requirement value from Eq. (10), without waterlogging. The appropriate equations presented by Bernstein (Bernstein, 1967; Rhoades and Bernstein, 1971) are

$$LF = 1 - \frac{E \cdot t_c}{I \cdot t_i} \tag{11}$$

where drainage is nonlimiting, and

$$LF = \frac{O \cdot t_c}{E \cdot t_c + O \cdot t_c} = \frac{O}{E + O} \tag{12}$$

where drainage is limiting. In these relations, t_c is the irrigation frequency, t_i is the irrigation duration, and O is the net downward drainage rate below the plant root zone. The case where drainage is limiting occurs when $O \cdot t_c$ (or D_{dw}) is less than the quantity $(I \cdot t_i - E \cdot t_c)$. Whenever LR [calculated from Eq. (10)] is greater than LF [calculated either from Eq. (11) or Eq. (12)], then salinity cannot be controlled under the specified operating conditions. It is then necessary either to choose a more salt-tolerant crop, or to accept a certain amount of yield reduction due to salinity. Salinity effects may also be lessened by adoption of appropriate water management practices to minimize root-zone salinity levels during crucial crop growth stages, or by using additional off-season leaching to minimize long-term salinity accumulations.

A major error in common usage of the LR concept has been a tendency to substitute saturation extract data (chemical data obtained for extracts from pastes of saturated soil) for drainage water data in Eq. (10). Because most soils have a water content at saturation that is two to three times the water content at "field capacity" (an approximation of the field water content following initial free drainage of water after irrigation), use of saturation extract data for LR calculations can result in calculated LR values that are two to three times higher than the values actually necessary for salinity control. As saturation extract data are more commonly available than drainage water data (with respect to crop tolerance), attempts to utilize such data accurately have included use of saturation extract EC values producing a 50% yield reduction as an approximation to ED_{dw} values producing only a 10 to 20% yield reduction for many crops (U.S. Salinity Laboratory Staff, 1954; Rhoades, 1968).

When using data relating crop tolerance to various salinity levels, it is important to keep in mind the conditions under which the crop tolerance measurements were made. Many such measurements have been made using small plots or greenhouse-scale containers maintained at nearly constant salinity or toxic element concentrations (U.S. Salinity Laboratory Staff, 1954). Under field conditions, the plant root zone usually contains some portions that are relatively free of soluble salts or toxic elements, and other portions that contain rather high concentrations of these same materials. Recent work by Bernstein and Francois (1973) has demonstrated that most plant root activity under such conditions is in the low-salt portion of the soil. Thus, if the low-salt zone is sufficiently large to provide adequate water and nutrients for plant growth, the plant can grow satisfactorily at drainage water salinity levels much higher than those that would be limiting for uniformly salinized soils. The same workers have demonstrated that alfalfa can be grown with essentially no growth reduction under nonuniform salinity distributions at drainage water salt concentrations 2.5 to 4 times higher than required to produce a 50% decrease in yield for uniformly salinized soils. They have also demonstrated under such conditions a greater apparent response to EC_{iw} than to EC_{dw}, because most alfalfa root activity and water uptake is associated with the more salt-free portions of the soil.

Although reduced values of the leaching requirement for salinity control are now considered to be valid, salinity control under field conditions must also take into account nonuniformities of water application and infiltration. The additional amounts of water required to assure adequate irrigation on *all* parts of a field will lead to greater water applications, in many instances, than required to satisfy crop water needs and provide for salinity control under average conditions for the field.

In managing sodium, it must be remembered that the SAR of the soil solution increases as the salt concentration increases during evapotranspiration. Relatively high SAR and ESP values may exist relatively deep in the soil profile without producing measurable reductions in soil permeability (because of the simultaneous presence of high salt concentrations). However, the subsequent passage of low-salt water through the soil (as in climates characterized by significant winter rainfall inputs) may lower the permeability drastically as excess salt is removed from the high-sodium zone. This is a hazard that must be kept in mind when managing irrigation agriculture for minimal leaching, with resultant production of high salinity and SAR levels deep in the soil profile. Decreases in permeability deep in the soil profile are particularly serious, because of the difficulty in mechanically increasing the size of conducting pores through the soil mass after the soil has been restored to a favorable chemical status. Surface applications of gypsum prior to the winter rainfall period have proved successful in Israel to alleviate the adverse physical effects of irrigation waters having moderate salinity and high SAR levels, by providing a source of soluble calcium to accompany the percolating rainwater (Dr. Isaac Shainberg, Volcani Institute, personal communication).

The situation with respect to specific ion toxicities is less thoroughly understood. The basic leaching requirement approach should still be applicable. However, if the zone of high salt accumulation deep in the soil profile is not sufficiently concentrated, the uptake of water from that zone may lead to the uptake of toxic elements as well, with substantial reductions in crop yield.

When water quality is viewed from the standpoint of available management practices, rather than from the standpoint of intrinsic properties of the water itself, the options available become extremely diverse. The problems of water quality degradation during multiple intrabasin use, of management practices available to growers in different portions of the basin, and of equitable allotments of waters acquired from interbasin transport become joint problems in soil chemistry, plant physiology, agricultural engineering, and agricultural economics. It is to the interrelationship of such considerations that much of the remainder of this volume are devoted.

Summary

Several aspects of the salinity problem as related to water resource use and development have been discussed. Appropriate water quality parameters that are needed for a thorough analysis, or that might be encountered in relevant reports, are dealt with. The former group includes the electrical conductivity, SAR, Langelier index, and specific toxic ion concentrations, whereas the latter includes total dissolved solids, other individual ion concentrations, residual sodium carbonate, and osmotic properties.

The viewpoint is reiterated that water quality is not an inherent property of the water itself, but rather a property intimately related to the management practices available to the growers of the agricultural region in question. Bernstein's integrated usage of the leaching requirement and potential leaching fraction concepts is pointed out as a viable means of quantifying water quality recommendations. The complex interrelationship of soil chemistry, plant physiology, agricultural engineering, and agricultural economics in dealing with water quality problems is stressed.

References

Bernstein, L. (1961). Salt tolerance of plants and the potential use of saline waters for irrigation. Nat. Acad. Sci., Nat. Res. Coun. Publ. *942*: 273-283.

Bernstein, L. (1967). Quantitative assessment of irrigation water quality. Amer. Soc. Testing Materials Spec.Tech. Publ. *416*: 51-65.

Bernstein, L., and Francois, L. E. (1973). Leaching requirement studies: Sensitivity of alfalfa to salinity of irrigation and drainage waters. Soil Sci. Soc. Amer. Proc. *37*: 931-943.

Bower, C. A. (1959). Cation-exchange equilibria in soils affected by sodium salts. Soil Sci. *88*: 32-35.

Bower, C. A. (1972). Colorimetric, semi-quantitative test for soil salinity. Soil Sci. Soc. Amer. Proc. *36*: 527-528.

Bower, C. A. (1974). Salinity of drainage waters. In J. van Schilfgaarde (ed.); *Drainage for Agriculture*, Agronomy *17*, pp. 471-487.

Bower, C. A., Ogata, G., and Tucker, J. M. (1968). Sodium hazard of irrigation waters as influenced by leaching fraction and by precipitation or solution of calcium carbonate. Soil Sci. *106*: 29-34.

Bower, C. A., and Wilcox, L. V. (1965). Soluble salts. In C. A. Black (ed.), *Methods of Soil Analysis*, Agronomy *9*, pp. 933-951.

Bower, C. A., Wilcox, L. V., Akin, G. W., and Keyes, M. G. (1965). An index of the tendency of $CaCO_3$ to precipitate from irrigation waters. Soil Sci. Soc. Amer. Proc. *29*: 91-92.

Bradford, G. R. (1963). Lithium survey of California's water resources. Soil Sci. *96*: 77-81.

Carter, D. L., Bondurant, J. A., and Robbins, C. W. (1971). Water-soluble NO_3-nitrogen, PO_4-phosphorus, and total salt balances on a large irrigation tract. Soil Sci. Soc. Amer. Proc. *35*: 331-335.

Dutt, G. R., Shaffer, M. J., and Moore, W. J. (1972). Computer simulation model of dynamic bio-physicochemical processes in soils. Univ. Arizona Agr. Exp. Sta. Tech. Bull. *196*, 101 pp.

Griffin, R.A., and Burau, R.G. (1974). Kinetic and equilibrium studies of boron desorption from soil. Soil Sci. Soc. Amer. Proc. *38*: 892-897.

Griffin, R. A., and Jurinak, J. J. (1973). Estimation of activity coefficients from the electrical conductivity of natural aquatic systems and soil extracts. Soil Sci. *116*: 26-30.

Halvorson, A. L., and Rhoades, J. D. (1974). Assessing soil salinity and identifying potential saline-seep areas with field soil resistance measurements. Soil Sci. Soc. Amer. Proc. *38*: 576-581.

Hatcher, J. T., Bower, C. A., and Clark, M. (1967). Adsorption of boron by soils as influenced by hydroxy aluminum and surface area. Soil Sci. *104*: 422-426.

McNeal, B. L. (1974). Soil salts and their effects on water movement. In J. van Schilfgaarde (ed.), *Drainage for Agriculture*, Agronomy *17*, pp. 409-431.

McNeal, B. L., Oster, J. D., and Hatcher, J. T. (1970). Calculation of electrical conductivity from solution composition data as an aid to in-situ estimation of soil salinity. Soil Sci. *110*: 405-414.

Nakayama, F. S. (1968). Calcium activity, complex and ion-pair in saturated $CaCO_3$ solutions. Soil Sci. *106*: 429-434.

Nelson, C. E., and Weaver, W. H. (1971). Salt balance for the Wapato project for 1970-71 compared with the salt balance for 1941-42. Wash. Agr. Exp. Sat. Bull. *743*, 12 pp.

Oster, J. D., and McNeal, B. L. (1971). Computation of soil solution composition variation with water content for desaturated soils. Soil Sci. Soc. Amer. Proc. *35*: 436-442.

Oster, J.D., and Rhoades, J. D. (1975). Calculated drainage water compositions and salt burdens resulting from irrigation with river waters in the western United States. J. Environmental Quality *4*: 73-79.

Oster, J. D., and Willardson, L. S. (1971). Reliability of salinity sensors for the management of soil salinity. Agronomy J. *63*: 695-698.

Pearson, G. A. (1960). Tolerance of plants to exchangeable sodium. U.S. Dept. Agr. Inform. Bull. *216*, 4 pp.

Reeve, R. C., Pillsbury, A. F., and Wilcox, L. V. (1955). Reclamation of a saline and high boron soil in the Coachella Valley of California. Hilgardia *24*: 69-91.

Rhoades, J. D. (1968). Mineral-weathering correction for estimating the sodium hazard of irrigation waters. Soil Sci. Soc. Amer. Proc. *32*: 648-652.

Rhoades, J. D., and Bernstein, L. (1971). Chemical, physical, and biological characteristics of irrigation and soil water. In L. L. Ciaccio (ed.), *Water* and *Water Pollution Handbook*, Marcel Dekker, New York, vol. I, pp. 141-222.

Rhoades, J. D., Ingvalson, R. D., Tucker, J. M., and Clark, M. (1973). Salts in irrigation drainage waters: I. Effects of irrigation water composition, leaching fraction, and time of year on the salt compositions of irrigation drainage waters. Soil Sci. Soc. Amer. Proc. *37*: 770-774.

Richards, L. A., Bower, C. A., and Fireman, M. (1956). Tests for salinity and sodium status of soil and of irrigation water. U.S. Dept. Agr. Circ. *982*, 19 pp.

Sinanuwong, S., and El-Swaify, S. A. (1974). Predicting exchangeable sodium ratios in irrigated tropical vertisols. Soil Sci. Soc. Amer. Proc. *38*: 732-737.

Tanji, K. K. (1969). Solubility of gypsum in aqueous electrolytes as affected by ion association and ionic strength up to $0.15\ M$ and at $25°C$. J. Environmental Sci. Tech. *3*: 656-661.

Tanji, K. K., and Biggar, J. W. (1972). Specific conductance model for natural waters and soil solutions of limited salinity levels. Water Resources Research *8*: 145-153.

Tanji, K. K., Doneen, L. D., and Paul, J. L. (1967). Quality of percolating waters: 3. Predictions on the quality of waters percolating through stratified substrata by computer analysis. Hilgardia *38*: 319-347.

U. S. Salinity Laboratory Staff (1954). *Diagnosis and Improvement of Saline and Alkali Soils*. U. S. Dept. Agr. Handbook No. 60, U.S. Govt. Printing Office, Washington, D.C., 160 pp.

Van Schilfgaarde, J., Bernstein, L., Rhoades, J. D., and Rawlins, S. L. (1974). Irrigation management for salt control. J. Irrigation and Drainage Division, Amer. Soc. Civil Engr., *100*: 321-338.

Wilcox, L. V. (1955). Classification and use of irrigation waters. U.S. Dept. Agr. Circ. *969*, 19 pp.

Wilcox, L. V. (1959). Effect of industrial wastes on water for irrigation use. Amer. Soc. Testing Materials Spec. Tech. Publ. *273*, 58-64.

Wilcox, L. V., Blair, G. Y., and Bower, C. A. (1954). Effect of bicarbonate on suitability of water for irrigation. Soil Sci. *77*: 259-266.

Wilcox, L. V., and Durum, W. H. (1967). Quality of irrigation water. In R. M. Hagan, H. R. Haise, and J. W. Edminster (eds.), *Irrigation of Agricultural Lands*, Agronomy *11*, pp. 104-122.

Wilcox, L. V., and Resch, W. F. (1963). Salt balance and leaching requirement in irrigated lands. U.S. Dept. Agr. Tech. Bull. *1290*, 23 pp.

3

Effects of Salinity and Soil Water Regime on Crop Yields

Leon Bernstein

U.S. Salinity Laboratory, Riverside, California

Introduction

Soluble inorganic salts are normal constituents of soils and provide the mineral elements essential for plant growth. Low total nutrient concentrations of 10 meq/liter provide adequate nutrition, but higher concentrations of nutrients or other salts may inhibit growth. Saline soil waters by definition contain 80 meq or more of salts per liter, although sensitive crops may be affected at half this salt concentration and tolerant ones only at twice this concentration. In most cases, the total salt concentration of the root medium, measured as osmotic potential, rather than the specific salts present, determines its inhibitory activity, although toxicities and nutritional effects may predominate in some cases (Bernstein, 1964a, 1975). The ions that may be present in high concentration are Na^+, Ca^{2+}, Mg^{2+}, Cl^-, SO_4^{2-}, and HCO_3^-. In rare cases, K^+ or NO_3^- concentrations may also be high. The ionic proportions and concentrations vary greatly from place to place. Although the salts may be derived from weathering of soil minerals in situ, most salts are brought into an area in irrigation waters or in groundwater draining into the area from surrounding locations. Poorly developed surface drainage in arid and semiarid zones restricts salt discharge into the sea and causes salt to accumulate in inland basins.

Since it is the concentration of salts in solution rather than their absolute amounts in the root medium that affects plant growth, the water content as well as the salt content of a soil determine its salinity. Sandy soils that retain less water become saline at lower salt contents on a dry soil basis than fine-textured soils that retain more water (U.S. Salinity Laboratory Staff, 1954).

Plants absorb water and dissolved salts independently. In saline soils, relatively little of the salts dissolved in the water is taken up as water is absorbed. At a given soil salinity, salt concentration tends to vary inversely with water content. When water cultures or sand cultures are used to study the effects of salinity on plants, water supply and, therefore, salt concentration may be essentially constant. In soils, fluctuating water contents usually result in inversely fluctuating salinities. The soil water regime may significantly influence plant response to salinity. In soils, unlike water cultures, salinity may also vary greatly within a given root zone, further complicating the determination of effective soil salinity. High saline water tables tend to cause accumulation of salts at the soil surface because of the upward movement of soil water under the influence of surface evaporation. In the absence of high water tables, salinity commonly increases with depth in the root zone, reflecting the increased extraction of water from increments of water applied in earlier irrigations. We discuss first the tolerance of plants to essentially uniform root zone salinity, and then the modifications in response caused by graded and fluctuating salinities.

Salt Tolerance of Plants

Experimental Methods

Some indication of salt tolerance may be obtained by observing crop responses in the field and their relation to ambient salinities. Salt tolerances so inferred may be subject to considerable error because salinities change with time, and the observed salinity may be quite different from that which damaged the crop at an earlier date. It is therefore preferable to determine the effects of salinity under carefully controlled and monitored salinity conditions. For some purposes, water or sand cultures may be advantageous, since they permit careful control of specific factors of the root environment, such as osmotic potential, specific ion concentrations, or nutrient levels. Salt tolerance data so obtained, although valid for the conditions studied, may not indicate salt tolerance under the field conditions used in the commercial production of the crop. Such conditions can be closely approximated in artificially salinized field plots that combine salinity control and commercial growing conditions. Briefly, the method involves the use of a series of carefully prepared plots on a permeable soil in which salinity can be controlled. The plot size commonly used at the U.S. Salinity Laboratory in Riverside, California, is 4.27 m × 4.27 m. Prior soil mixing within the plot area has minimized soil variability, and the small plot size facilitates highly uniform irrigation and salination (U.S. Salinity Laboratory Staff, 1954).

To assure a full stand of plants, the crop is planted in nonsaline soil that is salinized as soon as the seedlings are established. [Tolerance to salinity during germination may be studied in separate laboratory experiments using presalinized soil cultures (Ayers and Hayward, 1948).] Salination of the field plots is accomplished by irrigation of a series of plots (usually four) with waters of graded salt concentrations. The salts used may vary, depending on the specific salinity conditions of interest for the crop in question, but usually 50:50 mixtures by weight of NaCl and $CaCl_2$ have been employed. The high solubility of the chloride salts permits study of high salinity levels unaffected by salt precipitation. In some experiments, comparisons between chloride and sulfate salinities or various proportions of cations have been made, but care must be exercised in designing such experiments to take into account the precipitation of sparingly soluble salts such as $CaSO_4$.

To avoid osmotic shock, salination is achieved by stepwise increases in the salinity of the irrigation water, especially when salinities as high as 5,000 or 10,000 mg of salts per liter are employed. The first three salinizing irrigations are light applications at one-third, two-thirds, and three-thirds of the desired final levels applied at intervals of a few days so that salination is achieved rapidly while the crop is young. Depending on the salt tolerance of the crop studied, the amount of salt added to the low-salinity water (the Riverside water supply contains 450 mg. total salts per liter) may be 0, 500, 1,000, and 1,500 mg/liter for very sensitive crops up to 0, 5,000, 10,000, and 15,000 mg/liter for highly tolerant ones. Once the irrigation waters are at the desired salinity levels, the concentrations are maintained in all subsequent irrigations. After the first few irrigations, the soil salinity levels in the plots become essentially stable, provided that the soil is permeable and well drained. Irrigations must be adequate so that the leaching rate is high, to maintain stable salinity levels. The artificially salinized plots differ from most saline fields in that the salinity of the soil water approximates that of the irrigation water for each plot instead of greatly exceeding it because of progressive salt accumulation as in the field.

Three or four times during the growing season, soil samples are taken for determination of soil salinity by measuring the electrical conductivity of the saturation extract (EC_e). EC_e has been the standard measure of soil salinity rather than osmotic potential of the soil water because the former is more easily measured and the saturation extracts are reproducible dilutions of the soil salts that do not vary with fluctuating soil water contents (U.S. Salinity Laboratory Staff, 1954). Soil samples should be taken at several depths that include the major root zone of the crop. Localized zones of salt accumulation such as occur in the surface of furrow-irrigated ridges should be discarded or sampled separately from the underlying less saline soil. Root activity in such zones of high salt accumulation is nil.

Crop yields, quality, and any other crop property of interest are measured and related to EC_e graphically to determine the soil salinity that affects the crop adversely.

All plots in an experiment are fertilized at recommended levels for maximum yields, although in special studies fertility levels may be varied to determine the interaction of salinity and fertility. Irrigations are applied at recommended frequencies for the crop in question, all plots in an experiment being irrigated at the same frequency and with the same quantity of water, that is, enough to maintain maximum yields in the nonsaline control plot(s). This practice results in excess water application for the saline plots in which plant growth and water utilization are reduced. The excess water causes progressively greater leaching fractions with increasing salinity, preventing the accumulation of salt in the root zone with time. Tensiometers at two or more depths in each plot help time irrigations and provide data on soil water status. Frequently irrigated crops, such as potatoes, were irrigated daily in alternate furrows; crops requiring less frequent irrigation were irrigated every 5 to 25 days or so, depending on crop, weather conditions, and stage of growth. Cropping periods were selected to match the climatic requirements of the crop. Cool-season crops were grown during the winter at Riverside, California, and warm-weather crops during the summer. Since growing conditions varied for different crops, the salt tolerance data indicate crop response under the specific conditions that were optimal for each crop at Riverside, rather than under the same conditions for all crops.

Salt Tolerance Data

Salt tolerance data for field, forage, and vegetable crops are summarized in Table 1. The values reported are the EC_e's at which yield was reduced by 10, 25, and 50% compared to the nonsaline controls, determined by interpolation on graphs of yield versus EC_e. Values have been rounded off to the nearest 0.5 mmho/cm. Repeated experiments in different years with some crops (cotton, alfalfa) indicate the yield decrement data to be generally reliable within ±10%. In many experiments, four or five varieties of a crop were tested simultaneously. In most cases, differences were either nonsignificant or so small as to be of little practical importance (e.g., varieties of upland cotton, sugar cane, green beans, lettuce, etc.). Notably large differences occurred among Bermuda grasses, as noted in Table 1. Large varietal differences have also been claimed for soybean varieties, in which salt tolerance varied inversely with level of chloride accumulation, controlled by a single gene with low Cl^- accumulation dominant over high Cl^- accumulation (Abel and MacKenzie, 1964; Abel, 1969). The data for soybeans in Table 1 are from a field plot study at Riverside of the Lee variety, a tolerant soybean, according to Abel and MacKenzie (1964). Grass

Table 1 Salt Tolerance of Plants[a]

Crop	$EC_e \times 10^3$ (millimhos/cm at 25°C) at which yield decreased by[b]		
	10%	25%	50%

Forage Crops			
Bermuda grass[c] [*Cynodon dactylon* (L) Pers)]	13	16	18
Tall wheatgrass [*Agropyron elongatum* (Host) Beauv.]	11	15	18
Crested wheatgrass [*Agropyron desertorum* (Fisch. ex Link) Schult]	6	11	18
Tall fescue (*Festuca arundinacea* Schreb.)	7	10.5	14.5
Barley, hay[d] (*Hordeum vulgare* L.)	8	11	13.5
Perennial ryegrass (*Lolium perenne* L.)	8	10	13
Hardinggrass (*Phalaris stenoptera* Hack.)	8	10	13
Narrow-leaf birdsfoot trefoil [*Lotus tenuifolius* (L.) Reich]	6	8	10
Beardless wildrye (*Elymus triticoides* Buckley)	4	7	11
Alfalfa (*Medicago sativa* L.)	3	5	8
Orchardgrass (*Dactylis glomerata* L.)	2.5	4.5	8
Meadow foxtail (*Alopecurus pratensis* L.)	2	3.5	6.5
Alsike and red clovers (*Trifolium hybridum* L. and *T. pratense* L.)	2	2.5	4
Field Crops			
Barley, grain[d] (*Hordeum vulgare* L.)	12	16	18
Sugarbeet[e] (*Beta vulgaris* L.)	10	13	16
Cotton (*Gossypium hirsutum* L.)	10	12	16
Safflower (*Carthamus tinctorius* L.)	8	11	12
Wheat[d] (*Triticum aestivum* L.)	7	10	14
Sorghum (*Sorghum vulgare* Pers.)	6	9	12
Soybean [*Glycine max* (L.) Merr.]	5.5	7	9
Sesbania[d] (*Sesbania macrocarpa* Muhl.)	4	5.5	9
Sugarcane (*Saccharum officinarum* (L.)	3	5	8.5
Rice, paddy[d] (*Oryza sativa* L.)	5	6	8
Corn (*Zea mays* L.)	5	6	7
Broadbean (*Vicia faba* L.)	3.5	4.5	6.5
Flax (*Linum usitatissimum* L.)	3	4.5	6.5
Field bean (*Phaseolus vulgaris* L.)	1.5	2	3.5

Table 1 (Continued)

	$EC_e \times 10^3$ (millimhos/cm at 25°C) at which yield decreased by[b]		
Crop	10%	25%	50%

Vegetable Crops

Crop	10%	25%	50%
Beets[e] (*Beta vulgaris* L.)	8	10	12
Spinach (*Spinacia oleracea* (L.)	5.5	7	8
Tomato (*Lycopersicon esculentum* Mill.)	4	6.5	8
Broccoli (*Brassica oleracea* var. italica L.)	4	6	8
Cabbage (*Brassica oleracea* var. capitata L.)	2.5	4	7
Potato (*Solanum tuberosum* L.)	2.5	4	6
Sweet corn (*Zea mays* L.)	2.5	4	6
Sweet potato [*Ipomoea batatas* (L.) Lam.]	2.5	3.5	6
Lettuce (*Lactuca sativa* L.)	2	3	5
Bell pepper (*Capsicum annuum* L.)	2	3	5
Onion (*Allium cepa* L.)	2	3.5	4
Carrot (*Daucus carota* L.)	1.5	2.5	4
Greenbean (*Phaseolus vulgaris* L.)	1.5	2	3.5

[a]Reproduced from Bernstein (1974), by courtesy of American Society of Agronomy.
[b]In gypsiferous soils, EC_e's causing equivalent yield reductions will be about 2 mmho/cm greater than the listed values.
[c]Average for different varieties. Suwannee and Coastal Bermuda grasses are about 20% more tolerant and Common and Greenfield are about 20% less tolerant than the average. For most crops, varietal differences are relatively insignificant.
[d]Less tolerant during seedling stage. Salinity at this stage should not exceed 4 or 5 mmho/cm, EC_e.
[e]Sensitive during germination. Salinity should not exceed 3 mmho/cm during germination.

ecotypes have recently been shown to vary in salt tolerance. Tideland ecotypes transport less Cl^- and Na^+ to the shoots and are more salt-tolerant than upland ecotypes (Hannon and Barber, 1972).

When salinity affects plant nutrition, major varietal differences in salt tolerance may occur, associated with varietal differences in nutrition. Thus, some varieties of lettuce are more susceptible than others to internal browning, a calcium-deficiency disease that may be caused by high sulfate concentrations. Varietal differences in cation nutrition affect the salt tolerance of carrot varieties (Bernstein, 1964a). Foliar sprays are effective in correcting blossom-end rot of tomatoes and bell peppers and blackheart of celery, when these calcium-deficiency diseases have been caused or aggravated by salinity (Geraldson, 1957).

Salt-sensitive species produce maximum yields over only a very narrow range of soil salinities. Most vegetable crops, for example, decrease in yield by 10% at EC_e's of only 1.5 to 2.5 mmho/cm. Tolerant species such as cotton, barley, and sugar beets maintain maximum yields over a broad range of salinities, generally equal to two-thirds of the salinity that causes a 10% reduction in yield. Since salinity of the soil water will not be less than that of the irrigation water, the range of irrigation water salinity that permits maximum yield is restricted for sensitive crops to about 1 mmho/cm but may be as high as 6 to 8 mmho/cm for the most tolerant crops (see Table 6, below). Of course, when poor soil conditions prevent adequate leaching, or when irrigation is poorly managed, much lower irrigation water salinities can eventually cause harmful salt accumulation in the soil.

With increases in salinity, growth of most crops, like yields, is progressively reduced. Salinities that stunt growth decrease the yields of most sensitive and moderately tolerant crops. This is not true, however, for some highly tolerant crops. The vegetative growth of barley, wheat, and cotton may be drastically reduced by salinity without any effect on grain or fiber yields, respectively. Corn, however, being only moderately tolerant, shows equivalent effects on vegetative (stover) and grain yields (Bernstein, 1964b).

Salt Tolerance During Germination and Other Growth Stages

It was formerly thought that crops were more salt-sensitive during germination than at later growth stages. This belief was based on the frequent occurrence of germination failures in the field, even when plants that did germinate appeared to make normal growth. Comparison of responses during germination and later growth stages indicates comparable salt tolerances for most crops studied (Bernstein and Hayward, 1958). A notable exception are beet plants, which are much more sensitive during germination than at later stages. With most crops, germination failures are caused by the accumulation of high concentrations of salts near the soil surface as a result of salt transport by water moving up to the seed from furrows or by concentration of salts near the surface by evaporation of water in basin or flood-irrigated plantings. As a result, seed may be exposed to salinities several to many times greater than the average salinity of the root zone or even the plow layer (Bernstein et al., 1955; Bernstein and Fireman, 1957). Modified planting beds, such as the sloping seed bed, prevent salt accumulation around the seed.

Some crops are especially sensitive to salinity during the seedling stage. These include the grains—barley, wheat, and rice—and *Sesbania macrocarpa*. Tolerable salinity during the seedling stages of these crops is limited to about 4 mmho/cm (EC_e), despite the greater tolerances at later growth stages (Table 1).

Salt Tolerance of Fruit Crops

Fruit crops have not been studied as thoroughly as other crops with respect to salt tolerance. This is due partly to the greater complexity of their reaction to salinity and partly to the greater difficulty of studying tree crops that require years to mature and to show the cumulative effects of salinity. It is well established that fruit crops are generally specifically sensitive to chloride and sodium. When the leaves of most woody plants (trees, vines, etc.) accumulate 0.25 to 0.50% Na or 0.5 to 1.0% Cl on a dry-weight basis, they develop characteristic leaf injuries, mostly marginal or tip burns. Consequently, fruit crop response will be conditioned by the specific composition of soil salts (especially Cl^- and Na^+ concentrations) as well as by total salinity. Large differences in uptake of Cl^- by different varieties or rootstocks of a given crop further complicate the picture. Varieties or rootstocks that absorb less Cl^- can tolerate higher Cl^- concentrations in the root media. Despite the lack of complete information on fruit crop response to salinity, general limits may be stated for tolerable total salinity (Table 2) and for Cl^- tolerance of different varieties and rootstocks of important fruit crops (Table 3). Most fruit crops are salt-sensitive even when toxic accumulations of Cl^- do not occur, and most

Table 2 Salt Tolerance of Fruit Crops[a]

Crop	$EC_e \times 10^3$ (millimhos/cm) at which yield may decrease by about 10%[b]
Date (*Phoenix dactylifera* L.)	8
Pomegranate (*Punica granatum* L.), fig (*Ficus carica* L.), olive (*Olea europaea* L.)	4-6 (estimated)
Grape (*Vitis* sp.)	4
Muskmelon (*Cucumis melo* L.)	3.5
Grapefruit (*Citrus paradisi* Macfad.)	3.5
Orange (*Citrus sinensis* (L.) Osbeck)	3
Lemon [*Citrus limon* (L.) Burm. f.]	2.5
Apple, pear, apricot, plums, prunes, almond, peach (*Malus, Pyrus* and *Prunus* spp.)	2.5
Blackberry, boysenberry (*Rubus* spp.)	2.5
Avocado (*Persea americana* Mill.)	2
Red raspberry (*Rubus idaeus* L.)	1.5
Strawberry (*Fragaria* sp.)	1.5

[a]Reproduced from Bernstein (1974), by courtesy of American Society of Agronomy.
[b]These data are applicable when rootstocks are employed that do not accumulate Na or Cl rapidly, or when these ions do not predominate in the substrate.

Table 3 Chloride Tolerances in the Saturation Extract of Soil for Fruit Crop Rootstocks and Varieties if Leaf Injury Is to Be Avoided[a]

Crop	Rootstock or variety	Maximum permissible Cl in saturation extract (meq/liter)
	Rootstocks	
Citrus (*Citrus* spp.)	Rangpur lime, Cleopatra mandarin	25
	Rough lemon, tangelo, sour orange	15
	Sweet orange, citrange	10
Stone fruit (*Prunus* spp.)	Marianna	25
	Lovell, Shalil	10
	Yunnan	7
Avocado (*Persea*	West Indian	8
americana Mill.)	Mexican	5
	Varieties	
Grape (*Vitis* spp.)	Thompson Seedless, Perlette	25
	Cardinal, Black Rose	10
Berries[b]	Boysenberry	10
(*Rubus* spp.)	Olallie blackberry	10
	Indian Summer raspberry	5
Strawberry	Lassen	8
(*Fragaria* spp.)	Shasta	5

[a]Reproduced from Bernstein (1974), by courtesy of American Society of Agronomy.
[b]Data available for single variety of each crop only.

(citrus, stone fruits, avocado) are highly sensitive to Na^+ as well, accumulating damaging foliar concentrations when exchangeable sodium percentages are as low as 5%, or when soluble Na^+ concentrations exceed 7 to 10 meq/liter in the soil saturation extract. Some fruit crops (grapes, strawberries) do not transport Na^+ to their leaves and are not injured by even higher concentrations of Na^+ in the root medium (Bernstein, 1974).

Sprinkler irrigation of fruit crops is often hazardous because leaves may absorb Cl^- and Na^+ directly and accumulate these ions much more rapidly than these ions are transported from the roots. Consequently, irrigation waters that are not injurious when applied to the soil only may be highly injurious when they wet the leaves. As little as 2 to 3 meq Cl^- or Na^+ per liter in such cases may cause toxic accumulations of these ions in leaves of citrus and stone fruit trees. The leaves of some species, such as avocado and strawberry, are relatively impermeable and do not absorb harmful quantities of salts foliarly (Bernstein, 1965, 1975).

Although annual (nonwoody) crop plants are usually not injured by sprinkling with most irrigation waters, they may be damaged by sprinkling with saline waters containing 10 to 20 meq or more per liter of Na^+ or Cl^- (Bernstein and Francois, 1975). Even highly salt-tolerant crops such as cotton may be damaged by sprinkling with brackish waters (Bernstein, 1975). Again, not all species are susceptible since some, such as sugar cane (Bernstein, 1975), are not damaged by sprinkling with saline waters.

Other Toxicities

Since Na^+ and Cl^- are major saline constituents, toxic reactions to these ions occur usually on saline soils, except as noted above for foliarly absorbed salts. Elements that are toxic at very low concentrations may occur independently of soil salinity. Boron, an essential nutrient required in the root medium in concentrations of 0.2 to 0.5 mg/liter, becomes toxic to sensitive species at concentrations of only 1 mg/liter. Lithium is toxic to citrus at the very low concentration of 0.1 mg/liter, and selenium is toxic to sensitive species at concentrations above 0.2 mg/liter (Bernstein, 1974). These and other substances not normally found in toxic amounts in irrigation waters are important in evaluating irrigation water quality, with regard to toxic effects on animals consuming the plants as well as direct toxic effects on plants.

Tolerance to Exchangeable Sodium

Sodic soils (containing more than 10 to 20% exchangeable sodium, depending on soil texture) constitute a special and major class of salt-affected soils, especially with respect to their reclamation (U.S. Salinity Laboratory Staff, 1954). When nonsaline, sodic soils inhibit plant growth because of unfavorable physical conditions (poor soil structure) and potential deficiencies of calcium and magnesium. Plant species vary widely in their tolerance to low levels of Ca and Mg, but all species studied except paddy (flooded rice) are affected by poor soil structure. In saline soils, both the structural and nutritional effects of sodicity are lessened (Bernstein, 1964a, 1975). High soluble salt concentrations improve soil structure by promoting flocculation and tend to provide nutritionally adequate levels of Ca and Mg even though the percentages of exchangeable Ca and Mg are low. Crops specifically sensitive to Na toxicity are affected at low levels of sodicity and cannot tolerate sodic soils, whether nonsaline or saline.

Factors That Modify the Salt Tolerance of Plants

Plant nutrition, climate, and irrigation method and management may affect the salt tolerance of specific crops. As noted previously, even varietal nutritional differences may modify response to certain saline conditions. The examples cited included sulfate-induced calcium deficiency in some lettuce varieties, salinity-induced calcium deficiencies of tomato, bell peppers, and celery, and cationic imbalance in carrot varieties. Studies on the effects of variable soil fertility levels (N, P, and K) on salt tolerance have yielded conflicting results. Ravikovitch and Porath (1967) and Ravikovitch and Yoles (1971) reported increased salt tolerance of several crops when N or P was increased to levels greater than those optimum under nonsaline conditions. Bernstein et al. (1974) found, however, that the effects of salinity and fertility tend to be additive, especially when the inhibitory influence of each is relatively moderate (about 25% reduction in yield by either factor). According to these findings, given *relative* yield decreases (percent decrease in yield relative to nonsaline controls) tend to occur at the same salinity for both optimally fertilized treatments and partially deficient ones. High levels of fertility did not increase salt tolerance. Extreme nutrient deficiency or salinity tends to control yield so that changes in the other factor are ineffective.

Climatic factors may modify plant response to both toxic and osmotic effects of salinity. Toxic reactions (leaf burns) occur under hot, dry conditions and may not appear during cool, humid weather even though leaves contain toxic concentrations of Cl^- or Na^+ (Bernstein, 1974). Salt-sensitive species such as onions and beans are also affected more by salinity under dry, hot conditions even though osmotic, rather than toxic, effects predominate. The tolerance of salt-tolerant species such as beets is affected much less by climatic factors (Bernstein, 1974).

Some effects of irrigation methods on salinity responses have already been described in considering the effects of furrow irrigation on salt distribution and germination, and the effects of sprinkling on foliar salt absorption and toxicity. The important influence of irrigation frequency on salt tolerance will now be considered.

Irrigation Frequency and Salt Tolerance

Wadleigh and co-workers some 30 years ago studied the interactive effects of irrigation frequency and soil salinity on the yields of beans (Wadleigh and Ayers, 1945) and guayule (Wadleigh et al., 1946). Soil cultures were adjusted to salinity levels that covered the salt tolerance range of each species. Irrigation frequency treatments at each salinity level were also wide-range, with matric potentials from about -1 to -15 bars. In the nonsaline treatments, matric potential was the main component of total water potential; in the saline, frequently irrigated treatments, the main component was osmotic potential, and in other treatments, both matric and osmotic components contributed significantly to total water potential. For both species, yield was related to the mean integrated total water stress, or total water potential, regardless of the relative importance of the matric and osmotic components. The effects on yields of matric and osmotic potentials were thus equivalent and additive. When a saline treatment was infrequently irrigated, it developed not only low matric potentials prior to irrigation but also lower osmotic potentials because of the decrease in soil water and the resultant concentration of the residual soil solution. It was thus clearly demonstrated that yields on a saline soil were reduced more by a given water deficit than would be the case for the same soil if nonsaline. It was concluded that saline soils should be irrigated at a higher residual water content than comparable nonsaline soils. To some degree, this recommendation tends to be put into practice automatically when salinity varies within a field because water use decreases with increasing salinity and decreased plant growth. Thus the saline parts of a field tend to be irrigated at higher residual water contents than the nonsaline parts, which, by reason of earlier water depletion, indicate the need for irrigation. In a uniformly saline field, however, the irrigator does not have the nonsaline plants as a guide for irrigation and must irrigate when all of the field is still at relatively high matric potentials. The crop serves as a poor guide in such cases, since wilting does not, as a rule, occur even when total water potential is low if matric potentials are still high. It is a rapidly decreasing water potential that causes wilting, and in saline soils, potentials tend to be consistently low and to change more gradually with time because of lower rates of water utilization. Only when there are very abrupt and large increases in salinity will most crops wilt because of low osmotic potential. There is a very real hazard, therefore, of overextending the irrigation cycle on saline soils and of causing further decreases in yield because of lower osmotic potentials.

Irrigations for most crops are scheduled to minimize water stress so as to achieve maximum yields. (Water stress may be desirable for some crops, especially to promote maturation.) The salt tolerance data in Table 1 were in fact obtained on plots irrigated to maintain water tensions in the tensiometer range (tensions of 75 cbar or less). Tensiometer depth varied from as little as 15 cm to 60 cm or more, depending on the rooting depth of the crop, making possible the wide range of irrigation frequencies recommended for the different crops studied.

Having recognized the importance of keeping saline soils from becoming as dry as may be permissible for comparable nonsaline soils, we should now inquire to what extent the inhibitory effects of salinity can be overcome by irrigating at a greater frequency than is optimum for the crop on nonsaline soils. Recent attention to daily irrigation by drip (trickle) systems has demonstrated the feasibility of daily irrigation of row and orchard crops [For literature see Bernstein and Francois (1973a, 1975)].

To examine the possibilities of reducing yield losses by frequent irrigation of saline soils, let us consider the effects of high-frequency irrigation of a moderately tolerant crop. Salt tolerance data for broccoli (Table 1) will be used, but the results of the analysis are generally applicable. The EC_e's associated with 10, 25, and 50% decreases in yield with a conventional irrigation frequency provide the starting point for our analysis. In all salt tolerance studies, water tensions just prior to irrigation decrease progressively with increasing soil salinity in the range that affects growth. We shall assume constant water use efficiency so that water use is proportional to growth and yield. [This is valid for moderate yield depressions, up to 25% for alfalfa. With higher salinities and greater yield depressions, water use efficiency may decrease markedly (Bernstein and Francois, 1973b)]. In the salt tolerance experiments, it is safe to assume strict salt conservation since only highly soluble salts were employed. The EC of the soil water immediately after irrigation is calculated by taking the ratio of water volume at saturation:water volume after irrigation as 2. The salinity of the soil water prior to irrigation is calculated from the ratio of water volumes just after and just prior to irrigation with the assumption of strict salt conservation. To determine the mean integrated salinity ($\overline{EC}_{sw}$) over the irrigation cycle in plots in which the given fractional water depletions have occurred, we use the relationship (Bernstein and Francois, 1973b):

$$\overline{EC}_{sw} = \frac{EC_i v_i}{v_i - v_f} \ln \frac{v_i}{v_f}$$

EC_i is the EC of the soil water immediately after irrigation ($2EC_e$), v is the volumetric soil water content, and i and f refer to levels immediately after and prior

Table 4 Initial, Final, and Integrated Mean Salinities of Soil Water in Salinized Plots on Pachappa Sandy Loam Soil at Which 0 to 50% Yield Decreases Would Occur for a Moderately Salt-Tolerant Crop

| Relative yields (% of control) | Maximum tension (cbar) | Fractional water depletion, $\frac{v_i - v_f}{v_i}$ | EC_{sw}, immediately After irrigation | | Before irrigation | Mean integrated EC_{sw} |
			EC_e (mmho/cm)	($2EC_e$) (mmho/cm)	(mmho/cm)	(mmho/cm)
100	50	0.50	0.6	1.2	2.4	1.7
90	40	0.45	4.0	8.0	14.5	10.6
75	32	0.375	6.0	12.0	19.2	15.1
50	25	0.25	8.0	16.0	21.3	18.3

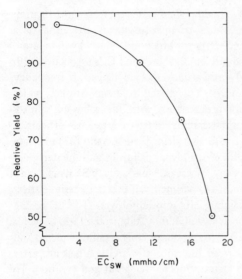

Fig. 1 Relation of relative yield of a moderately salt tolerant crop to mean integrated soil water salinities.

to irrigation, respectively. We now plot yields against the mean integrated soil water salinities ($\overline{EC_{sw}}$), as given in Table 4, for the four salinity levels (Fig. 1), instead of against EC_e as is customary. To determine the effect of maintaining EC_{sw} at the minimum possible level by high-frequency irrigation, we read the yields for the minimum EC_{sw}'s from the curve of Fig. 1. Comparing these values with the yield decrements for the conventional irrigation regime (irrigation when the control plot is at a tension of 50 cbar), we find that maintaining minimum salinity by high-frequency irrigation causes progressively greater increases in yield with increasing soil salinity despite the progressively smaller degrees of water depletion (Table 5). The decreases in EC by high-frequency irrigation tend to be similar for all saline treatments. The progressively smaller *proportional* increase in EC, because of lower fractional water depletion, is balanced by the higher EC_i's. The greater effect on yield at higher salinities is thus seen to depend on the progressive steepening of the yield-salinity curve (Fig. 1). A given change in EC_{sw} has a greater effect at higher than at lower salinity.

These calculated benefits of reducing mean integrated soil water salinity by high-frequency irrigation agree well with experimental findings in bell peppers (Bernstein and Francois, 1973a). Furrow irrigation with a saline water at a conventional frequency (controlled by tensiometer at 15 cm depth in the nonsaline plot) reduced yield by about 18%, but daily drip irrigation with this water decreased yield by 14% [cf. also Bernstein and Francois (1975)]. *Partial*

Table 5 Comparison of Yields for Conventional Irrigation and for High-Frequency Irrigation[a]

EC_e	Frequency	Maximum EC_{sw}	Relative Yield %	Yields for high frequency as percent of conventional frequency
0.6	High	1.2	100	100
	Conventional	2.4	100	
4.0	High	8.0	95	106
	Conventional	14.5	90	
6.0	High	12.0	86	114
	Conventional	19.2	75	
8.0	High	16.0	70	140
	Conventional	21.3	50	

[a]See Table 4 for EC_{sw} data and text for method of estimating yields from Fig. 1 for high-frequency irrigation.

restoration of salinity-caused yield losses by high-frequency irrigation of the magnitude shown in Table 5 may therefore be expected. Yield restoration is limited because the salinity of the irrigation water is, itself, inhibitory, and the salinity of the soil water does not increase greatly over the conventional irrigation cycle.

In nonsaline soil, the yields of bell pepper were not measurably affected by high-frequency irrigation, indicating that matric potentials were maintained above the level that affects bell peppers, even with conventional irrigation frequencies (Bernstein and Francois, 1973a). Disregarding the matric component of water potential in relating yield to salinity only (Fig. 1) is therefore justified. The yields of some crops may be improved by moderate stress, especially during maturation (Wadleigh et al., 1946). For such crops, the maintenance of high water potential may actually decrease yield.

Sprinkling with brackish water may cause severe leaf injury, and early reports on drip irrigation with saline water appeared to exaggerate the benefits by contrast to the nearly total loss of yield with sprinkling. Compared to surface irrigation at conventional frequencies, drip irrigation does decrease yield reduction by salinity but only by about one-half or less (Table 5). The major benefits of drip irrigation are related to factors other than salinity. Significant water savings, especially for annual crops, may be realized (Bernstein and Francois, 1973a), and, most important, the method has the *potential* of assuring uniform water supplies for all plants in a field population to perhaps a greater degree than other irrigation methods. This is of particular importance in connection with recent findings that leaching requirements may be quite small for most crops and irrigation waters, and that leaching may be deferred for most or all of a crop season (Bernstein and Francois, 1973b). When water applications are designed to meet only immediate crop requirements, the tolerance to nonuniformities in water application *or* infiltration becomes very low. Low rates of application that do not exceed infiltration anywhere in the field become essential to assure uniform infiltration. When application rates are low, high application frequencies may be required to meet the water requirements of crops. Although drip irrigation may reduce the inhibitory effects of saline irrigation waters by providing a steady supply of water at the minimum salinity of the irrigation water (not concentrated by evapotranspiration during extended periods between irrigations), the method of application does leave zones of salt accumulation midway between drip orifices and in the periphery of the wetted zone that may necessitate some leaching before the next crop can be started (Bernstein and Francois, 1973a). Even high salinities of 50 to 100 mmho/cm (EC_e) in the surface soil at the wetted periphery do not inhibit growth as long as the salt remains there. Rain during the cropping period may leach the salts from the surface soil into underlying root zones, causing injury inless the salts are promptly leached out by irrigation.

Effects of Depth-Variable Soil Salinities

The salt tolerance data describe crop response in essentially uniformly saline soils. Often, however, soil salinity in a given root zone may vary 10-fold, or more. It was formerly thought that plants respond to the average salinity of the root zone, and that the occurrence of high salinity anywhere in the active root zone is harmful. Recently, however, a salinity of 8 mmho/cm (EC_e) that causes a 50% decrease in yield of alfalfa when present throughout the root zone was shown to have no effect when restricted to the lower part of the root zone (Bernstein and Francois, 1973b). Twice this salinity in the lower root zone decreased yield by about 15%. In contrast, an increase of only 1 mmho/cm in the EC of the irrigation water and therefore in the soil water of the upper root zone affected alfalfa yields as much as about a 20 mmho/cm increase in the soil water of the lower root zone. These results are consistent with the greater water uptake from the upper root zone where salinity is low compared to that from the highly saline lower root zone. In a later study (Bernstein et al., 1975), the effects of salinity increments in the upper and lower root zones were confirmed. Obviously, tolerance limits for soil salinity need to be specified separately for the upper (minimal salinity) root zone and lower or maximal salinity zone. The former is usually governed by the salinity of the irrigation water and the latter by the leaching fraction (LF), which determines the salinity of the drainage water, neglecting soil mineral solubilization (Rhoades et al., 1968) and salt

Table 6 Crop Tolerances to Salinity of Irrigation Water (EC_{iw}) and to Drainage Water Salinity (EC_{dw})

	EC_{iw}[a] (mmho/cm)	EC_{dw}[b] (mmho/cm)
Tolerant crops: Bermuda grass, tall wheatgrass, barley, sugar beets, cotton	6.5 – 8.5	40 – 45
Moderately tolerant crops: alfalfa, soybeans, rice, tomato	2 – 3.5	32 – 35
Sensitive crops: clovers, beans, onions, carrots	1 – 1.3	14 – 16

[a]Limits for full yields when EC_e in upper root zone does not exceed EC_{iw}.
[b]Yields of 85 to 100% of maximum are obtained when these EC_{dw} values are used to calculate LR.

precipitation (Bower et al., 1968). Since $LF = v_{dw}/v_{iw}$ (U.S. Salinity Laboratory Staff, 1954) and with strict salt conservation, $v_{dw}/v_{iw} = EC_{iw}/EC_{dw}$,

$$LF = \frac{EC_{iw}}{EC_{dw}} \qquad (2)$$

where dw and iw refer to drainage and irrigation water, respectively, and v is the volume of water. The LF becomes the leaching requirement (LR) when the maximum permissible EC_{dw} for a crop is specified and used to calculate the minimum LF or LR for a given irrigation water by Eq. (2). Tolerances to upper root zone and lower root zone salinities therefore indicate tolerable irrigation water and drainage water salinities, respectively (Table 6). Tolerable salinities of irrigation waters specify only their *potential* for crop production. Inadequate irrigation can cause salt to accumulate in the upper root zone so that minimum soil salinities in the profile greatly exceed EC_{iw}. Irrigation deficiencies may occur in parts of a field that are less permeable or that are underirrigated because of poor land leveling, improper slope, or poor water distribution.

The EC_{dw} values in Table 6 refer to the salinity of the soil water at the bottom of the root zone. Corresponding EC_e values may range from $1/2\ EC_{dw}$ to EC_{dw}, depending on drainage and the proximity of the water table to the bottom of the root zone.

Ion uptake is also affected differently by concentrations in the upper and lower root zones. The uptake of Cl^- and Na^+ by alfalfa was sensitive to small changes in the concentrations of these ions in the irrigation water and therefore in the upper root zone, but quite insensitive to much larger changes of concentrations in the lower root zone (Bernstein and Francois, 1973b). Thus, the already quite complex reactions of crops to salinity in uniform root media become even more complex when the variable salinity profiles encountered in the field are considered. With better understanding of the controlling salinity parameters and with improved irrigation methods, it should be possible to realize the full potentialities of irrigation waters for crop production with maximum irrigation efficiency.

References

Abel, G. H. (1969). Crop Sci. *9*: 697.
Abel, G. H., and MacKenzie, A. J. (1964). Crop Sci. *4*: 157.
Ayers, A. D., and Hayward, H. E. (1948). Soil Sci. Soc. Amer. Proc. *13*: 224.
Bernstein, L. (1964a). Plant Anal. and Fertilizer Problems *4*: 25.
Bernstein, L. (1964b). U.S. Dept. Agr. Inform. Bull. *283*.
Bernstein, L. (1965). U.S. Dept. Agr. Inform. Bull. *292*.
Bernstein, L. (1974). In J. van Schilfgaarde (ed.), *Drainage for Agriculture*, Agronomy *17*: 39.
Bernstein, L. (1975). Ann. Rev. Plant Pathol. *13*: 295.
Bernstein, L., and Fireman, M. (1957). Soil Sci. *83*: 249.
Bernstein, L., and Francois, L. E. (1973a). Soil Sci. *115:* 73.
Bernstein, L., and Francois, L. E. (1973b). Soil Sci. Soc. Amer. Proc. *37*: 931.
Bernstein, L., and Francois, L. E. (1975). Argon. J. *67*: 185.
Bernstein, L., Francois, L. E., and Clark, R. A. (1974). Argon. J. *66*: 412.
Bernstein, L., Francois, L. E., and Clark, R. A. (1975). Soil Sci. Soc. Amer. Proc. *39*: 112.
Bernstein, L., and Hayward, H. E. (1958). Ann. Rev. Plant Physiol. *9*: 25.
Bernstein, L., MacKenzie, A. J., and Krantz, B. A. (1955). Soil Sci. Soc. Amer. Proc: *19*: 240.
Bower, C. A., Ogata, G., and Tucker, J. M. (1968). Soil Sci. *106*: 29.
Geraldson, C. M. (1957). Proc. Amer. Soc. Horticultural Sci. *69*: 309.
Hannon, N. J. and Barber, H. N. (1972). Search *3*: 259.
Ravikovitch, S., and Porath, A. (1967). Plant and Soil *26*: 49.
Ravikovitch, S., and Yoles, D. (1971). Plant and Soil *35*: 555.
Rhoades, J. D., Kruger, D. B., and Reed, M. J. (1968). Soil Sci. Soc. Amer. Proc. *32*: 643.
U.S. Salinity Laboratory Staff (1954). *U.S. Dept. Agr. Handbook No. 60*, U.S. Govt. Printing Office, Washington, D.C.
Wadleigh, C. H., and Ayers, A. D. (1945). Plant Physiol. *20*: 106.
Wadleigh, C. H., Gauch, H. G., and Magistad, O. C. (1946). U.S. Dept. Agr. Tech. Bull. *925*.

4

Irrigation and Soil Salinity

Eshel Bresler

Hebrew University of Jerusalem, Rehovot, Israel

Introduction

Soil salinity is an important factor of the environment in which plants grow. In agriculture, soil salinity is defined in terms of accumulation of salts in the root zone to an extent that may cause a certain crop yield reduction. Salinity problems are known to exist in many soils throughout the world, particularly in arid and semiarid regions, where irrigation is necessary for successful agriculture.

All irrigation waters contain hundreds and, in extreme cases, thousands of parts per million (ppm) of dissolved salts, as compared with 10 ppm in rainwater. The salts most commonly present in irrigation waters are chloride (Cl^-), sulfate (SO_4^{2-}), and bicarbonate (HCO_3^-) of calcium (CA^{2+}), sodium (Na^+), and magnesium (Mg^{2+}). In irrigated lands, the concentration and composition of the soil solution are derived from the salinity of the irrigation water. The salt concentration of the soil solution is always greater than or equal to the concentration of the irrigation waters. The solution concentration is increased as water evaporates into the air and plants extract water from the soil for growth, both of which processes concentrate the salt in the soil solution. Thus, even with relatively good-quality water, the permanent irrigation practice causes the irrigated soil to be affected by excess soluble salts and to be subjected to the deleterious effect of adsorbed sodium on the physical properties of the soil, as a result of the increase in the soil solution concentration. Moreover, world prospects indicate that the quality of the irrigation water tends to deteriorate, so that it becomes necessary to utilize water of even poorer quality for irrigation. Removal of excess salts from the soil and reclaiming saline and sodic soils by proper irrigation management is therefore necessary for attaining optimum crop production and preventing transformation of highly productive soils into nonproductive salt-affected soils.

Control of saline and sodic soils by irrigation, with or without chemical amendments, is therefore essential to the operation of a permanent successful irrigated agriculture. It involves reclamation of saline soils already affected by improper irrigation, and maintenance or improvement of nonsaline or reclaimed irrigated soils. A combination of crop yield-salinity data and an understanding of soil chemistry and physics with applications of their principles lead to a management irrigation practice that can result in economic optimization of water quality-quantity in irrigated lands. Thus, applications of existing theories and knowledge on salinity-irrigation relationships can play a major role in advancing water resources use for irrigation while meeting the challenge of minimizing deterioration of soils and degradation in groundwater quality.

The discussion here is an analysis of the chemical and physical aspects of soils that are relevant to water management for controlling soil salinity by irrigation. We discuss first the salt status of the soil solution that is in equilibrium with the adsorbed phase of the soil, and then the dynamics of the soil solution under irrigation conditions. Estimation procedures for leaching and accumulation of salts in irrigated soils and effects of the irrigation method on the control of soil salinity are also discussed. The study is concerned with the physicochemical aspects of the control problem. No attempt is made to discuss economic approaches to soil salinity and its control.

Salts in Soil Solution

Experimental data indicate that plants are strongly affected by the salt concentration of the soil solution within the root zone (e.g., Wadleigh et al., 1951; Hayward, 1954; Bernstein, 1965; Bierhuizen, 1969; Shalhevet et al., 1969). Salt-affected soils are therefore those soils that contain excessive concentrations of soluble salts, and the extent to which soil salinity has to be controlled is dictated by the response of agricultural plants to salt concentration and to the composition of the soil solution. Soluble salts also produce harmful effects on soils when the soil solution composition in contact with the solid surface results in a high percentage of sodium ions in the exchange complex of the soil. The exchangeable sodium present in high-sodium soils may have a negative effect on the physical and chemical properties of the soil. The type of exchangeable ions and the concentration of the soil solution have very marked effects on the macroscopic properties of the soil, because of their effects on the electrical phenomena at the solid-solution interface.

Ionic Distribution in the Vicinity of Charged Solid Surfaces

The solid part of the soil system is composed of various sizes of small solid particles, the specific surface areas of which become larger as the particles become smaller. Thus the particles of the soil clay fraction may have a specific surface area as large as 800 m^2 per gram of montmorillonite clay, which is typical for arid and semiarid soils.

The soil clay particles have a plate shape. Their relative dimensions (100 to 1,000 times larger and wider than they are thick) and general form when not compressed resemble a crumpled piece of tissue paper. The crystal of the clay particles is made up of aluminosilicates, with a certain degree of substitution of other metallic ions for the trivalent aluminum and tetravalent silicon ions. Substitution of trivalent aluminum for the tetravalent silicon and of bivalent magnesium or ferrous ions for trivalent aluminum result in a decrease in positive charge or an increase in electrons (negative charge) in the clay crystal lattice. The phenomenon responsible for the origin of charge in this manner is called isomorphous substitution. The charge due to this isomorphous substitution is considered to be distributed evenly in the surface of the clay particle.

The negative particle charge is neutralized by accumulation of an equivalent amount of cations in the solution surrounding the charged particle. This amount is known as the cation exchange capacity (CEC) of the soil and is expressed as milliequivalent (meq) exchangeable cations per 100 g of dry soil. Many physical properties of soils are sensitive to the relative composition of the exchangeable cations, which in turn depend on concentration and composition of the soil solution in equilibrium with the charged solid surface. The divalent cations are in favor of the physical properties such as fluid conduction, aggregate stability, and aeration. Even amounts as small as 20% or less of monovalent cations in the equilibrium soil solution or in the exchange complex are sufficient to deteriorate the soil and to interfere with the growth of most crop plants. However, high soil-solution concentration prevents the unfavorable physical effects of the adsorbed monovalent sodium ions.

The counter ions (cations) that compensate for the negative charges on the particle form a diffuse layer. Two forces acting on ions in the diffuse layer cause opposite tendencies. The electrostatic field pulls the cations toward the negatively charged solid surface and repulses the anions from it. At the same time thermal motion tends to cause equal distribution of the ions in the solution phase according to Fick's first law of diffusion. The result of these two opposing forces (Fig. 1) is the distribution of the ions in such a way that the concentration of the counter ions (cations) is very high close to the solid surface and gradually decreases as the distance from the surface increases. Conversely, the concentration of the coions (anions) is the lowest at the solid surface, increases gradually with distance, and reaches its maximum midway

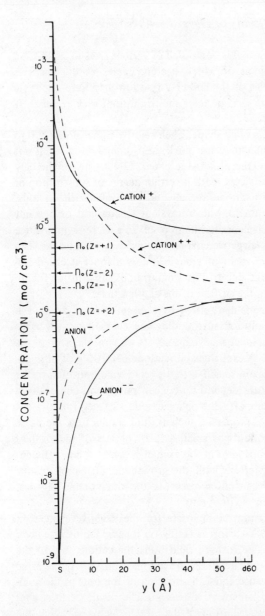

Fig. 1 Ion distribution between two negatively charged soil particles 100 Å apart in a mixed ionic system with two non-symmetrical electrolytes (see also Bresler, 1970). Note that n_0 denotes ion concentration in the equilibrium solution (moles/cm^3), z its valency, and $s \leq y \leq d$ is the coordinate of distance between the particles where s stands for the particle "surface" and d is the midplane distance between the particles.

between two parallel clay platelets. The effect of these forces on the configuration of the diffuse layer is demonstrated in Fig. 1, where a mixed system with two nonsymmetrical salts is represented. One salt has a monovalent anion ($z = -1$), a bivalent cation ($z = {^+}2$), and an equilibrium solution concentration of 0.001 M; the other one has a bivalent anion ($z = -2$), a monovalent cation ($z = +1$), and a concentration of 0.003 M. The mathematical description leading to the data of Fig. 1 is given in detail elsewhere (Bresler, 1970; Overbeek, 1952).

The electrostatic attraction of the charged surface for the divalent cations (or repulsion for the divalent anion) is double the attraction for the monovalent cation (or repulsion for the monovalent anion). However, the electrolyte concentration in the bulk (external equilibrium solution is larger for the monovalent cation ($n_0 = 6 \times 10^{-6}$ mole/cm^3, $z = +1$, Fig. 1) than for the divalent cation ($n_0 = 10^{-6}$, $z = +2$, Fig. 1). Thus the tendency for the monovalent cation to diffuse away from the vicinity of the surface to the bulk solution is less than the diffusion tendency for the divalent cation. The diffusion tendency to the bulk solution diminishes as the concentration of the equilibrium solution (n_0) increases. The net result of equilibrating the electrostatic attraction of cations (and repulsion of anions) with diffusion away from the surface to the bulk solution for cations, and from bulk solution to surface for anion, is that the diffuse layer of the divalent cation is more compressed (and that of the divalent anion is less compressed) than the diffuse layer of the monovalent cation. This tendency is somewhat compensated by the higher bulk solution concentration of the monovalent cation, which tends to compress the diffuse cations atmosphere when the equilibrium solution concentration increases.

Phenomena Affected by Solution Concentration and Composition

Swelling and Hydraulic Conductivity

When two particles of the clay fraction of the soil come close to each other, their diffuse-ion atmospheres overlap and reduce the free energy of water in the midplain between the clay platelets, relative to the equilibrium soil solution. If the clay particles are confined, the water molecule diffusing into the space between the clay particle causes hydraulic pressure between the particles. This pressure builds up until the increase in free energy of water due to the swelling pressure is equal to the decrease in free energy of the water caused by the greater concentration of ions between the clay particles as compared with the external solution. This swelling pressure, which results from the repulsive force between the particles, decreases with an increase in salt concentration and an increase in the valency of the cations in the equilibrium soil solution (Warkentin et al., 1957).

As mentioned before, Ca and Na ions are most commonly contained in soil and irrigation water, with Ca being the dominant ion. It is therefore important to consider a mixed sodium-calcium system. The swelling retention curves for mixtures of mono- and divalent ions can be calculated on the basis of two models: (1) The adsorbed ions are randomly distributed in the exchange soil phase (Bresler, 1972); and (2) there is "demixing" of the two ions, whereby calcium is concentrated on internal surfaces and sodium ions concentrate on the external surfaces of packets of clays called tactoids (Shainberg et al., 1971). Both models predict that when the exchangeable sodium percentage (ESP) is greater than about 20%, the amount of solution retained by the soil because of swelling is essentially the same. This result is important when one considers the effect of ion composition and concentration on the hydraulic conductivity of saturated and unsaturated soil.

The most severe deleterious effects of the sodium ions on the physical properties of field soils are observed by the changes in the hydraulic conductivity of the soil [Eq. 14]. The hydraulic conductivity is a function of size of the water-filled pores, so that any solution composition that causes a decrease in the size of these water conduction pores may have a marked effect on the hydraulic conductivity of the soil. The size of the water conduction pores is decreased by swelling of clay particles and by dispersion of the colloidal soil material. It may be deduced from the diffuse layer theory (Bresler, 1972) that both swelling and particle dispersion increase as the concentration of salts in the soil decreases, and the sodium-to-calcium ratio of the soluion increases. Experimental results (Fig. 2) confirmed that the saturated, as well as the unsaturated, hydraulic conductivity behaves accordingly, that is, higher hydraulic conductivity in more concentrated soil solution, or lower Na/Ca ratio than in dilute solution, and a high sodium-to-calcium ratio. The decrease in hydraulic conductivity with decreasing soil solution concentration and increasing Na/Ca ratio was pronounced for soils with a high percentage of montmorillonite-type clay (McNeal and Coleman, 1966). Soils high in kaolinite clay and sesquioxides are insensitive to concentration and composition of the soil solution. The most sensitive soils are therefore those that contain clay minerals capable of swelling and dispersion.

Negative Adsorption of Anions (Anion Exclusion)

Whereas cations are attracted to the negatively charged particle surface by an electrostatic force, anions are repelled from the surface by the same field force. A concentration gradient is established, with low concentration of anions near the surface and the diffusion of anions opposite in direction to the movement due to electrostatic repulsion, creating the distribution of anions shown in Fig. 1. The reduction in concentration of anions near negatively charged solid

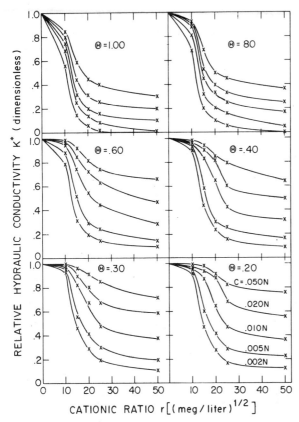

Fig. 2 Relative hydraulic conductivity (K^*) as a function of equivalent $Na/\sqrt{Ca}$ ratio (r) for five different equilibrium soil solution concentrations and six different effective degrees of water saturation (Θ). Note that C is the concentration of the equilibrium soil solution, K^* is given by the ratio $K(\theta, C, r)$ to $K(\theta)$ of Ca saturated soil, that is, $K^* = K(\theta, r, C)/K(\theta, 0, C)$; r is defined by the equivalent ration $r = Na/\sqrt{Ca}$, and the quantity Θ is defined by $\Theta = (\theta - \theta)/(\theta_s - \theta)$. (See also Russo, 1976.)

surfaces is known as *negative adsorption* or *anion exclusion*. Negative adsorption may be calculated from the anion concentration near the particle (Fig. 1), and the equilibrium solution concentration. Negative adsorption, in equivalents per square centimeter, for an anion i (with a valence z) can be calculated by

$$\Gamma^- = z_{ia} \int_{y=s}^{y=d} (n_{iao} - n_{ai}) \, dy \qquad (1)$$

where the subscript a indicates an anion, s and d are the particle surface and the midplane, respectively, and n is concentration in mols per cubic centimeter. The integral in Eq. (1) represents the area under the curves bounded by $s \leq y \leq d$ and n_{iao} in Fig. 1, and can be evaluated numerically for the general case (Bresler, 1970) or analytically (de Hann, 1965) for many specific cases. Effects of negative adsorption on water and salt flow are discussed in Chap. 3.

Exchange Equilibria Between Mono- and Divalent Cations

Cation exchange equations describe the distribution of cations between the adsorbed soil phase and the soil solution phase. The preference of the clay particles for one of two cations at equilibrium can be described by the mass action law, which gives the apparent selectivity coefficient as (Shainberg, 1976)

$$K = \frac{(\bar{x}_{Na})^2 (C_{Ca})}{(\bar{x}_{Ca})(C_{Na})^2} \tag{2}$$

where $\bar{x}$ is the equivalent fraction of the adsorbed ion, C is the concentration in the solution phase in moles per liter, and K is the apparent constant estimated to be 0.25 for typical Israeli soils. Using this value for K, Shainberg (1976) shows that this is identical to the expression suggested by the U.S. Salinity Laboratory Staff (1954), that is,

$$\frac{\bar{x}_{Na}}{\bar{x}_{Ca}} = 0.0158 \frac{(Na)}{\sqrt{(Ca) + (Mg)}} \tag{3}$$

where the parentheses denote the concentration of the cations in the solution in millimoles per liter. This equation states that the cation of higher valency is preferentially adsorbed by the clay fraction of the soil and that the preference increases with the dilution of the soil solution. This effect is also observed if one calculates exchange equilibria between mono- and divalent ions by integrating the concentration of each ion as distributed according to the diffuse layer theory (Bresler, 1972). The ratio of the monovalent ions adsorbed $\bar{x}_{Na}$ to the total cation exchange capacity (CEC) of the soil is given similar to Erikson (1952) as (Bresler, 1972)

$$\frac{\bar{x}_{Na}}{CEC} = \frac{R}{\Gamma_s / \sqrt{C} \sqrt{\alpha(R+1)}} \sinh^{-1} \left[\frac{\Gamma_s}{\sqrt{C}} \frac{\sqrt{\alpha(R+1)}}{R + 4\cosh(e\psi_d/kT)} \right] \tag{4}$$

where Γ_s is the surface charge density of the soil (esu/cm^2), C is the solution concentration (moles/liter), R is the ratio between molar concentration of mono- and divalent cations in the solution (i.e., $C_{Na^+}/C_{Ca^{2+}}$), k is Boltzmann's constant, e is the charge of an electron (esu), T is the absolute temperature, ψ_d is the electrical potential at the midplane (y = d in Fig. 1), a function of cation exchange capacity, specific surface area of the soil, soil water content (θ), C, and R, and the constant α is approximately 1.3×10^{-8} at room temperature. The value of ψ_d for any mixed sodium-calcium soil system can be obtained from Figs. 1 to 5 of Bresler (1972). The increase in the affinity of the clay for the divalent ion and the increase in this preference with dilution of the soil solution are explained by Eq. (4), which is based entirely on the diffuse layer theory.

Dynamics of Water and Salts in Irrigated Soils

Transport of Solutes in Soils

Water-soluble salts entering the soil profile through the process of irrigation may accumulate in the root zone or may be leached out of this zone, depending on the convective diffusive transport processes and solute interactions within the soil. Understanding these processes is important in establishing management practices directed toward preventing the hazardous effect of salt on the plant and the soil. Theories based on macroscopic considerations provide a satisfactory means for describing such transport phenomena. In these theories, the soil is considered to be a continuous porous medium, and the governing equations are derived for a representative elementary volume that is sufficiently large for its properties to be expressed in terms of statistical averages.

In general, the macroscopic rate at which a given solute moves in the soil system depends on the average flow pattern, on the rate of molecular diffusion, and on the ability of the porous material to spread the solute as a result of local microscopic deviations from the average flow. For proper modeling and understanding of the manner of solute transport in a natural soil profile, these phenomena must be considered simultaneously. In dealing with the problem of simultaneous transfer of solute and water, one usually assumes that the transport of the solute is governed by convection (viscous movement of the soil solution) and diffusion (thermal motion within the soil solution).

Transport by Diffusion Without Solution Flow

In this case the solute transport is by molecular diffusion in response to a concentration gradient of a particular ion in the soil solution. Diffusion of a solute in a uniform medium of water is described macroscopically by Fick's fist law,

$$J_d = -D_o \frac{dc}{dx} \qquad (5)$$

where J_d is the amount (grams, moles, etc.) of solute diffusing across a unit area (cm^2) per unit time (sec), D_o is the diffusion coefficient of the solute in water (cm^2/sec). c is the solute concentration (g/cm^3 of solution, or $mole/cm^3$), and x is the distance (cm) along the direction of net movement of the solute.

Since the soil constitutes a complex charged porous medium, the effective diffusion coefficient (D_p) for any given ion is less than the equivalent coefficient (D_o) in a pure water system. The relationship between D_p and D_o may be expressed as

$$D_p = \theta \left(\frac{L}{L_e}\right)^2 \alpha \gamma D_o \qquad (6)$$

where θ is the volumetric water content of the soil (cm^3 water/cm^3 bulk soil). When a solute moves by diffusion in the water phase of the soil, the cross-sectional area available for flow is only a fraction, θ, of the macroscopic cross-sectional area. Therefore, θ represents the reduction in the area available for flow in a soil, as compared with the area available in a body of water. L is the macroscopic average path of diffusion, whereas L_e is the actual tortuous path along which a particle moves. Since the tortuous path is longer, dx must be multiplied by L_e/L. Furthermore, since the actual path is likely to form an angle with the macroscopic direction, the macroscopic cross-sectional area is decreased by L/L_e. The resulting tortuosity factor, arising from the porous nature of the soil, is therefore $(L/L_e)^2$; α accounts for reduction in water viscosity due to the presence of charge particles in the soil matrix, and γ takes into account the retarding effect of anion exclusion on flow in the vicinity of negatively charged soil particles.

The factors θ, $(L/L_e)^2$, α, and γ are less than unity, and therefore the effective diffusion coefficient (D_p) is less than the equivalent diffusion coefficient (D_o) in a free water solution. However, in a given unsaturated soil, reducing the soil water content results in a smaller cross-sectional area available for diffusion, a longer flow path, and more significant values of water viscosity and negative adsorption. Thus, one can express D_p in terms of D_o and θ. It has been shown (see Olsen and Kemper, 1968) that, in a clay-water system, the solute diffusion coefficient is a positive exponential function of water content (θ) and, for pratical purposes, is independent of salt concentration. This simply means that

$$D_p(\theta) = D_o a e^{b\theta} \qquad (7)$$

where a and b are empirical constants. Olsen and Kemper (1968) stated that data collected on soils fit Eq. (7) reasonably well with b = 10 and a ranging from 0.005 to 0.001, depending on the surface area of the soil studied (sandy loam to clay). This simplified $D_p(\theta)$ relationship was found to apply to water contents corresponding to a range of suctions between 0.30 and 15 bar, which is enough for most cases where diffusion may be important. Thus, taking into account the effect of water content in the soil on ion diffusion, Fick's modified first law can be written as

$$J_d = -D_o \theta \left(\frac{L}{L_e}\right)^2 \alpha\gamma \frac{dc}{dx} = -D_p(\theta) \frac{dc}{dx} \qquad (8)$$

where c is the solute concentration in the soil solution, and $D_p(\theta)$ may be calculated from Eq. (7). The diffusion process described by Eq. (8) may be important in some specific situations (Bresler, 1973a). However, the magnitude of D_p is such that fluxes of salts are usually very small when there is no significant movement of the bulk solution.

Transport of Solute by Convection

In general, molecular diffusion takes place together with convective transport, and each process contributes to the final dispersion of the solute. It is generally assumed that macroscopic transport by convection must take into account the average flow velocity as well as the mechanical, or hydrodynamic, dispersion. Solutes are transported by convection at the average velocity of the solution, and in addition they are dispersed about the mean position of the front. Mathematically the mechanical dispersion may be treated as a diffusion process if one replaces D_p with the mechanical dispersion coefficient D_h. Many experimental and theoretical works (Perkins and Johnston, 1963; Ogata, 1970; Passioura et al., 1970) have shown that the magnitude of the mechanical dispersion coefficient D_h in a given porous material depends on the average flow velocity. It has been shown that, under saturated and steady-flow conditions, D_h may often be taken to be proportional to the first power of the average flow velocity (e.g., Ogata, 1970),

$$D_h(V) = \lambda |V| \qquad (9)$$

where V, in centimeters per second, is the average interstitial flow velocity and λ, in centimeters, is an experimental constant depending on the characteristics of the porous medium. We assume here that similar relationships will also hold under our unsaturated transient conditions.

Assuming that average and local velocities are additive, the total amount of solute transported across a unit area, in the direction of flow (x) by convention, is obtained from

$$J_h = -D_h(V)\frac{dc}{dx} + V\theta c \tag{10}$$

where J_h represents the total amount of solute moving by convection across a macroscopic area, in unit time (g, or mole, cm^{-2} sec^{-1}). The first term on the right-hand side represents solute flow due to dispersion and the second term is the flow due to the average flow velocity.

Combined Diffusion and Convective Flows

The joint effect of diffusion and convection is derived by combining their mathematical expressions (8) and (10) to obtain

$$\begin{aligned} J &= -[D_h(V) + D_p(\theta)]\frac{\partial c}{\partial x} + V\theta c \\ &= -D(V, \theta)\frac{\partial c}{\partial x} + qc \end{aligned} \tag{11}$$

where
- J = total flux of solute
- c = solute concentration of the soil solution
- x = flow direction coordinate
- D = combined diffusion-dispersion coefficient (cm^2/sec)
- q = volumetric specific flux of solution (cm^3/cm^2 sec).

Although Eq. (11) is an approximation, it is useful in practice for prediction purposes.

In agricultural soils, changes in water content due to infiltration, redistribution, evaporation, and transpiration bring about the simultaneous movement of water and salt. Steady-state flow conditions occur very seldom under natural coditions. A mathematical expression for one-dimensional transient conditions is derived from a consideration of continuity, or mass conservation. This states that the rate of change of solute within a given soil element must be equal to the difference between the amounts of solute that enter and leave the element.

By equating the difference between outflow and inflow to the amount of salt that has accumulated in the soil element, and considering the case of one-dimensional vertical flow, one obtains the expression

$$\frac{\partial}{\partial t}(Q + \theta c) = \frac{\partial}{\partial z}\left[D(V, \theta)\frac{\partial c}{\partial z}\right] - \frac{\partial(qc)}{\partial z} + S' \tag{12}$$

where
- Q = local concentration (positive or negative) of solute in the "adsorbed" phase (meq/cm^3 soil), usually depends on both θ and c
- c = concentration in the solution phase (meq/cm^3 soil solution)
- S' = any sink or source rate term due to salt uptake, production, precipitation, or dissolution
- t = time
- z = vertical space coordinate (considered to be positive downard)

Equation (12) is the mathematical statement of trasient vertical diffusive-convective solute transport under the conditions described. It applies when the solute does not interact or interact chemically with the soil and when there is also gross loss or gain of salt inside the flow system. Mathematical solutions of Eq. (12) may be developed for problems involving salinity and fertility under natural soil conditions, when c represents either the total salt concentration or a spcific ion concentration, with or without dissolution, adsorption, or exclusion.

Nielsen and Biggar (1961) worked out a system with a steady-state horizontal flow of water. Under these conditions V and θ are constants and Eq. (12), ignoring Q and S', becomes

$$\frac{\partial c}{\partial t} = D\frac{\partial^2 c}{\partial z^2} - V\frac{\partial c}{\partial z} \tag{13}$$

where $D = (D_h + D_p)/\theta$ is the hydrodynamic dispersion coefficient obtained by fitting the breakthrough data to an analytical solution of (13) using the appropriate boundary conditions (Nielsen and Biggar, 1962). Biggar and Nielsen (1967) discussed the limitations of this analytical approach. They concluded, on the basis of their data and boundary conditions, that for purposes of predictions, "for some range of velocities, simple porous material and reasonable magnitude of error some satisfaction may be obtained" by the method described.

Soil Salinity and Water Flow

Macroscopic water flow in unsaturated soil under isothermal and isosainity conditions is described quite well by Darcy's equation, adapted to unsaturated soil,

$$q = V\theta = -K(\theta)\frac{dH}{dx} \tag{14}$$

where q is the volumetric specific water flux (cm^3 H_2O/cm^2 sec^2), $K(\theta)$ is the hydraulic conductivity of the soil (a function depending on θ alone), and H is the hydraulic head, which is the sum of the suction or pressure head (h) and gravity head. In this Darcy-type water flow equation it is assumed that the hydraulic gradient is the only driving force that causes water to flow. However, the dynamic changes of salt concentration, because of mass movement of salt and water content fluctuations, may create an additional driving force due to osmotic gradients. Also, variation in salt concentration and composition may affect the hydraulic conductivity function, $K(\theta)$, because of density, electrical changes, particle swelling, and dispersion. Thus, in applying Eqs. (12), (14), and (19) to a given salinity and irrigation situation, the mutual salt-water flow effects must be considered.

Effect of Osmotic Pressure Gradients

Kemper and Evans (1963) have shown that when a solute is completely restricted by the soil, water flow in response to an osmotic pressure gradient is the same as when a hydraulic pressure gradient is applied, provided that they are of equal magnitude. For this case of complete restriction, the hydraulic conductivity coefficient K can be used for both hydraulic and osmotic potential gradients. When the salt is only partially restricted by the soil, Darcy's law in its usual form does not apply, and the specific flux q (assuming uniform temperature) must now be expressed as

$$q = -K(\theta, C, R)\left[\frac{dH}{dz} - \frac{\sigma(\theta, c)}{\rho g}\left(\frac{d\pi}{dz}\right)\right] \tag{15}$$

Here σ is the osmotic efficiency coefficient (Kemper and Rollins, 1966) expressed as a function of c and θ, K is the hydraulic conductivity as a function of C, θ, and R, and π is the osmotic pressure. The osmotic pressure may be estimated by using van's Hoff's law,

$$\pi = \phi R'TC \tag{16}$$

where ϕ is the osmotic coefficient of the electrolyte, R' is the universal gas constant, T is the absolute temperature, and C is the sum of molar concentrations of

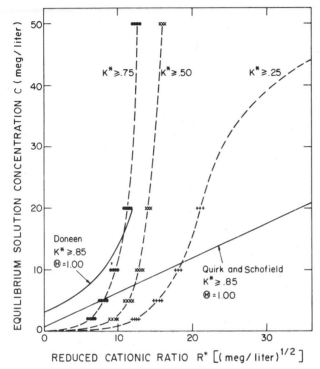

Fig. 3 The combination of soil solution concentration (C), cationic ratio (r), and effective degree of water saturation (Θ) that causes a given reduction in the unsaturated hydraulic conductivity. Reductions in K* of 15, 25, 50, and 75% are given by the lines labeled K* ⩾ 0.85, K* ⩾ 0.75, K* ⩾ 0.50, and K* ⩾ 0.25, respectively. Note that R* is defined by the relationship R* = rΘK* and that r, K*, and Θ are the same as in Fig. 2. (See also Russo, 1976.)

all cations and anions in the equilibrium solution. Note that a nonunique relationship between K and θ has been assumed in Eq. (15). This is a reasonably good assumption as long as the total electrolyte concentration is not greater than that given by Fig. 3 for any Na/Ca relationship.

Kemper and Rollins (1966) were able to express the osmotic efficiency coefficient in terms of the distribution of adsorbed ions and the thickness of a single thin film. We can use a similar approach in estimating the values of $\sigma(\theta, c)$.

A macroscopic estimate of the film thickness can be obtained from unsaturated hydraulic conductivity data, $K(\theta)$, using the formula

$$b(\theta) = \left[\frac{3\eta K(\theta)}{\alpha\gamma(L/L_e)^2 \theta \rho g} \right]^{1/2} \tag{17}$$

which is based on Poiseuille's law (Kamper, 1961). Here, 2b represents the approximate average effective thickness of the films, η and ρ are the average dynamic viscosity and density, respectively, g is acceleration due to gravity, θ is the volumetric water content, $K(\theta)$ is the hydraulic conductivity of the soil, and $\alpha\gamma(L/L_e)^2$ represents soil-water interactions and tortuousity as before. The last mentioned can be evaluated approximately for various water contents using values obtained by interpolation and extrapolation of the data published by Kemper and van Schaik (1966) [Eqs. (6) and (7)], or Porter et al. (1960).

If we denote $c'(y)$ as the anion concentration at any plane y (Fig. 1), then $\sigma(\theta, c)$ in Eq. (15) is obtained similar to Kemper and Rollins (1966) from

$$\sigma(\theta, c) = \frac{U}{U_{ex}} = \frac{\int_0^b \left\{ \int_0^y dy \int_y^b [1 - c'(y)/c] \, dy \right\} dy}{\int_0^b (2by - y^2) \, dy} \tag{18}$$

where U_{ex} and U are the average flow velocities of the solution across each film for the complete and partial salt exclusion, espectively. The term $[1 - c'(y)/c]$ is the salt exclusion factor at y, and can be obtained as a function of $b(\theta)$ and c from the diffuse layer theory (Fig. 1). If the salt is completely excluded [all $c'(y) = 0$], then the system becomes a perfect osmotic membrane ($\sigma = 1$). If, on the other hand, no exclusion takes place [$c'(y) = c$], then $\sigma = 0$ and Darcy's classical equation (14) holds in Eq. (15). Usually, however, $0 \leq \sigma \leq 1$. To calculate $\sigma(\theta, c)$ from Eq. (18), one must first estimate the salt exclusion factor and the effective thickness of the film, 2b. These are obtained from the diffuse double-layer theory (Fig. 1) and Eq. (17), respectively.

Effects of Soil Hydraulic Conductivity

As mentioned before, swelling of soils containing clay and dispersion of the soil colloidal material alter the geometry of the soil pores and thus affect the intrinsic permeability of the soil. It was deduced from the double-layer theory that both swelling and particle dispersion increase as soil solution concentration and Ca/Na ratio decrease. The hydraulic conductivity is affected not only by the intrinsic permeability, but also by the properties of the soil solution, such as fluid density and viscosity, which are also affected by the composition and concentration of solutes. Investigations have confirmed that the hydraulic conductivity behaves accordingly, that is, higher in concentrated solutions or high Ca/Na ratios and lower in dilute solutions or low Ca/Na ratio (Fig. 2). However, it is possible to find for any given $r = Na^+/(Ca^{2+})^{1/2}$ ratio a threshold value of soil solution concentration required to prevent an appreciable decrease in soil permeability. Such data are given in Fig. 3 and have been summarized by Russo (1976). In Fig. 3; Θ is defined by $\Theta = (\theta - \theta_o) / (\theta_s - \theta_o)$ where θ_o is the volumetric water content of an air-dry soil and θ_s is the value of θ at saturation. K^* is the value of the unsaturated hydraulic conductivity relative to its value in calcium saturated soil as defined by $K^* = K(\theta, C, r)/K(\theta \mid Ca\text{-soil})$ and R^* is a reduced cationic ratio defined by $R^* = r\theta K^*$. Note that the data of Quirk and Schofield (1955) and of Doneen (see McNeal and Coleman, 1966) are for saturated soils, whereas those of Russo were obtained for unsaturated conditions in Gilat loam soil.

Effects of salt concentration and composition on the soil hydraulic conductivities are clearly demonstrated in Fig. 3. An appreciable decrease in soil hydraulic conductivity may be expected in soils with combinations of R^* and C that are below the pertinent curves as given in Fig. 3. For a combination of solution concentration and composition in which values of C are larger and values of R^* are smaller than the values given by these pertinent lines, a unique relationship between K and θ can be assumed independent of the sodium-to-calcium ratio and of the concentration of the flowing soil solution. In other cases the $K(\theta)$ function must depend on both C and r. Because high electrolyte concentration affects soil permeability in sodic soils, it is suggested that a reclamation of sodic soils should start with high-salt-concentration water to increase permeability and facilitate leaching (Reeve and Bower, 1960).

Governing Water Flow Equations

The one-dimensional vertical flow of water in an incompressible unsaturated soil with water uptake by plant roots can be described by

$$\frac{\partial \theta}{\partial t} = -\frac{\partial q}{\partial z} - S(z, t) \qquad (19)$$

where θ is the volumetric water content of the soil (cm^3 H$_2$O/cm^3 soil), t is the time (sec), z is the vertical distance from reference level (cm), q is the specific water flux in the soil (cm^3 H$_2$O/cm^2 soil sec) as given by Eq. (15), and S is the water uptake by roots (cm^3 H^{2}O/cm^3 soil) considered positive from the soil into the roots. A major difficulty in solving Eq. (19) for θ (z, t) stems from the unknown form of function S(z, t).

As a result of the lack of physical understanding of flow within the plant itself, there has been a tendency to describe water uptake by macroscopic equations which assume that the rate of uptake is proportional to soil hydraulic conductivity K and to the difference between h_r and $h + \pi$. According to this approach the sink term entering into Eq. (19) can be expressed as

$$S(z) = \frac{\Delta q(z)}{\Delta z} = \frac{-K(\theta)[h_r(z) - \pi(z) - h(z)]}{b(z)} \tag{20}$$

where S is the volumetric rate of water uptake per unit bulk volume of the soil, π is the osmotic pressure of the soil solution as given by Eq. (16), and h_r is the total (osmotic + matric) pressure head at the root-soil interface. The coefficient of proportionality 1/b is allowed varying degrees of freedom by various authors, and owing to our ignorance of the physics involved it must presently be considered merely as an empirical entity. It follows from theoretical considerations (e.g., Gardner, 1960) that 1/b represents the geometry of the flow, and if one is willing to assume analogy to Darcy's law, then one might reason that 1/b is proportional to the specific area of the root-soil interface (total surface area of roots per unit bulk volume of soil), and inversely proportional to the impedance (thickness divided by the hydraulic conductivity) of the soil-root interface. This, however, must presently be viewed as a purely speculative conjunĉture. Notwithstanding, the physical meaning of b(z) indicates that it must have the units of square centimeters. Information about the function b(z) is very scarce because direct experimental determination of the factors involved is extremely difficult. Feddes (1971) tried to overcome this problem by deriving b(z, t) per layers of 1-cm thickness from direct field measurements of vertical flow q under the assumption of a quasi-steady state, together with the assumption that the rate of flow from the roots through the plant to the atmosphere is linearly proportional to the corresponding pressure differences between these media. Nimah and Hanks (1973a, 1973b) do not determine b(z, t) but determine a root distribution function $RDF(z)/(\Delta x \Delta z)$ as the weight fraction of the roots in the depth interval Δz relative to the total weight of the roots, and Δx is arbitrarily set equal to unity. Values of b(z) for various field crops growing on different soils are given by Feddes et al. (1974).

Boundary Conditions

Equations (12) and (19) with (15) and (20) must be supplemented by appropriate initial and boundary conditions. When hysteresis is considered and h is not a single-valued function of θ, the initial conditions are simply

$$h(z, 0) = h_n(z) \quad \text{or} \quad \theta(z, 0) = \theta_n(z) \qquad 0 \leqslant z \leqslant Z \quad t = 0 \tag{21a}$$

$$c(z, 0) = c_n(z) \qquad\qquad\qquad\qquad 0 \leqslant z \leqslant Z \quad t = 0 \tag{22a}$$

where c_n, h_n, and θ_n are predetermined initial conditions, prescribed functions of z, and Z is the lower geometric boundary, that is, the groundwater table depth or an arbitrary depth always below the wetting front. The boundary conditions at Z at any time $t \geqslant 0$ are therefore

$$\frac{\partial h}{\partial z} = 0 \quad \text{or} \quad h = 0 \qquad z = Z \quad t \geqslant 0 \tag{21b}$$

$$\frac{\partial c}{\partial z} = 0 \quad \text{or} \quad c = c_{gw} \qquad z = Z \quad t \geqslant 0 \tag{22b}$$

where c_{gw} is constant or a prescribed function of t.

Along soil-air interfaces (in the absence of ponding), the soil can lose water to the atmosphere by evaporation or gain water by infiltration. Whereas the potential (i.e., maximum possible) rate of evaporation from a given soil depends on atmospheric conditions only, the actual flux across the soil surface is limited by the ability of the porous medium to transmit water from below. Similarly, if the potential rate of infiltration (e.g., the rain or irrigation intensity) exceeds the infiltration capacity of the soil, some of the water may be lost by surface runoff. Here, again, the potential rate of infiltration is controlled by atmospheric or other external conditions, whereas the actual flux depends on antecedent moisture conditions in the soil. Thus, the exact boundary conditions to be assigned for water flow at the soil surface (z = 0) cannot be predicted a priori, and a solution must be sought by maximizing the absolute value of the water flux (while maintaining the correct sign) subject to the requirements (see Hanks et al., 1969)

$$|q(0,t)| \leqslant |R(t)| \quad \theta_d \leqslant \theta(0,t) \leqslant \theta_s \quad h_d \leqslant h(0,t) \leqslant h_o \quad z=0 \quad t>0 \tag{21c}$$

For the solute, $c(z, t)$, one has the following boundary conditions that must be satisfied at the soil surface:

$$J(0, t) = -[D_p[\theta(0, t)] + D_h[V(0, t)]]\frac{\partial c}{\partial z} + q(0, t)c(0, t) \tag{22c}$$

$z = 0 \quad t > 0$

Here [Eqs. (21c), (22c)], $q(0, t)$ is the specific flux of water at the soil surface as given by Eq. (15); $h(0, t)$ is the pressure head at the surface; h_d and θ_d are the pressure and water content, respectively, in an air-dry soil under given evaporation conditions; h_o and θ_s are the highest pressure and soil water content, (usually having values of zero and saturations), respectively; $J(0, t)$ is the solute flux at the inlet; D_p is the diffusion coefficient in the soil solution, D_h is the mechanical dispersion coefficient, and $R(t)$ is the prescribed potential surface water flux at the soil surface, which may vary with time. It should be noted that

$R(t) > 0$ and $h(0, t) \leqslant h_o$, or $\theta(0, t) \leqslant \theta_s$ during the irrigation period (infiltration), $R(t) = 0$ during drainage or redistribution, $R(t) < 0$ and $h(0, t) \geqslant h_d$, or $\theta(0, t) \geqslant \theta_d$ during evaporation and the period of water extraction by plant roots. Note also that $J(0, t) = q(0, t)C_o(t)$ during infiltration, $J(0, t) = 0$, and $q(o, t) = 0$ during redistribution or drainage and therefore $(\partial c/xz)_{z=0} = 0$; and $J(0, t) = 0$ during evaporation or the water extraction period, where C_o is the solute concentration at the inlet of the irrigation water.

Methods for calculating $R(t)$ and $h_o(t)$ during the evapotranspiration period on the basis of atmospheric data are presented by Feddes et al. (1974). They suggest that the total potential evapotranspiration from both soil and crop, E^*, be calculated according to

$$E^* = \frac{\delta(R_n - G) + \rho_a c_p (e_z^* - e_z)/r_a}{(\delta + \gamma)L} \qquad (23)$$

where δ is slope of saturation vapor pressure curve, R_n is net radiation flux, G is heat flux into soil, ρ_a is density of moist air, c_p is specific heat of air at constant pressure, e_z^* is saturated vapor pressure, and e_z is unsaturated vapor pressure at elevation z and ambient temperature, r_a is resistance to vapor diffusion through air layer around leaves (for table of values see Feddes, 1971), γ is psychometric constant, and L is latent heat of water vaporization. The potential evaporation $R(t)$ is calculated from a simplified form of Eq. (23) by neglecting the aerodynamic term and taking into account only that fraction of R_n which reaches the soil surface (see Ritchie, 1972):

$$R(t) = \frac{\delta}{(\delta + \gamma)L]\,R_n} \exp[-0.39(LAI)] \,. \qquad (24)$$

where LAI is the leaf-area index.

The value of h_d in Eq. (21c) can be determined by equilibrium conditions between soil water and atmospheric vapor. Feddes et al. (1974) suggest using the formula

$$h_d = \left(\frac{R'T}{Mg}\right) \ln(f) \qquad (25)$$

where R' is universal gas constant, T is absolute temperature, M is molecular weight of water, g is acceleration due to gravity, and f is relative humidity of air.

Irrigation and Soil Salinity

The potential rate of transpiration by plants, E_{pl}^*, is assumed to be equal to the maximum possible rate of water extraction by roots per unit horizontal area of the soil and is also dependent on atmospheric conditions. This quantity can be calculated direction from Eqs. (23) and (24) according to

$$E_{pl}^* = E^* - R(t) \tag{26}$$

The quantity E_{pl}^* is required to calculate the rate of transpiration (E_{pl}) and water uptake by plants (Nimah and Hanks, 1973; Neuman et al., 1975). For this the purpose is to maximize the value of E_{pl} subject to the requirement that it does not exceed the potential rate of transpiration [Eq. (26)], that is,

$$|E_{pl}| \leqslant |E_{pl}^*| \tag{27}$$

In addition, the pressure head in the roots should not be allowed to drop below the wilting point, h_w (which is usually taken as $-15{,}000$ cm of water), that is,

$$h_r \geqslant h_w \tag{28}$$

This constrained maximization process is accomplished by iterative procedure.

Estimation of Leaching, Accumulation, and Distribution of Salts in Irrigated Soils

Estimating the distribution of salt in depth and time is important in many problems involving irrigated agriculture. Among these are problems concerning soil salinity and alkalinity, placement and leaching of plant nutrients, soil pollution, and so on. Estimation procedures to be described here are applicable to most salinity control situations under irrigation conditions, when the effect of vegetation is considered or being neglected. The description of the procedures includes estimation of total salt concentration when neither salt precipitation nor uptake by plants take place [S' of Eq. (12) is dropped]. The procedure may also be useful to estimate a specific ion distribution when its exclusion or adsorption is taken into consideration or being ignored.

The problem is defined by the nonlinear partial differential equations (12) and (15) with (19) and (20), with the boundary conditions (21) through (28) and, at present, can be solved by numerical methods only with the aid of a computer. Solution to many applied specific cases may be obtained (e.g., Bresler and Hanks, 1969; Hanks, et al., 1969; Bresler, 1973a, 1973b; Nimah and Hanks, 1973; Neuman et al., 1975; Bresler, 1975). These methods of solution resulted in a numerical technique that is simple, efficient, unconditioned stable, and second-order accurate.

Fallow (No Vegetation) Isothermal Conditions

Noninteracting Solute

In this case the solute is assumed to be inert with respect to its environment, so that interaction with the soil matrix is ignored and therefore the effect of salt fluctuation on water flow, adsorption, exclusion, and gravity segregation effects are neglected. From a salinity and irrigation point of view this estimation is restricted to the case of total salt concentration when neither salt precipitation nor dissolution takes place. This may also be a good approximation for anions (such as chlorides) under a wide range of natural conditions where precipitation of carbonate and sulfate minerals in the soil (Rhoades et al., 1973; Oster and Rhoades, 1974); which may decrease the salt burden of drainage water, are irrelevant.

Under these conditions the governing equation for transient vertical diffusive convective flow of an inert solute under isothermal conditions is Eq. (12) with zero values for Q and S'. Similarly, the differential equation that governs vertical solution (water) flow in this soil is Eq. (19) plus Eq. (15) with both S and σ having a value of zero and $K(\theta, c, R) = K(\theta)$ a unique function of θ. As mentioned before, numerical methods must be used to solve these differential equations subjected to the pertinent boundary conditions (21) and (22). A finite difference solution is given by Hanks et al. (1969) to calculate the values of θ (z, t), h(z, t,), and q(z, t) for the processes of infiltration, redistribution, drainage, and evaporation. The calculated values can then be used to obtain c(z, t) from Eq. (12). To solve Eq. (12), the values of $D_h(z, t)$ and $D_p(z, t)$ must be known. Since $\theta(z, t)$ is now known, the value of the diffusion coefficient $D_p(z, t)$ may be calculated from Eq. (7) by adapting the values of a and b from Olsen and Kemper (1968) or from Porter et al. (1960). Since q(z, t) is also known, one calculates $V(z, t) = q(z, t)/\theta(z, t)$ and from it determines the mechanical dispersion coefficient D_h by using $D_h(V)$ in Eq. (9). The coefficient λ in (9) may be approximated with the aid of a best-fit breakthrough curve to the solution of Eq. (13) for a constant value of V and θ (Passioura et al., 1970) or from theoretical considerations (Perkins and Johnston, 1963).

The finite difference solution of Eq. (12) may cause considerable smearing of the concentration profile, a phenomenon known as *numerical dispersion.* This numerical dispersion may limit the applicability of the solution. Its value usually depends on the flow velocities as well as on the space and time increments. Following Lantz (1971), Bresler (1973a) used an approach similar to Chaudhari (1971) that minimizes the effect of numerical dispersion and eliminates almost completely the numerical smearing, and the solution is efficient and highly accurate.

Any estimation technique must be compared with experimental results

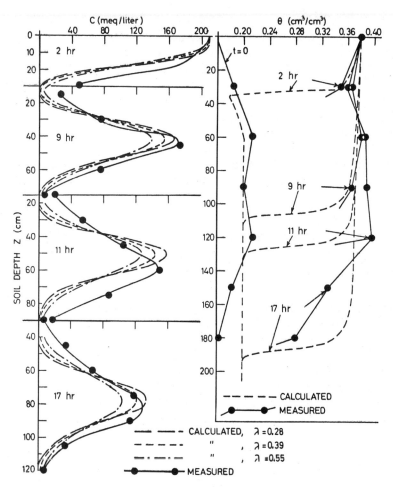

Fig. 4 Field distribution of chloride and water (Warrick et al., 1971, Figs. 5, 6, and 8) (solid line connecting the black circles), compared with estimated profiles (dashed lines) for four infiltration times and three values of λ. (See also Bresler, 1973b.)

before one can gain confidence in its reliability. The field infiltration experiment reported by Warrick et al. (1971) provides a good opportunity for such a comparison in salt concentration and water content profiles. The comparison between the calculated results and the experimental data of Warrick et al. for 2, 9, 11, and 17 hr of infiltration is presented in Fig. 4. Note that the origin ($z = 0$) on the left-hand side of Fig. 4 is shifted and the data are accordingly translated along the z axis to distinguish among the different sets of times. Also note that there is a small discrepancy in the total salt balance between the measured and the calculated data. Since the calculated salt balance in the system appeared to be correct, the small discrepancy may result from minor deviations from the assumptions. However, good agreement generally exists between the calculated and measured salt concentration and water content profiles, especially if one considers the nonhomogeneity of the soil profile with depth and also the variability and the precision involved in the measurements of water and salt in soils. Figure 4 also shows that a narrow range of λ of 0.28 to 0.55 properly describes the field data and the proper value of λ is generally not affected by infiltration time or position in the soil profile.

The good agreement between the calculated and measured water and salt profiles suggests that macroscopic-scale theoretical approaches are generally satisfactory for estimation purposes.

Solute Interacting with Soil

To take into account the physicochemical interactions between the ions and the soil matrix, the effects on the flow of osmotic gradients and of anion exclusion or cation adsorption must be considered. Here the estimation procedure must use the governing equation (12) with zero value for S' only and Eq. (15) with zero S values. It is clear that in evluating the value of $Q = Q(c, \theta)$ in Eq. (12), a distinction must be made between anions and cations.

Cations. For cations the value of $Q(\theta, c)$ is always positive ($Q > 0$) and can be estimated from Eqs. (2) and (3), or (4). Magdoff and Bresler (1973) used an equation similar to (3) in order to estimate the distribution of exchangeable sodium percentage in sodic soil (ESP = 22) irrigated with various concentrations of $CaCl_2$ in the irrigated water. They then used these estimates to evaluate various methods for reclamation of sodic soils by $CaCl_2$ solution. They concluded that the estimation procedure is "able to distinguish between treatments and rank them in the same order of efficiency of Na replacement by Ca as the experimental results. In some cases there was a fairly good quantitative agreement between calculated and experimental ESP profiles. Thus, even with the simple assumptions underlying it, the procedure was found to be sufficiently good for screening and general prediction purposes."

Anions. For anions the value of $Q(c, \theta)$ is always negative ($Q < 0$) because of anion exclusion from the negatively charged soil particles. When local equilibrium between the soil solution phase and the exchange phase can be assumed, the amount of excluded anions per centimeter of the soil may be estimated from (Bresler, 1973b)

$$Q(c, \theta) = \Gamma - [b(\theta), c] \, A_{ex} \rho_b = \theta_{ex}(\theta, c)c \tag{29}$$

Here $\Gamma-$ (meq/cm^2) is the calculated negative adsorption [a function of $b(\theta)$

and c], A_{ex} (cm^2/g) is the specific exclusion surface area of the soil (Edwards et al., 1965), ρ_b (g/cm^3) is the bulk density of the soil, and θ_{ex} is the volume of anion-free solution per unit volume of the soil (cm^3/cm^3). The value of $Q(c, \theta)$ may be obtained by experimentation or may be estimated from Eq. (29) if Γ^- is calculated, for external surfaces from the double-layer theory, assuming a symmetric or a nonsymmetric mixed system (de Hann, 1965; Bresler, 1970).

The irrigation experiment conducted in the field by Warrick et al. enables us also to evaluate the effect of possible interactions between anions and the soil matrix on the estimation of accumulation and leaching of salts in soils. Such interactions are considered by Bresler (1973b) by including data on $\sigma(\theta, c)$ from Eq. (18) and on $Q(\theta, c)$ from Eq. (29) in the numerical solution to the governing equation (12). A comparison between calculated results and the data of Warrick et al. for 2, 9, 11, and 17 hr of infiltration is given in Fig. 5. The dispersion para-

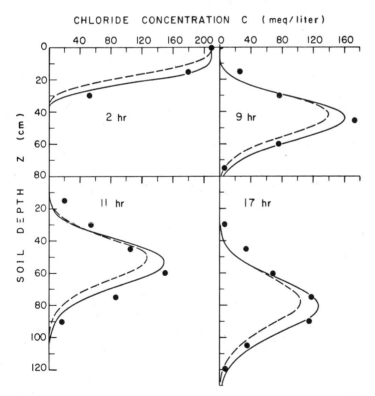

Fig. 5 Computed chloride distribution (continuous lines) as compared with the field data (black circles) of Warrick et al. (1971) for four infiltration times. Dashed lines denote data of Fig. 4 for noninteracting solute with λ = 0.55; solid lines denote a solution in which anion exclusion was considered. (See also Bresler, 1973a.)

meter $D(V, \theta)$, with $\lambda = 0.55$, is the same as in Fig. 4. Equations (18) and (29) were used to calculate $\sigma(c, \theta)$ and $\theta_{ex}(c, \theta)$, respectively. The function $\sigma(\theta, c)$ was computed from a table of values of the $CaCl_2$ system. The results using a dispersivity value of $\lambda = 0.55$, which were similar to those obtained with the two other values tested (0.28, 0.39), are given in Fig. 5. Some slight improvement in the agreement between the theoretical results and the measured chloride distribution data can be seen. On the other hand, the water content distribution data remain the same as in Fig. 4.

In analyzing the effect of ion-soil interaction (Fig. 5), it appears that salt concentration gradients were not an important factor in the movement of the soil solution. In this case the water flux due to osmotic gradients exceeded 15% of the total flux at several points along the soil profile. However, this was not characteristic, and the contribution of osmotically induced flow to the total flow was negligibly small in most parts of the system. On the average over the whole flow system, the osmotic part of the flow was calculated to be less than 5% of the total solution flow. It remains, therefore, that the slight difference between the two calculated lines shown in Fig. 5 may be a consequence of the effect of anion exclusion and not of σ on the flow.

The effect of anion exclusion on the rate of chloride movement in the soil may explain part of the small discrepancy in total salt balance between measured and calculated data that was noted in Fig. 4. This small discrepancy was related to minor deviations from the experimental assumptions, to soil nonhomogeneity, and to measurement errors. The data in Fig. 5 suggest that anion exclusion may explain some of this discrepancy. However, only about 10% of the chlorides in the profile were excluded. Thus, the agreement between the estimated and the measured data is only slightly improved under the present conditions.

Crop-Growing Conditions

The assumption that salt uptake by plants might be neglected sould also be applied in the presence of plants. However, water uptake by plants is a dominant factor and must be considered when salt distribution under crop-growing conditions is of interest. Under these conditions, a nonzero macroscopic water-extraction term must be introduced in Eq. (19) to account for water loss resulting from transpiration.

Introducing either empirical (Shimshi et al., 1975) or a theoretical [Eq. (20)] extraction term into Eq. (19) and solving it numerically subject to the boundary conditions (21) through (28) (Nimah and Hanks, 1973; Feddes et al., 1974; Neuman et al., 1975) makes it possible to estimate the salt distribution as before from the solution of Eq. (12). Such an approach was taken by Child and

Hanks (1965) to estimate the effect of soil salinity on crop growth. Assuming that the only salinity effect on plants is osmotic potential, they could predict the effect of irrigation management and system on crop yield.s

The degree of accuracy that one obtains with the methods in estimating the soil salinity vairables, $c(z, t)$, $\theta(z, t)$, and $h(z, t)$, depends primarily on the following factors: (1) proper choice of the mathematical model to give an adequate description of the physical system at hand; (2) proper specification of boundary conditions; (3) accuracy and stability of the numerical procedure; and (4) accuracy of the soil and plant parameters used in the computation. Of these factors, the most critical one is the last, that is, accuracy of the soil and plant parameters.

Because of the difficulties involved in obtaining all the necessary soil-water-plant-salt parameters required to estimate the salt distribution by these methods, a simplified estimation procedure seems to be of practical usefulness. Such a simplified procedure to estimate salt distribution in the soil profile under crop-growing field conditions was described by Bresler (1967). The method is essentially a numerical solution of the mass balance equation (12) when $D(\theta, V) = 0$, $S' = 0$, and the time increment Δt was set equal to the time interval between two successive water applications (rains or irrigation). Furthermore, only the downward flow was considered and was assumed to take place in the range of water content (θ) between saturation and the assumed "field capacity" of the soil. In addition, the amount of water passing the depth $z \neq 0$ in Δt [i.e., $q^j(z) \Delta t^j$] was estimated by the difference between the amount of water applied [$\bar{Q}^j = \bar{q}(0, j) \Delta t^j$] and the water consumption by the crop from the soil surface down to the depth z in Δt^{j-1}. With these approximations, Eq. (12) becomes

$$\int_0^z [c^j(z) - c^{j-1}(z)] \bar{\theta}(z) \, dz = \bar{Q}^j C_o^j - [Q^j - \int_0^z E^{j-1}(z) \, dz] c^{j-\frac{1}{2}}(z) \qquad (30)$$

where $\bar{Q}$ is the amount of water applied to the soil surface by irrigation or rain (cm), j is the index of water applied by irrigation or rainstorm, and E is the amount of water per unit volume of soil (cm^3 H$_2$O/cm^3 soil) consumed by the crop during the the time interval Δt^{j-1}. Estimation of the salt profile from the solution of Eq. (30) is possible if $\bar{\theta}(z)$ and $E(z)^{j-1}$ can be evaluated, provided that the initial salt concentration [$c_n(z)$], the amount of water applied, and its salt concentrations are measured. Evaluation of $E^{j-1}(z)$ may be obtained from the large number of water-requirement experiments that have been conducted in many countries (particularly in Israel). The applicability of this estimation method was tested by D. Yaron (personal communication) and Yaron et al. (1972), who compared many irrigation experiment data with the calculated reults, and found the former to be applicable in many practical cases.

Irrigation Method and Soil Salinity Control

The problem of irrigation method from the point of view of control of soil salinity may have short-run or long-run implications. In the short run, the direct effect of irrigation water and salinity on the yield of a specific crop, or of seasonal irrigation water on soil salinity within a relatively short time period, is considered. The effects of salt accumulation over relatively long periods are not taken into account. The short period involved is usually of the order of one year or one growing or irrigation season. The short-run problems include irrigations for the reclamation of initially saline soil to optimize the yield of the following crop, and water use for the control of soil salinity throughout the crop-growing season.

In the long run, the perennial soil salinity effect must be considered. Usually the long-run effects are the time integral of the short-run effects. Thus, when the short-run effects can be estimated or predicted, it is also possible to predict the long-run effects simply by integrating all the short-period effects over the longer period of interest. An alternative method is to approximate the system by taking average values over relatively large time and space intervals, but these effects will not be discussed here.

Reclamation of Saline Soils by Flood, Sprinkle, and Trickle Irrigation

The reclamation process is aimed at removing the salinity-limiting factor so that the crop which follows it will not suffer yield reduction. Reclamation of saline soil is essentially a process in which the high-concentration soil solution is displaced by a less concentrated solution of irrigation water. Theoretical results suggested what was found in several laboratory experiments and field trials—that reclaiming saline soil by leaching with irrigation water is more efficient when the soil is maintained unsaturated and the flow rates are relatively slow (Bigger and Nielsen, 1967; Mokady and Bresler, 1968). That is the reason why water for leaching of initially saline soil can be utilized more efficiently by applying sprinkler or trickle-drip irrigation methods.

The salt concentration (meq/liter solution) and salt content (meq/gm soil) profiles in flooded and sprinkler-irrigated soil are demonstrated in Fig. 6. In both irrigation methods the upper part of the soil profile, at the end of the irrigation, has a salt concentration identical to the low concentration of the applied water. The salt profile increases to a maximum value close to the wetting front and drops to its initial value below the wetting depth. Because of the slower wetting rate under sprinkling, the zone of complete leaching at the end of irrigation (infiltration) extends more deeply into the profile than under flood irrigation.

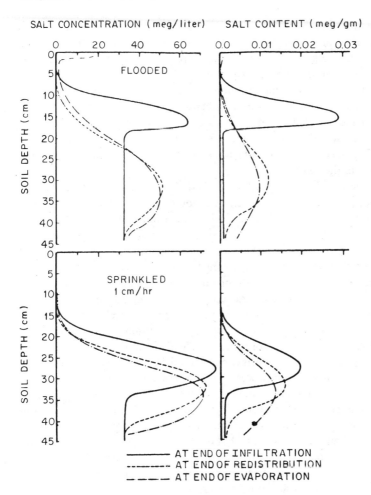

Fig. 6 An example of the effect of irrigation method on salt concentration (in the soil solution) and salt content (per gram soil) as a function of soil depth. (See also Bresler and Hanks, 1969.)

From the end of irrigation until the end of redistribution there is a wide distribution of salt concentrations. The salt added by irrigation water and the salt already present in the soil are thoroughly mixed and dispersed along the entire profile. The dispersal zone is narrower in the sprinkler-irrigated soil during redistribution than in the flooded soil. When the soil is subjected to evaporation, flow occurs upward and downard, simultaneously, in different parts of the soil profile. Thus, some salt continues to move down with the redistributed water and, at the same time, salt near the surface moves upward toward the surface, where it accumulates. The amount of salt that moves to the top depends largely on the salt quantity present in that part of the soil which is shallower than the depths from which water flows upward. Thus, only negligibly small amounts of salt move up during evaporation from soil previously irrigated by sprinklers and none to the region above a depth of 12 cm. In the flooded soil, on the other hand, more salt moves upward and accumulates at the top. The data in Fig. 6 clearly demonstrate the superiority of sprinkler irrigation over flooding, with respect to water utilization, when leaching of hazardous salts initially present in the soil is considered.

Use of the *trickle-drip irrigation method for* irrigating saline fields with relatively saline water has become quite common in recent years. The success of the surface approach, as opposed to underground application of trickling, is attributed partly to the formation of a low-salt root zone in the vicinity of the trickle source. In this approach the trickle emitter is placed directly on the soil surface, so that the area across which irrigation takes place is small compared with the total soil surface. As a result, one has a case of three-dimensional flow, which differs from the usual one-dimensional flow of flood and sprinkler irrigation, where the area across which flow takes place is identical to the total area of the wetted field. The general pattern of salt distribution in two soils and for two trickle discharge rates is demonstrated in Fig. 7. From this example one can evaluate the applicability of this irrigation method to the reclamation of saline soil by leaching. This leaching process involves the case of low-concentration irrigation water that miscibly displaces a high-concentration soil solution originally oresent in the field. Figure 7 presents the salt content field data in terms of salt content per unit bulk volume of soil (as milliequivalents of solute per liter of bulk soil). It demonstrates the effect of trickle discharge and soil properties on the shape of salt distribution. In both soils with both discharge rates there is a leached zone close to the position of the emitter, and an accumulation zone close to the wetting front. Here the salt quantities from the leached part of the soil are accumulated and reach a maximum at a certain distance from the irrigation source. The location of this maximum and the accumulation zone is largely dependent on the soil properties and discharge rates of the emitter.

The general pattern of salt distribution and its dependence on initial salinity, salinity of the irrigation water, rate of trickle discharge, and the hydraulic characteristics of the soil, as demonstrated in Fig. 7, is of practical interest for problems connected with water utilization for reclamation of saline soils by irrigation in order to control soil salinity in the wetted root zone. Consider, for instance, a saline field being irrigated by a set of symmetrical trickle emitters. Suppose that it is important to evaluate the leaching effectiveness of removing salts from a given root volume in which the main plant roots function.

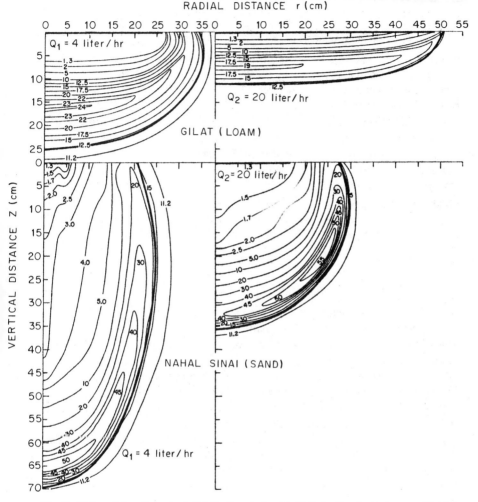

Fig. 7 Volumetric salt content distribution during trickle irrigation for two cases of trickle discharges (Q_1 and Q_2) and two soils (Gilat and Hahal Sinai). The numbers labeling the curves indicate volumetric salt content ($c\theta$) in milliequivalents per liter of bulk soil. (See also Bresler, 1975.)

With a sufficiently long irrigation duration, depending on soil and discharge rate, this leached zone may be sufficient for most of the roots to concentrate and function without disturbance. On the other hand, a large quantity of salt may accumulate to an appreciable level at a certain distance from the source close to the wetting fronts (Fig. 7). It is apparent that this limitation to plant growth can be overcome by changing the discharge rate or position of the source, or both.

Knowing the initial soil conditions and the final desired soil salinity (as determined by the crop to be grown), the quantity-quality relationships of water used in the leaching process can be determined for each irrigation method. Leaching efficiency may be improved by regulating the conditions at the soil surface boundary as, for example, by adapting the proper irrigation method or by artificially decreasing the rates of water applied by sprinkler. In some cases it is also possible to increase the leaching efficiency by controlling the bottom boundary conditions by means of artificial drainage. In addition, the reclamation process can be improved by regulating the $K(\theta, C, R)$ function simply by changing the salt composition and/or concentration of the water applied. Changing the soil hydraulic conductivity by means of soil-conditioning chemicals is also possible. At present, however, these materials appear to be too costly and there is insufficient information on the persistence of their effect in the field.

Salinity Control by Irrigation During the Crop-Growing Period

Control of soil salinity during the growing period of a given crop depends on knowledge of the crop-response pattern to an appropriate soil salinity index. The salinity index chosen may be the specific concentration of an ion or the electrical conductivity of the soil solution (or of the extract obtained from the saturated soil paste). It may also be defined in terms of the total water suction $[s = \pi(c) + h]$ when the soil matrix suction (h) is being considered.

Consider first the conception maintained by certain workers (e.g., Bernstein, 1965; Bierhuizen, 1969) that significant yield response to salinity occurs only above a certain critical threshold concentration. Below this threhold value, the salinity effect is negligible. This critical salinity index is unique for a given crop and growing conditions. With this conception in mind, Eqs. (12), (15), and (19) through (29) or (30) can be applied for purposes of salinity control during the crop-growing period.

The systems of these equations can be classified into four major groups: (1) predetermined functions, such as $K(\theta, C. R)$, $h(\theta)$, $D(\theta, V)$, $S(t, z)$, $\sigma(\theta, c)$, and $Q(\theta, c)$; (2) predetermined variables (initial conditions) or postreclamation variables $\theta(z,0)$ and $c(z,0)$; (3) man-controlled variables $C_o(t)$, $\bar{Q}(t) = \int_t q(0,t)dt$; and (4) dependent variables, $c(z,t)$, $\theta(z,t)$, and $h(z,t)$. Of particular interest for irrigation and salinity-control purposes is the dependence of $c(z,t)$ and $h(z,t)$ on the man-controlled variables $\bar{Q}(t)$ and $C_o(t)$, which determine uniquely the salinity regime of the root zone throughout the growing period [subject to Eqs. (12), (15), and (19) through (29) or (30), the predetermined functions, and the initial conditions]. In other words, control of soil salinity during the crop-growing period by irrigation may be obtained by regulating the quantity ($\bar{Q}$) and a salinity (C_o) of the water applied, as well as rate of application, irrigation method, time of application, and so on.

An application of linear programming and computer simulation modeling to the estimation of the "best" combination of the controllable variables quantity ($\bar{Q}$) and salinity (C_o) of the irrigation water was described by Yaron and Bresler (1970). Their methods enable one to consider changes in time and space of the critical theshold values. Thus, variation in the response of the crop during different stages of growth can be taken into account. The problem may be formulated in many different ways. To illustrate, let (1) the amount of water applied be predetermined at some specific level, $\bar{Q}(t)$; and (2) the soil salinity index at any time and depth, $c(z, t)$, be restricted not to exceed prespecified time and space threshold values, that is, $c(z, t) < C_r(z, t)$. The problem is to find the maximum level of C_o that will not violate the restricted threshold values $C_r(z,t)$ subject to any set of the finite difference linear form of (12), (19) through (29) or (30). The linear programming formulation consists of a linear objective function to be maximized subject to a set of linear equalities (the salt distribution estimation methods) and inequalities (restrictions on the critical threshold levels). By successively varying the predetermined $\bar{Q}(t)$ and solving for C_o, a set of $\bar{Q}$ - C_o combinations is obtained that maintains the soil salinity index below the prespecified critical values. Another method of solving the same problem is a computer simulation model. Here the "best" combination of the man-controlled variables $\bar{Q}$ - C_o subject to the same conditions is found by computerized trial and error (Yaron and Bresler, 1970).

The threshold concentration concept may, in many situations, be of limited use since it represents only a point on the crop response-salinity function. This point is usally varied when the water regime is changed. Yaron et al. (1972) obtained an empirical relationship between crop yield and the total suction $[s = (h + \pi)]$ or the actual soil salinity index, which was either measured or estimated. Available data on crop response function to the combined effect of soil water regime and salinity add an alternative for controlling soil salinity throughout the irrigation season (Bresler and Yaron, 1972). Regulating the soil water regime by varying the frequency of water application and the wetting depth, within the root zone, will affect the total water suction ($s = h + \pi$) by changing both the actual soil solution concentration and therefore π, and the soil matrix suction, h. In regions with a Mediterranean-type climate (i.e., semiannual alternation of irrigation in dry summers with rainy winter seasons), it might be more efficient to control salinity by regulating the soil water regime rather than by leaching. This is demonstrated in Fig. 8, which is based on estimates using irrigation experiments in combination with mathematical-physical models (Bresler and Yaron, 1972). Figure 8 shows that under certain conditions and assumptions, a given amount of irrigation water for salinity control is used more

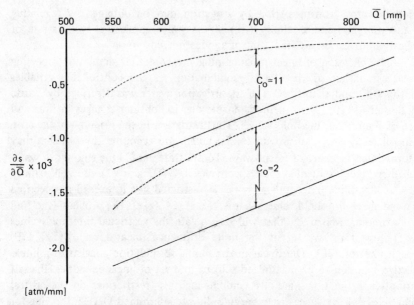

Fig. 8 Rate of changes of total water potential (s = π + h) with water quantity ($\bar{Q}$) as a function of amount of irrigation water ($\bar{Q}$) for two cases of irrigation water salinity (C_0). (See also Bresler and Yaron, 1972.)

efficiently for changing the soil-water regime than for leaching during the irrigation season. Irrigation frequency to change the soil-water regime is therefore an additional man-controlled variable by which the salinity factor may be managed more efficiently (with total soil water suction taken into consideration) by irrigation. Thus, assuming that soil water regime may modify the plant response to salinity, and on the basis of data from an irrigation experiment, Bresler and Yaron (1972) showed (Fig. 8) that situations exist in which increasing the quantity of irrigation water, shortening the time interval between irrigations, and increasing the depth of wetting lead to increasing salt concentration in the soil solution while at the same time decreasing s. They also show that under certain situations it is more efficient to use a given quantity of irrigation water for changing the soil water regime rather than for leaching.

References

Bernstein, L. (1965). Salt tolerance of fruit crops. U.S. Dept. Agr. Inform. Bull. 292.
Bierhuizen, J. F. (1969). *Water Quality and Yield Depression.* Institute of Land and Water Management Resources, Wageningen, The Netherlands.

Biggar, J. W., and Nielsen, D. R. (1967). Miscible displacement and leaching phenomena. In *Irrigation of Agricultural Lands*, Agronomy *11*, pp. 254-274.

Bresler, E. (1967). A model for tracing salt distribution in the soil profile and estimating the efficient combination of water quality and quantity under varying field conditions. Soil Sci. *104*: 227-233.

Bresler, E. (1970). Numerical solution of the equation for interacting diffuse layers in mixed ion systems with non-symmetrical electrolytes. J. Colloid Interface Sci. *33*: 278-283.

Bresler, E. (1972). Interacting diffuse layers in mixed mono-divalent ionic systems. Soil Sci. Soc. Amer. Proc. *36*: 891-896.

Bresler, E. (1973a). Anion exclusion and coupling effects in nonsteady transport through unsaturated soils: I. Theory. Soil Sci. Soc. Amer. Proc. *37*: 663-669.

Bresler, E. (1973b). Simultaneous transport of solute and water under transient unsaturated flow conditions. Water Resources Res. *9*: 975-986.

Bresler, E. (1975). Two-dimensional transport of solutes during nonsteady infiltration from a trickle source. Soil Sci. Soc. Amer. Proc. *39*: 604-613.

Bresler, E., and Hanks, R. J. (1969). Numerical method for estimating simultaneous flow of water and salt in unsaturated soils. Soil Sci. Soc. Amer. Proc. *33*: 827-832.

Bresler, E., and Yaron, D. (1972). Soil water regime in economical evaluation of salinity in irrigation. Water Resources Res. *8:* 791-800.

Chaudhari, N. M. (1971). An improved numerical technique for solving multi-dimensional miscible displacement equations. Soc. Petrol. Eng. J. *11*: 277-284.

Childs, S. W., and Hanks, R. J. (1965). Model of soil salinity effects on crop growth. Soil Sci. Soc. Amer. Proc. *39*: 617-623.

de Haan, F. A. M., (1965). The interaction of certain inorganic anions with clays and soils. Centrum voor landbouwpublikaties en landbouwdocumentatie, Wageningen, The Netherlands, p. 167.

Edwards, D. G., Posner, A. M., and Quirk, J. P. (1965). Repulsion of chloride ions by negatively charged clay surfaces. Part 3. Di- and trivalent cation clays. Trans. Faraday Soc. *61*: 2820-2823,

Erikson, E. (1952). Cation-exchange equilibria on clay minerals. Soil Sci. *74*: 103-113.

Feddes, R. A. (1971). Water, heat and crop growth. Thesis, Comm. Agr. Univ. Wageningen 71-12, The Netherlands, 184 pp.

Feddes, R. A., Bresler, E., and Neuman, S. P. (1974). Field test of a modified numerical model for water uptake by root systems. Water Resources Res. *10(6)*: 1199-1206.

Gardner, W. R. (1960). Dynamic aspects of water availability to plants. Soil Sci. 89: 63-73.

Hanks, R. J., Klute, A., and Bresler, E. (1969). A numeric method for estimating infiltration, redistribution, drainage and evaporation of water from soil. Water Resrouces Res. 5: 1064-1069.

Hayward, H. E. (1954). Plant growth under saline conditions. UNESCO Arid Zone Res. 4: 37-71.

Kemper, W. D. (1961). Movement of water as affected by free energy and pressure gradients. I. Application of classic equations for viscous and diffusive movements to the liquid phase in finely porous media. Soil Sci. Soc. Amer. Proc. 25: 260-265.

Kemper. W. D., and Evans, N. A. (1963). Movement of water as affected by free energy and pressure gradients. III. Restriction of solute by membranes. Soil Sci. Soc. Amer. Proc. 27: 485-490.

Kemper, W. D., and Rollins, J. B. (1966). Osmotic efficiency coefficients across compacted clays. Soil Sci. Soc. Amer. Proc. 30: 529-534.

Kemper, W. D., and van Schaik, J. C. (1966). Diffusion of salts in clay-water systems. Soil Sci. Soc. Amer. Proc. 30: 534-540.

Lantz, R. B. (1971). Quantitative evaluation of numerical diffusion (truncation error). Soc. Petrol. Eng. J. 11: 315-320.

Magdoff, F., and Bresler, E. (1973). Evaluation of methods for reclaiming sodic soils with $CaCl_2$. A. Hadas et al. (eds.), *Ecological Studies. Analysis and Synthesis,* Springer-Verlag, Berlin-Heidelberg-New York, vol. 4, pp. 441-452.

McNeal, B. L., and Coleman, N. T. (1966). Effect of solution composition on soil hydraulic conductivity. Soil Sci. Soc. Amer. Proc. 30: 308-312.

Mokady, R.S., and Bresler, E. (1968). Reduced sodium exchange capacity in unsaturated flow. Soil Sci. Soc. Amer. Proc. 32: 463-467.

Neuman, S. P., Feddes, R. A., and Bresler, E. (1975). Finite element analysis of two-dimensional flow in soils considering water uptake by roots: I. Theory. Soil Sci. Soc. Amer. Proc. 39: 224-230.

Nielsen, D. R., and Biggar, J. W. (1961). Miscible displacement in soils: I. Experimental information. Soil Sci. Soc. Amer. Proc. 25: 1-5.

Nielsen, D. R., and Biggar, J. W. (1962). Miscible displacement: III. Theoretical considerations. Soil Sci. Soc. Amer. Proc. 26: 216-221.

Nimah, M. N., and Hanks, R. J. (1973a). Model for estimating soil water, plant and atmospheric interrelations, 1, Description and sensitivity. Soil Sci. Soc. Amer. Proc. 37: 522-527.

Nimah, M. N., and Hanks, R. J. (1973b). Model for estimating soil water, plant and atmospheric interrelations, 2, Field test of model. Soil Sci. Soc. Amer. Proc. 37: 528-532.

Ogata, A. (1970). Theory of dispersion in a granular medium. Geological Survey Professional Paper 411-I, U.S. Govt. Printing Office, Washington, D.C.

Olsen, S. R., and Kemper, W. D. (1968). Movement of nutrients to plant roots. Advan. Agron. *30*: 91-151.

Oster, J. D., and Rhoades, J. D. (1974). Calculated Drainage Water Compositions and Salt Burdens Resulting from Irrigation with Western U.S. River Waters. Journal of Environmental Quality.

Overbeek, J. Th. (1952). The interaction between colloidal particles. In H. R. Kruit (ed.), *Colloid Science I*, Elsevier, Amsterdam, p. 251.

Passioura, J. B., Rose, D. A., and Haszler, K. (1970). Lognorm: A program for analysing experiments on hydrodynamic dispersion. Div. Land Res. Rech. Memo. 70/5, Commonwealth Sci. and Ind. Res. Organ., Canberra, Australia, 9 pp.

Perkins, T. K., and Johnston, O. C. (1963). A review of diffusion and dispersion in porous media. Soc. Petrol. Eng. J. *3*: 70-84.

Porter, L. K., Kemper, W. D., Jackson, R. D., and Steward, B. A. (1960). Chloride diffusion in soils as influenced by moisture content. Soil Sci. Soc. Amer. Proc. *24*: 460-463.

Quirk, J. P., and Schofield, R. K. (1955). The effect of electrolyte concentration on soil permeability. J. Soil Sci. *6*: 163-178.

Reeve, R. C., and Bower, C. A. (1960). Use of high salt water as a flocculant and source of divalent cations for reclaiming sodic soils. Soil Sci. *90*: 139-144.

Rhoades, J. D., et al. (1973). Salts in irrigation drainage waters I. Effects of irrigation water composition, leaching fraction, and time of year on the salt compositions of irrigation drainage waters. Soil Sci. Soc. Amer. Proc. *37(5)*: 770-774.

Ritchie, J. T. (1972). A model for predicting evaporation from a row crop with incomplete cover. Water Resources Res. *8(5)*: 1204 1213.

Russo, D. (1976). Hydraulic properties of an unsaturated soil as affected by Na-Ca composition and soil soltuion concentration. Unpublished M.Sc. thesis (in Hebrew), Hebrew University, Jerusalem.

Shainberg, I. (1976). Cation and anion exchange reactions. In Chesters and Brenner (eds.), *Soil Chemistry*, Marcel Dekker, New York.

Shainberg, I., Bresler, E., and Klausner, Y. (1971). Studies of Na/Ca montmarillonite systems. 1. The Swelling pressure. Soil Sci. *111*: 214-219.

Shalhevet, J., Reiniger, P., and Shimshi, D. (1969). Peanut response to uniform and non-uniform soil salinity. Agron. J. *61*: 584-587.

Shimshi, D., Yaron, D., Bresler, E., Strateener, G., and Weisbrod, M. (1975). A simulation model for evapotranspiration of wheat: Empirical approach. J. Irrigation and Drainage Div. ASCE *101 (IR1)*, Proc. paper 11170: 1-12.

U.S. Salinity Laboratory Staff (1954). *Diagnosis and Improvement of Saline and*

Alkali Soils, U.S. Dept. Agr. Handbook No. 60, U.S. Govt. Printing Office, Washington, D.C.

Wadleigh, C. H., Gandi, H. G., and Kolisch, M. (1951). Mineral composition of orchard grass grown on Pachappa loam salinized with various salts. Soil Sci. *72*: 275-282.

Warkentin, R. P., Bolt, G. H., and Miller, R. D. (1957). Swelling pressures of montmarillonite. Soil Sci. Soc. Amer. Proc. *21*: 495-499.

Warrick, A. W., Biggar, J. W., and Nielsen, D. R. (1971). Simultaneous solute and water transfer for an unsaturated soil. Water Resources Res. *7*: 1216-1225.

Yaron, D., Bielorai, H., Shalhevet, J., and Gavish, Y. (1972). Estimation procedures of response functions of crops to soil water content and salinity. Water Resources Res. *8*: 291-300.

Yaron, D., and Bresler, E. (1970). A model for the economic evaluation of water quality in irrigation. Aust. J. Agr. Econ. *14*: 53-62.

5

Fertilization and Salinity

J. J. Jurinak and Robert J. Wagenet

Utah State University, Logan, Utah

Introduction

The interactions among soil fertility, salinity, and yield are of major concern to those interested in maximizing crop production in arid or semiarid areas. Classical approaches to salinity-fertility relationships have involved salt effects on crops growing under optimal conditions of soil fertility and soil moisture status. These experiments established certain criteria of salt tolerance for plants (U.S. Salinity Laboratory Staff, 1954; Bernstein, 1964a; Maas and Hoffman, 1976). These criteria are, however, useful only within very broad limits, as they relate only minimally to the variable soil fertility factors such as available nitrogen or phosphorus that can greatly influence crop response in any particular situation. It is therefore necessary to consider crop response to fertilization in the presence of variable amounts of soil salts, a consideration of primary concern as agricultural enterprises are expanded to include additional salt-affected soils and saline irrigation waters.

The limited literature on the question of maximizing productivity of salt-affected soils through fertilizer application generally indicates that under a particular soil fertility level crop yields decrease with increases in salinity but for a given salinity level there is an increase in yield with fertilizer application (Luken, 1962; Amer et al., 1964; Lunin and Gallatin, 1965a). The factors contributing to such tendencies are not well understood, but center around two mechanisms. First, the osmotic potential of a solution increases with additions of salt, making a given volume of soil water less available to plants as salt concentration in solution is increased. This osmotic effect, as discussed in other chapters, will therefore influence crop yield by limiting the amount of soil water available to the plant. Second, several investigators have noted an influence of salinity on the uptake of both macro and micro nutrients (Lunin and Gallatin, 1965a; Hassan et al., 1970a, 1970b) and hence an influence on plant nutrition. This effect depends greatly on the particular crop and nutrient and is manifested not only in the mass yield of the crop but also in the elemental composition of the plant's component parts. Both these concepts are discussed below at greater length.

Fertilizers are usually applied as inorganic salts of nitrogen, phosphorus, and potassium, either separately or in mixtures, known as N-P-K fertilizers. Although organic sources of nitrogen, such as urea, are becoming more popular, the investigation of salinity and fertilizer effects has been limited primarily to inorganic sources of N, P, and K. Aside from providing a source of available plant nutrients, these compounds have a distinct effect on soil chemistry and salinity that is manifested in the yield-influencing mechanisms of osmotic potential and nutrient uptake previously mentioned.

Fertility-Salinity Effects of Fertilizer Form

The influence of fertilizer form on soil salinity is best understood by separate consideration of the influence of N, P, and K in soils. Each has an influence determined by its basic chemical properties. In general, these influences can be summarized in the following manner. Soil salinity can be increased significantly by addition of nitrogen, because nitrogen salts are generally quite soluble with little resultant effect on the solubility of other salts that may already be present in the soil solution. Dissolved nitrogen fertilizers are therefore additive to the total salt in solution. In fact, certain forms of nitrogen, such as NH_4NO_3, tend to reduce soil pH, resulting in increases of dissolved Ca^{2+}, Mg^{2+}, and K^+ in solution, a factor that will serve to increase the measured electrical conductivity of the saturation extract (EC_e). Phosphorus fertilization has the effect of reducing soil salinity slightly, as insoluble Ca^{2+} and Mg^{2+} phosphates will form, removing these ions from solution, decreasing the total dissolved salts, which decreases the EC_e. Potassium, in contrast to nitrogen, has no major effect on pH and does not form insoluble precipitates as phosphorus does. Additions of potassium fertilizer alone therefore act to raise the soil's EC_e, but the increase is not as great as with some forms of nitrogen. All three of these nutrients are often added in combination. The effects discussed above are thereupon interactive, as has been presented by Lunin and Gallatin (1965a) and reproduced in Table 1.

The soil used to develop the data of Table 1 was a silty clay loam with an exchange capacity of 10.9 meq/100 g which has been limed at the rate of 3 tons/acre to produce a resultant pH of 5.5. Prior to salt or fertilizer treatment, it

Table 1 Effects of Added Fertilizers and Salinity on Selected Soil Parameters[a,b]

Salinity level	Fertilizer added							
	None	N	P	K	NP	NK	PK	NPK
				Soil pH				
A	5.42	5.13	5.32	5.29	5.13	5.09	5.18	5.04
B	5.28	4.99	5.16	5.04	4.95	4.96	5.11	4.98
C	5.06	4.89	5.06	5.04	4.93	4.88	5.09	4.94
D	5.00	4.85	5.01	4.98	4.81	4.84	5.02	4.88
				EC_e (mmho/cm)				
A	1.7	2.0	1.2	1.8	1.3	2.4	1.4	1.5
B	3.5	4.3	3.4	3.7	3.8	5.1	3.8	4.2
C	5.5	6.4	5.7	5.6	6.8	7.0	6.5	7.2
D	7.8	9.2	7.8	8.0	8.4	10.3	7.9	8.8
				Ca_e^{2+} (meq/100 g)				
A	0.41	0.61	0.31	0.52	0.36	0.77	0.39	0.45
B	0.67	0.96	0.62	0.79	0.75	1.15	0.74	0.86
C	0.87	1.18	0.92	1.11	1.07	1.38	1.03	1.40
D	1.35	1.71	1.29	1.29	1.47	1.64	1.23	1.46
				Mg_e^{2+} (meq/100 g)				
A	0.04	0.05	0.04	0.05	0.04	0.07	0.04	0.03
B	0.09	0.15	0.08	0.09	0.10	0.13	0.10	0.11
C	0.16	0.20	0.16	0.20	0.18	0.24	0.18	0.22
D	0.27	0.32	0.25	0.23	0.27	0.31	0.24	0.29
	Na			Na_e^+ (meq/100 g)				
A	0.16	0.11	0.12	0.11	0.11	0.13	0.13	0.13
B	0.53	0.59	0.54	0.55	0.62	0.63	0.60	0.78
C	1.04	1.02	1.12	1.15	1.21	1.23	1.20	1.30
D	1.76	1.85	1.87	1.64	1.80	1.78	1.62	1.69
				K_e^+ (meq/100g)				
A	0.03	0.04	0.02	0.11	0.03	0.15	0.05	0.05
B	0.05	0.05	0.03	0.12	0.03	0.18	0.10	0.07
C	0.05	0.07	0.05	0.17	0.05	0.23	0.14	0.15
D	0.11	0.11	0.07	0.26	0.12	0.32	0.17	0.19

[a]From Lunin and Gallatin (1965a).
[b]Subscript "e" indicates composition of saturation extracts.

contained 5.5, 0.8, 0.3, and 0.2 meq/100 g of exchangeable Ca^{2+}, Mg^{2+}, Na^+, and K^+, respectively. The four salinity levels consisted of (A) demineralized water and three treatments (B), (C), (D) irrigated with synthetic waters of EC values of 2, 4, and 6 mmho/cm, respectively. Four irrigations of each treatment were accomplished in a 3-week period. Each fertilizer element was added at the rate of 150 lb/acre.

Table 1 demonstrates that at any given salinity of irrigation water, there are independent effects on soil chemistry resulting from the type of fertilizer added. It also presents the interactive effects of various combinations of these three nutrients. It is obvious that pH, electrical conductivity, and availability of Ca^{2+}, Mg^{2+}, Na^+, and K^+ are greatly affected by the content of the added fertilizer.

Except under unusual conditions, the osmotic potential of the soil solution should never become high enough to injure the crop when the fertilizer is uniformly broadcast. However, when the fertilizer is applied by a method that localizes it in high concentrations in a small volume of soil, severe osmotic effects may result in injury to the plant. Such injury can be reduced by locating the fertilizer zone farther from the row, by reducing the amount of fertilizer, and by reducing the tendency of the fertilizer to injure the plant. In general, the most efficient fertilizer use is achieved when the zone of placement is close to the row, as less fertilizer is consumed in weed growth, and the roots of the new seedling quickly reach the fertilizer, a positive consequence in soils requiring fertilization for plant establishment and growth. Reducing the quantity of fertilizer is a risky management technique, as the possibility exists of going below the optimum amount of fertilizer necessary for plant growth. Therefore, there is good reason to consider the tendency of the fertilizer to create adverse salt effects, in order that the optimum amount of chemical may be placed in the best soil position without adversely affecting growth.

One method of quantifying the effects of different fertilizer forms on the osmotic potential of the soil solution is the salt index (Rader et al., 1943). This parameter provides a means of predicting the relative effect of any mixed fertilizer on the osmotic potential of the soil solution. It is defined as "the ratio of the increase in osmotic potential produced by the fertilizer in question to that produced by the same weight of sodium nitrate, multiplied by 100." In equation form,

$$\text{Salt index} = \frac{p}{p'} \times 100 \qquad (1)$$

where p is the increase in osmotic potential of the soil soltuion as a result of application of a definite weight of a fertilizer mixture, and p' is the increase as a result of application of the same weight of sodium nitrate under identical conditions. The selection of sodium nitrate as the reference compound, as well as other considerations of this index, are presented well by Rader et al. (1943).

It should be noted that the salt index does not predict the exact amount of fertilizer that will produce injury on a particular soil, but it does classify the

Table 2 Salt indexes of Selected Fertilizers[a]

Material	Analysis (%)[b]	Salt index
Nitrogen carriers		
Ammonia	82.2	47.1
Ammonium nitrate	35.0	104.7
Monoammonium phosphate	12.2	29.9
Diammonium phosphate	21.2	34.2
Ammonium sulfate	21.2	69.0
Calcium nitrate ($4H_2O$)	11.9	52.5
Potassium nitrate	13.8	73.6
Sodium nitrate	16.5	100.0
Urea	46.4	75.4
Phosphate carriers		
Monoammonium phosphate	61.7	29.9
Diammonium phosphate	53.8	34.2
Monocalcium phosphate (H_2O)	56.3	15.4
Monopotassium phosphate	52.2	8.4
Monosodium phosphate (H_2O)	51.4	36.2
Superphosphate	16.0	7.8
Superphosphate	20.0	7.8
Superphosphate	48.0	10.1
Potassium carriers		
Potassium chloride	63.2	114.3
Potassium nitrate	46.6	73.6
Potassium sulfate	54.0	46.1
Miscellaneous		
Calcium carbonate	56.0	4.7
Calcium sulfate	32.6	8.1
Sodium chloride	53.0	153.8
Sodium sulfate	43.6	74.2

[a]From Rader et al., (1943).
[b]Percentage analysis refers to the proportion of each fertilizer material consisting of the primary component, i.e., nitrogen, phosphorus, potassium, calcium, or sodium.

fertilizer with respect to other fertilizers with regard to their osmotic effects. The salt index scale is arranged such that the effect of sodium nitrate is a value of 100 with the effects of other materials expressed relative to 100.

Table 2 summarizes salt indexes for several commonly applied fertilizers. It shows that chloride salts of potassium and sodium have very high salt indexes, a reflection of their high solubility and therefore great effect on soil solution osmotic potential. In general, phosphorus fertilizers have low salt indexes, a result of the low solubility of phosphorus compounds. Nitrogen fertilizers vary in their effect, depending on the combined factors of chemical solubility and soil pH effects.

The values presented in Table 2 were developed for one soil type, a Norfolk sand (4.2% clay, 10.6% silt, 85.2% sand). As a test of the applicability of the salt index concept to other soils, its developers tested it on six other soils varying in texture from sandy loam through clay. It was found that in all seven situations, the value of the salt index remained relatively constant, despite widely varying moisture, silt, and clay contents. This illustrates the convenience of using the salt index as a measure of the relative tendency of a fertilizer to injure crops rather than actual osmotic pressure values, which vary widely from soil to soil. The salt index arranges fertilizers in order of their relative effects on soil solution osmotic pressure and allows prediction of the effects of these chemicals on crop growth.

Crops grown on sandy soils are more susceptible to injury through osmotic effects than are those grown on more clayey soils. This is due to the greater effect of applied soluble salts (applied fertilizer) on osmotic potential in cases where there is lower soil water content. Therefore, it is particularly appropriate to use the salt index for evaluation of potential osmotic effects problems in connection with the management of sandy soils.

Fertility-Salinity Effects of Irrigation Water Quality

It is important to recognize not only the effects of soil salts on nutrient status, but also the effects of salt applied in irrigation water on nutrient status. Table 1 presents, in addition to previously described relationships, a comparison of the effects on soil nutrient status (as established by fertilizer treatments) of the addition of irrigation water of varying salinity. The water used in the three treatments B, C, and D consisted of dilutions of synthetic seawater containing 473 meq/liter NaCl, 102 meq/liter $MgCl_2$, 20 meq/liter $CaSO_4$, and 12 meq/liter K_2SO_4. As can be seen, levels of soil-exchangeable cations rise appreciably, but this apparent increase in fertility status of the soil is more than offset by increases in EC_e, which act to limit water availability to the plant through the previously mentioned osmotic effect. It is important to note the concentration effect of soil-plant processes on the EC_e value. Although the waters added were of EC 0, 2, 4, and 6 mmho/cm, the observed EC_e was in all cases greater. This is due partially to the differential absorption of water and salts by the plant root, which tends to concentrate remaining salts in the soil solution as water is removed by the plant. Additional salt is also contributed to solution from the dissolution of precipitated soil salts. The general tendencies of the chemical effects of N, P, and K fertilizers are also apparent in these trends, as the EC_e is most increased in the case of added nitrogen and least increased by the addition of phosphorus.

Evaluation of situations other than those presented in Table 1 must be undertaken carefully. Results are presented for only one soil type and one composition of irrigation water (albeit several dilutions of this water). As the relative proportions of the major cations and anions (Ca^{2+}, Mg^{2+}, K^+, Na^+, HCO_3^-, CO_3^{2-}, SO_4^{2-}, and Cl^-) in the irrigation water change, so too will the effects of repeated application of the water. However, the effect of any particular combination of N, P, and K will be manifested according to the same general principles explained previously.

It is therefore apparent that independent of plant effects that are reflected as increased or decreased productivity, there are greatly interactive effects of soil salinity, irrigation water salinity, and fertilizer chemistry. These interactions result in alteration of the salt status of soils to which fertilizers are added, depending on the nitrogen, phosphorus, or potassium content of the added chemical. There are also alterations in plant availability of cations and anions comprising the system.

Salinity Effects on Fertilizer Transformations

Many nitrogen fertilizers are applied in forms chemically unavailable to plants, such as urea, or of limited availability because of immobility in the soil, such as ammonium salts, which depend on soil microbiological processes to be converted to mobile, plant-available forms, such as nitrate ion. This conversion process, known as nitrification, is influenced by soil salinity. According to Westerman and Tucker (1974), this effect is dependent on type and concentration of salt. At the end of 49-day incubations, 59, 80, and 75% of added nitrogen remained as NH_4^+ in soils amended with 1.0 M sodium, copper, and calcium salts of chloride, respectively. The nitrification process was decreased with any of these salts as the concentration of the salt increased, with the effects of sodium being less detrimental than the effects of copper or calcium. Complete inhibition of nitrification occurred with high concentrations of $CuCl_2$ and $CaCl_2$ after 49 days. This increasing salt effect on nitrogen transformations helps explain partially the observed slower response times of crops to nitrogen fertilization when grown on salt-affected soils. The effect could be minimized by utilization of a nitrogen fertilizer form not requiring nitrification to be plant available, such as $Ca(NO_3)_2$.

When fertilizer nitrogen is applied in conjuction with saline irrigation water, there is an important effect of the salt in the water on utilization of soil nitrogen. Soil nitrogen is that nitrogen contained in soil organic matter plus adsorbed ammonium and nitrate, which is made available to the plant by mineralization to ammonium with subsequent conversion to nitrate (nitrification). This can be an important supplemental source of nitrogen to the plant requiring little or no cost to the producer. It has been found (Broadbent and Nakashima, 1971; Westerman and Tucker, 1974) that dilute salt concentrations and added nitrogen enhance mineralization of soil nitrogen. This is hypothesized to be either through increased solubility of organic nitrogen and hence easier mineralization of soil nitrogen, or through the stimulatory effect of added nitrogen on microbial decomposition of soil organic matter—a priming effect. As salt concentration is increased, these beneficial effects disappear, and the osmotic effects of excess salts in solution inhibit microbial activity (Johnson and Guenzi, 1963).

Salinity Effects on Nitrogen Fixation

The conversion of atmospheric nitrogen to organic nitrogen, a process known as nitrogen fixation, is in many cases an important source of soil nitrogen to be utilized as a plant nutrient. Such nitrogen either supplements or substitutes for fertilizer-applied nitrogen. In symbiotic nitrogen fixation, bacteria (*Rhizobium*) cause the formation of root nodules in certain host plants (mostly legumes) and then inhabit these growths where they convert gaseous nitrogen (N_2) to forms useful to the plant. This process has been determined to be sensitive to soil salinity, although the exact mechanisms are still subject to debate, as are the fundamental mechanisms of nitrogen fixation. There are generally two methods whereby salt effects are manifested upon fixation. First, a general osmotic effect is hypothesized, in which total soil water potential decreases such that nodules lose water to the soil solution, with resultant damage to their metabolic processes. Second, it has been experimentally suggested (Sprent, 1972) that high levels of soil salts interfere with the metabolism of nodule cortical cells, which prohibits delivery of materials of respiratory origin to nitrogen-fixing regions of the plant. This latter theory explains the observed rapidity of salt effects, as well as including the fact that salinity effects are manifested at or via the nodule surface.

Work with soybeans (Sprent, 1972) showed nitrogen fixation was depressed both with time of exposure and concentration of NaCl up to 0.12 M, the greatest concentration studied. Calcium and alkaline ions are also inhibitory at high concentrations (Vincent, 1974); $CaCl_2$, 0.14 M; KCl, 0.3 to 0.6 M; $NaNO_3$, 0.15 to 0.3 M; Na_2SO_4, 0.3 to 1.2 M. Additional work remains to be done, particularly with reference to agricultural crops, before the inhibition of nitrogen fixation by soil salinity is well understood. However, there is little doubt that success in salt-affected areas in the use of legumes as supplemental nitrogen sources (to replace or minimize use of inorganic fertilizers) will depend greatly on soil salinity levels.

Interaction Effects of Fertilizer and Salinity on Crop Parameters

The literature on the effects of fertilizing saline soils appears contradictory. This is due primarily to the wide variety of crop types and experimental conditions used to study the question. Not only is crop response a function of the type of fertilizer used, it is also influenced by the composition of the nutrient or saline solution used in the experiments. In the following discussion various observations of a number of researchers are presented, with the predicative statement that a wide range of responses to fertilization is possible under saline conditions. Caution should therefore be employed in the extrapolation of any results presented beyond the conditions of the parent study from which they were developed.

Potassium Effects on Yield

The three major fertilizer-applied nutrients, N, P, and K, influence plant growth under saline conditions in differing manners and to different degrees. Of particular importance are N and P, which are discussed in detail below. The third nutrient, K, can be quickly summarized. Potassium absorption for most crops is dependent on a highly specific low-concentration mechanism (Epstein et al., 1963), which for most crops can supply adequate K even in the presence of high concentrations of other cations. Khalil et al. (1967) have suggested K fertilization in saline soils to counteract high cation levels for a few crops such as some varieties of carrots (Bernstein and Ayers, 1953). However, according to Bernstein et al. (1974), this seems to be a special case rather than a rule, as these researchers found that reducing applied K to a few tenths of a milliequivalent per liter had no effect on yield or K content of corn under saline conditions. There are some data (Bernstein et al., 1974) to suggest that K content of leaves is dependent on N and P treatments, but the effect of K fertilizer on uptake and distribution of N and P within the plant is unresolved. Generally, it would be correct to state that most crops can supply themselves with adequate K under saline conditions, minimizing the value of applying K fertilizer. In such cases the effect of adding K alone may be to increase total soluble salts in the soil to such a degree that slight decreases in yield may be detected, such as have been reported for beans by Lunin and Gallatin (1965a).

Nitrogen and Phosphorus Effects on Yield

Plant growth requires the movement of essential nutrients from the soil system into the plant. The effects of excess amounts of soluble salts on this process is extremely variable and can be generalized only with the statement that plant yield is eventually depressed at high salinity levels. The exact amount of depression, and the EC_e to produce the effect are dependent on crop and soil.

These variable factors can be elaborated. Studies of corn and cotton (Bernstein and Ayers, 1953; Broadbent and Nakashima, 1971) have shown that dry matter yields decrease with increasing salinity and increase with nitrogen application. At low levels of soil fertility, nitrogen is limiting to crop growth, with variations due to salinity becoming more evident as N stress is relieved. These effects become particularly pronounced above an EC_e of 8 mmho/cm, with dry matter production by stem decreased more than leaves, which in turn is decreased more than tassels. It was also observed that the percent N in these two crops increased with N applied and with increasing salinity. This apparently indicates that cotton and corn tend to continue to accumulate nitrogen under saline conditions despite reduced dry matter yield. A reduction in N fertilization efficiency may then be observed under such conditions as a result of decreased plant production from salt effects rather than from decreased nitrogen uptake. On the other hand, P percentage was observed to remain constant in the plant with both P applied and salt level. The observed failure of P to increase with salinity as did nitrogen may be explained with the concept of nutrient mobility. Nitrogen,

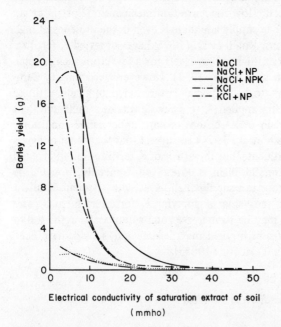

Fig. 1 Barley yields as related to added salt and fertilizer. (From Dregne and Mojallall, 1969.)

when applied as soluble, mobile nitrate ion is relatively unaffected by soil salinity and is capable of being present at the root interface in approximately the same amount regardless of salt levels. Phosphorus, on the other hand, has limited mobility largely because of its relative insolubility. When calcium and magnesium salt concentrations increase (as is the case in most increases in salinity), not only will more P be precipitated to unavailable forms, but the development of the plant's root system will be retarded as a result of increased salinity (Bernstein et al., 1974), decreasing the amount of surface area available to contact the immobile P.

It is not sufficient simply to state that increasing salinity results in decreased yield, for the actual decrease in yield may not become apparent until appreciable increases in salt have occurred. Barley has been shown to respond to salt in the soil by increasing yield slightly, up to an EC_e of 12 mmho/cm (Hassan et al., 1970a). Dregne and Mojallall (1969) have shown improvement in barley yield (Fig. 1) by addition of N, P, and K in soils moderately salinized with Na and K salts. Their findings support the contention that fertilizers can reduce the adverse effects of soil salinity but do so only when salinity is relatively low. Figure 1 also demonstrates the important fact that KCl is equally or more toxic to barley than are equivalent amounts of NaCl. Applications of large amounts of KCl to offset the adverse effects of high NaCl levels (such as would be present in some sodic soils) may actually be more harmful than beneficial.

Effects of Fertility-Salinity Treatments on Plant Mineral Composition

Measurements of plant mineral contents resulting from fertility-salinity treatments have shown this parameter of plant response to be extremely variable (Lunin and Gallatin, 1965a; Hassan et al., 1970a, 1970b; Bernstein et al., 1974) depending on the crop, the particular mineral, and the portion of the plant sampled. It is not therefore considered a reliable indicator of plant response to the interactive effects of soil N, P, and K and salinity. Only general statements concerning mineral nutrition of plants under adverse soil salinity conditions can be made.

In the most extensive studies (Bernstein et al., 1974), it has been found that in only a few cases did salinity per se induce or aggravate a nutrient deficiency. Salinity induced deficiencies of Ca; a few other similar effects have been observed (Bernstein, 1964b). Some decreases in contents of certain micronutrients in barley and corn (Cu, Fe in barley and Zn, Fe, Cu, Mn in corn) have been noted (Hassan et al., 1970a, 1970b) with increases in salinity. However, effects of fertilization were not included in the latter study. Generally, it is accepted that fertilization under saline conditions will promote adequate uptake

Table 3 Salinity Fertility Effects on Evapotranspiration and Yield in Beans[a]

Salinity level	N added				P added			
	0	75	150	300	0	75	150	300
	Water used (ml/g dry weight of leaves)							
A	53.1	58.2	56.9	58.8	103.4	71.7	61.0	58.9
B	53.6	54.2	52.7	54.0	85.1	74.5	54.4	54.0
C	42.7	45.1	45.2	42.2	59.9	49.8	49.0	42.2
D	34.2	25.3	42.1	32.1	–	–	–	–

[a]From Lunin and Gallatin (1965b).

of mineral nutrients from the soil solution, as long as the salinity is not so great as to severely restrict root growth. This generalization still remains to be completely resolved in light of research results into plant nutrition under salinized conditions.

Salinity Effects on Evapotranspiration

The effect of salinity on evapotranspiration is of considerable interest. Since increasing salinity increases the soil water osmotic potential (and therefore the total soil water potential), a resultant decreases in transpiration would be expected. Studies with beans (Lunin and Gallatin, 1965b), presented in Table 3 (treatments A, B, C, and D are identical to those of Table 1) have shown evapotranspiration per unit dry weight of leaves over a 48-hr period decreased with increasing salinity in response to treatment with phosphorus.

Quantifying and Interpreting the Results of Fertility-Salinity Interaction Studies

The analysis of data gained from studies concerning fertility-salinity interactions can take several forms. It is not always apparent from the presentation of such data whether interaction between different experimental treatments actually was indicated by the data gathered. In light of these possible ambiguities, it is important to discuss briefly the most common and useful methods of data presentation of interaction studies.

Two-dimensional Plots

The most widely used presentation involves the plotting of the dependent parameters (yield, nutrient content, evapotranspiration) on the ordinate axis

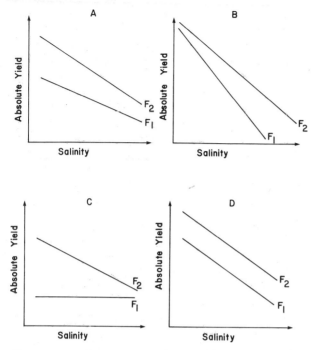

Fig. 2 Types of interaction of salinity and fertility.

with one of the independent variables (salinity or fertility level) plotted on the abscissa. The experimental results are then plotted on this xy axis as functions of the second independent variable utilized in the study. Bernstein et al. (1974) have discussed this type of presentation quite well. For example, Fig. 2 shows four possible responses to fertility treatments as salinity is increased. In each case of Fig. 2, the upper line (F_2) represents a theoretical yield response to salinity under conditions of optimum nutrition. In Fig. 2A, the lower curve (F_1) represents independent responses to salinity and a limiting fertility condition; that is, response to salinity is the same when nutrition is limiting as when it is adequate. The curves actually converge, and may show a statistically significant interaction because when nutrition limits yield, the effects of salinity on absolute yield cannot be as great as when yields are maximal with optimum nutrition. If relative yields (nonsaline treatment for each level of nutrition taken as 100%) instead of absolute yields are plotted, the two curves in Fig. 2A would coincide. Relative yield, often taken as a measure of salt tolerance, is not affected in this case. Figures 2B and 2C contrast yield responses to salinity under differing fertility levels. Comparing the F_1 curves of these two plots, a type B response indicates that at a particular salinity level, there is a greater decrease in yield when nutrition is limiting than when it is adequate, whereas a type C response indicates no salinity effects on yield at an F_1 fertility level. A type B response occurs whenever nutrition that is adequate under nonsaline conditions becomes limiting under saline, as with the previously mentioned salinity-induced Ca deficiency in some crops. Hot, dry climate and low relative humidity may increase the sensitivity of some crops to salinity, also producing a type B response. A type C response can occur when a crop is grown under a moderate (F_1) fertility regime with no salinity effect on it, but is affected by the same salt levels at higher (F_2) fertility, a condition perhaps induced by the salt effect of the applied fertilizer upon the soil solution. The curves in Fig. 2 are idealized to represent only the range over which yield is sensitive and linearly responsive to salinity. Actual salinity response plots may be curvilinear or include plateaus in addition to the sensitive response ranges depicted.

Figure 2D indicates no interaction between salinity and fertility. The parallel nature of the lines shows that yield is simply depressed by a limiting fertility condition, with effects of salinity imposed identically on each treatment. The importance or unimportance of interaction effects can be judged from such cases. When the lines are almost parallel, an interaction effect is present, but is only minimally exhibited, and can therefore be judged relatively unimportant. As the plotted lines begin to deviate substantially from parallel, the interaction effects become more apparent and therefore can be considered of more importance in the operation of the system under study.

Response Surfaces

A second commonly used means to represent data is the response surface. These surfaces, often referred to as regression surfaces, are three-dimensional representations of the relationships between the dependent and independent variables produced through multiple regression analysis of experimental data. A wide variety of approaches to regression modeling of data are available. This technique involves considerable expertise in statistics, but is quite useful and often more informative than the two-dimensional technique explained above.

A typical response surface for a fertility-salinity interaction study is presented in Fig. 3A (Langdale and Thomas, 1971). The crop in this case was Bermuda grass grown under greenhouse conditions. The exact numbers corresponding to the salinity and fertility levels are unimportant. Both salt and nitrogen applied increased with increasing subscript.

The utility of the response surface in presenting data of this type should be apparent from Fig. 3A. With a minimum of effort, the reader is able to grasp the two-way effects of the independent variables on the dependent variable, dry

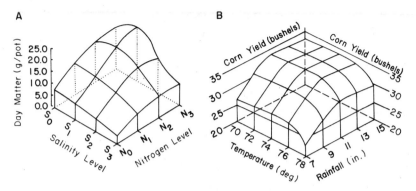

Fig. 3 Examples of response surfaces for interacting independent variables (A) and noninteracting independent variables (B). (fig. DA from Langdale and Thomas, 1971.)

matter yield. Absolute values of experimental results are not as easily ascertained as in the case of tabular presentation of the data. However, the interaction effects are readily apparent from the nonsymmetrical nature of the surface. If no interaction between independent variables had occurred, then all curves of nitrogen response at a chosen salinity (or conversely all curves of salt effect at a chosen fertility) would have had the same slope and differed only by a constant. An idealized representation of a response surface for noninteracting independent variables is presented in Fig. 3B. Note how all response curves are the same shape and differ only in the magnitude of yield detected. Figure 3B is hypothetical and used as an illustrative example. No such response surfaces have been reported for salinity-fertility studies.

Summary

The interactive effects of soil fertility and salinity are of major concern in situations often encountered in the agricultural utilization of salt-affected lands or saline irrigation waters. Although it is impossible to generalize the effects of salts and fertilizers on crop yield, it can be stated that in most cases moderate levels of soil salinity can be compensated by increased fertilization, so long as the salinity level is not excessively high nor the crop particularly salt-sensitive. The exact nature of crop response, however, depends greatly on crop type, fertilizer form and quantity, soil type, and irrigation water quality and soil salinity (see Chap. 4). At present, these interacting factors have been studied only superficially, with results often contradictory depending on the particular environmental conditions surrounding each study. It is therefore wise to consider the basic physical and chemical mechanisms operative in the soil system in assessing the potential for salt damage to a crop of any specific combination of applied fertilizer and salt. Such mechanisms include the solubility relationships of the applied fertilizer, osmotic effects of salts on water availability and microbiological activity, nutrient absorption by plants, and the influence of specific soil properties (i.e., cation exchange capacity, water-holding capacity). With these fundamental considerations in mind, each case of fertility-salinity interaction can and should be evaluated individually.

References

Amer, F., Elgabaly, M. M., and Balba, A. M. (1964). Cotton response to fertilization on two soils differing in salinity. Agron. J. *56*: 208-211.

Bernstein, L. (1964a). Salt tolerance of plants. U.S. Dept. Agr. Inform. Bull. 283, 23 p.

Bernstein, L. (1964b). Effects of salinity on mineral composition and growth of plants. Plant Anal. Fert. Probl. *4*: 25-45.

Bernstein, L. and Ayers, A.D. (1953). Salt tolerance of five varieties of carrots. Amer. Soc. Hort. Sci., Proc. *61*: 360-366.

Bernstein, L., Francois, L. E., and Clark, R. A. (1974). Interactive effects of salinity and fertility on yields of grains and vegetables. Agron. J. *66*: 412-421.

Broadbent, F. E., and Nakashima, T. (1971). Effect of added salts on nitrogen mineralization in three California soils. Soil Sci. Soc. Amer. Proc. *35*: 457-460.

Dregne, H. E., and Mojallall, H. (1969). Salt-fertilizer-specific ion interactions in soil. New Mexico Agr. Exp. Sta. Bull. 541, 16 pp.

Epstein, E., Rains, D. W., and Elzam, O.E. (1963). Resolution of dual mechanisms of potassium absorption by barley roots. Nat. Acad. Sci. Proc. *49*: 684-692.

Hassan, N. A. K., Drew, J. V., Knudsen, D., and Olson, R. A. (1970a). Influence of soil salinity on production of dry matter and uptake and distribution of nutrients in barley and corn: I. Barley (*Hordeum vulgare* L.). Agron. J. *62*: 43-45.

Hassan, N. A. K., Drew, J. V., Knudsen, D., and Olson, R. A. (1970b). Influence of soil salinity on production of dry matter and uptake and distribution of nutrients in barley and corn: II. Corn (*Zea mays* L.). Agron. J. *62*: 46-48.

Johnson, D. D., and Guenzi, W. D. (1963). Influence of salts on ammonium oxidation and carbon dioxide evolution from soils. Soil Sci. Soc. Amer. Proc. *27*: 633-666.

Khalil, M. A., Amer, F., and Elgabaly, M. M. (1967). A salinity-fertility interaction study on corn and cotton. Soil Sci. Soc. Amer. Proc. *31*: 683-686.

Langdale, G. W., and Thomas, J. R. (1971). Soil salinity effects on absorption of nitrogen, phosphorus, and protein synthesis by coastal bermudagrass. Agron. J. *63*: 708-711.

Luken, H. (1962). Saline soils under dryland agriculture in South Eastern Saskatchewan (Canada) and possibilities for their improvement. Part II. Evaluation of effect of various treatments on soil salinity and crop yield. Plant Soil *17*: 26-48.

Lunin, J., and Gallatin, M. H. (1965a). Salinity-fertility interaction in relation to growth and composition of beans: I. Effect of N, P, and K. Agron. J. *57*: 339-342.

Lunin, J., and Gallatin, M. H. (1965b). Salinity-fertility interactions in relation to the growth and composition of beans. II. Varying levels of N and P. Agron. J. *57*: 342-345.

Maas, E. V., and Hoffman, G. J. (1976). Crop salt tolerance: Evaluation of existing data. In H. E. Dregne (ed.), *Managing Saline Water for Irrigation. Proceedings of the International Conference on Managing Saline Water for Irrigation: Planning for the Future.* Center for Arid and Semi-Arid Land Studies, Texas Tech University, College Station, Tex.

Rader, L. F., White, L. M., and Whittaker, C. W. (1943). The salt index—A measure of the effect of fertilizers on the concentration of the soil solution. Soil Sci. *55*: 201-218.

Sprent, J. E. (1972). The effects of water stress on nitrogen fixing root nodules. III. Effects of osmotically applied stress. New Phytologist *71*: 451-460.

U.S. Salinity Laboratory Staff (1954). *Diagnosis and Improvement of Saline and Alkali Soils*, U.S. Dept. Agr. Handbook No. 60, U.S. Govt. Printing Office, Washington, D.C., 160 pp.

Vincent, J. M. (1974). Root nodule symbiosis with Rhizobia. In A. Quispel (ed.), *The Biology of Nitrogen Fixation*, American Elsevier, New York, pp. 265-341.

Westerman, R. L., and Tucker, T. C. (1974). Effect of salts and salts plus nitrogen-15 labeled ammonium chloride on mineralization of soil nitrogen, nitrification and immobilization. Soil Sci. Soc. Amer. Proc. *38*: 602-605.

6

Impact of Irrigation on the Quality of Groundwater and River Flows

Gaylord V. Skogerboe and Wynn R. Walker

Colorado State University, Fort Collins, Colorado

Introduction

Whether the goal is minimizing diversions to new croplands because of limited water supplies, reducing future diversions to irrigated agriculture to provide water supplies for new demands, minimizing water quality degradation in receiving streams resulting from irrigated agriculture, or maximizing agricultural production on existing croplands, the solutions are identical—improved water management practices.

Irrigation is one of the most important agricultural practices developed by man, with irrigation being practiced in some form since the earliest recorded history of agriculture. The economic base for many ancient civilizations was provided by irrigation. Irrigation farming not only increases productivity, but it also provides flexibility that allows shifting from the relatively few dryland crops to many other crops that may be in greater demand. Irrigation contributes to strengthening other facets of a region's economy in that it usually creates more employment opportunities than rain fed agriculture, through intensive and diversified cultivation, the stimulation of important agribusinesses and public institutions, the provision of products for export, and the creation of a healthy domestic market (Skogerboe, 1974).

History is replete with civilizations declining because of declining agricultural productivity, which has usually been the result of improper soil and water management practices. This potential danger is present today at many places throughout the world. This comes at a time of dangerous food shortages in a hungry world, when the goal is to increase per-acre crop yields.

To maintain agricultural productivity in irrigated agriculture—and we must do more than that today—salts applied onto the croplands, which are dissolved in the irrigation water supplies, must be moved below the root zone in order not to retard plant growth. Therefore, it is mandatory that water supplied to a crop must exceed the actual water requirement of the plants to include evapotranspiration needs, leaching requirements, seepage losses, and in most cases other transit or ditch losses which in most countries are substantial.

As is known, usually the quantity of irrigation water diverted from a river far exceeds the cropland water requirement. Data from many irrigation regions indicate that seepage losses from canals and laterals throughout the irrigation distribution systems are extremely high. Added to this problem is the excessive application of water on farm fields, which results in surface runoff from the lower end of the field (tailwater runoff) and/or large quantities of water moving below the root zone (deep percolation). The combination of seepage and deep percolation losses cause groundwater levels to rise (water-logging). In many irrigated regions, the groundwater levels have reached the vicinity of the root zone, which frequently results in the upward movement of groundwater as a result of capillary action. When upward-moving water reaches the soil surface and evaporates, the salts contained in the moisture are left behind on the ground surface. This process of salinization has not only resulted in declining agricultural production, but has caused many lands to become essentially barren.

The quality of water draining from irrigated areas is materially degraded from that of the irrigation water applied, as we have discovered along the Colorado and other rivers.

Agriculturists have viewed this phenomenon as a natural consequence of several complex processes, and strangely enough, even today little attention has been given to the possibility that progress could be made toward controlling or alleviating this degradation of water quality.

Historically, some degree of salt concentration as a result of irrigation has usually been accepted as the price for irrigation development. In some areas, however, there has been so much laxity that quality degradation has become a serious matter. As pressures on water resources become greater, as a result of increasing populations and the necessity to produce food in increased quantity and improved quality, there is a mounting concern for proper control of serious water quality deterioration and soil salinization. The need, then, for more precise information as a basis for wise policy action is a matter of critical importance.

The major problems resulting from irrigation are due to the basic fact that plants are large consumers of water resources. Growing plants extract water from the supply and leave salts behind, resulting in a concentration of the dissolved mineral salts that are present in all natural water resources. In addition to having a greater concentration of salts in the return flow resulting from evapotranspiration, irrigation also adds to the salt load by leaching natural salts arising from weathered minerals occurring in the soil profile, or deposited below. Irrigation return flows provide the vehicle for conveying the concentrated salts and other pollutants to a receiving stream or groundwater reservoir. It is necessary, then, to examine the waterlogging and salinity problems resulting from this process and to develop and implement measures to control or alleviate the detrimental effects.

Impaired crop production resulting from salinity is not limited to the western United States, but is a major problem in many areas of the world. In fact, the more one observes problems in other countries, the more one is convinced that this is indeed an international problem. Unfortunately, the portions of the world now facing the greatest population pressures are the same areas that have the least amount of additional land available for agriculture and in some cases where waterlogging and salinization are rampant. In such areas, increased food production must come from more intensive farming with consequent increased yields; and, as the economist reminds us, the modernization of agriculture requires that output continue to rise with decreasing per-unit input costs. On the one hand, there is a great need to increase the productivity of such lands; but on the other hand, agricultural production is continually being damaged because of rising groundwater tables (waterlogging) and increased salinity in the soils and groundwater supplies. Although there are estimates of the extent of these salt problems in the world, suffice it to say that it is a creeping plague of death in many places.

Whenever water is diverted from a river for irrigation use, the quality of the return flow becomes degraded. The degraded return flow then mixes with the natural flows in the river systems. This mixture is then available to downstream users to be diverted to satisfy their water demands. This process of diversion and return flow may be repeated many times along the course of a river. In the case of the original diversion, if the increase in pollutants contained in the return flow is small in comparison to the total flow in the river, the water quality would probably not be degraded to such an extent that it would be unfit for use by the next downstream user.

If the quantity of pollutants (e.g., salinity) in the return flow is large in relation to the river flow, then it is very likely that the water will not be suitable for the next user unless the water is treated to remove objectionable constituents. Since water is diverted many times from the major rivers, the river flows show a continual degradation of quality in the downstream direction. As the water resources become more fully developed and utilized, without controls, the quality in the lower reaches of the river will likely be degraded to such a point that the remaining flows will be unsuitable for many uses, or previous uses of the waters arriving at the lower river basin no longer will be possible.

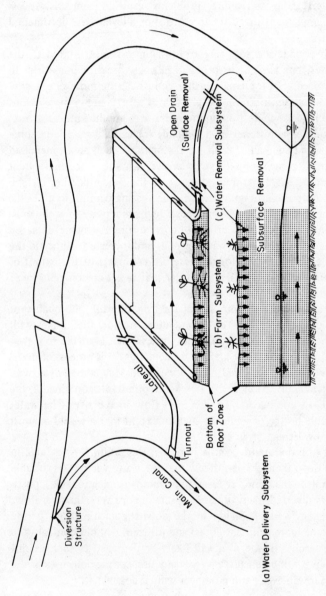

Fig. 1 The water delivery, farm, and water removal subsystems.

Definition of an Irrigation System

An irrigation system can be subdivided into three major subsystems: (1) the water delivery subsystem; (2) the farm subsystem; and (3) the water removal subsystem (Fig. 1). The water delivery subsystem can be further subdivided into two components, namely, (a) the transport of water and pollutants from the headwaters of the watershed to the cross section along the river where water is diverted to irrigate croplands, and (b) the transport of water and pollutants from the river diversion works to the individual farm. The farm subsystem begins at the point where water is delivered to the farm and continues to the point where surface water is removed from the farm. Also, the farm subsystem is defined vertically as beginning at the ground surface and terminating at the bottom of the root zone. The water removal subsystem consists of (a) the surface runoff from the tail end of the farm, and (b) water moving below the root zone.

Planning for Effective Water Management

The resource base for irrigated agriculture has not changed substantially since its inception thousands of years ago. Over the last century there has been little incentive for any major innovation to improve efficiency in the use of water, which now fortunately is considered a scarce resource. The provision of irrigation water since ancient times has been considered a governmental or collective responsibility, and the direct charges made for water have usually not been high enough to encourage proper budgeting of water or other necessary innovations. This custom of undercharging for water has continued to this very day; only a few regions charge the farmer for the real cost of water. However, there are a few examples of extremely water-scarce areas in the world where considerable ingenuity has been applied in utilizing water supplies effectively.

Aggravating this situation of so-called cheap water is the fact that the development of irrigated agriculture in most places, even in the last few decades, has focused almost entirely on the construction of water delivery subsystems. This preoccupation with the installation of "hardware" results from a naive single-discipline approach to water management (Wiener, 1972); let us be reminded that one discipline cannot begin to solve the complex physical, economic, political, and sociological problems involved. Probably the greatest deterrent to improved water management in most irrigation systems today is the inordinate focus on the water delivery subsystem and the almost complete neglect of other problems. How long can we afford to neglect other problems of the system, such as the need for improved soil-plant-water management techniques, improvements in cultural practices, farm machinery, agrarian structural reforms, roads, marketing systems, advisory services and input supply systems, administration of institutions, water laws, cooperatives and water users associations, and many other factors, all of which must fit together in all their interdependencies and complementarities to form a most complex system? In reality, especially in irrigation agriculture, we come face to face with the wide gap that frequently exists between "hardware development" and the development of all the other requisites for increased agricultural production.

The approach that has been applied to irrigated agricultural development in the past is characterized by separating the development of water resources from the management aspects of water resource utilization. Therefore, the record shows development being emphasized greatly while management is most often neglected. This orthodox approach has been used almost exclusively in the western United States with reasonable success. However, as the water resources become more fully utilized, the necessity for meeting new water demands (along with physical, socioeconomic, and political problems of water quality degradation) require that we reject much of the conventional wisdom of the past. It should also be obvious that the less developed countries (LDC's) today have neither the time nor the resources many Western irrigated regions had to utilize in their development. Pressures created by rapidly rising population rates alone will at some point force these countries to reevaluate their approaches.

In contrast to the mere development of water resources approach, the "management" approach attempts to achieve water development objectives by applying a variety of measures after studying the *entire system*, thereby attempting to modify the total system to meet new and changing demands as well as estimated future demands. Therefore, instead of constructing new engineering works to meet new demands, the focus should be on water resource management, with construction works being considered only as a tool when necessary to meet water management objectives (Wiener, 1972). Unfortunately, in most cases, water management and the many disciplines required to produce efficient management are relegated to the post analysis of engineering works (e.g., much future emphasis will be geared toward improving existing irrigation systems), which aggravates not only the implementation of technology, but really constrains or makes extremely difficult the implementation of a host of services requiring strong institutional measures.

Design of Irrigation System

The "heart" of an irrigation system is the farm subsystem. The purpose of an irrigation system is to grow food, and this "action" takes place in the root zone. The purpose of the water delivery subsystem and water removal subsystem is to *support* this "action." Therefore, the proper design of an irrigation system requires, first of all, that the farm subsystem be adequately designed so that the water delivery subsystem can be designed to provide the quantities of water at the times required by the plants. The most important constraint in the design procedure is the necessity for assuring adequate drainage through the root zone in order to maintain a root zone salt balance to ensure continued long-term agricultural productivity. A "balanced" design is achieved by circuitously designing and redesigning each subsystem until the "total" system will operate in such a manner as to optimize crop production while minimizing drainage requirements, and sustaining irrigated agriculture on a long-term basis.

Farm Subsystem

The first important variables in designing the farm subsystem are climate, soils, and crops. The interrelationships among these variables dictate the capability of the land resource for producing food and fiber. We need to understand the interactions and relationships between water and management factors such as optimum rates of fertilizer, proper pest control measures, seeding and tillage practices, and other improved cultural practices. Much of this research is adaptive and protective, which because of its rather low level of sophistication, is sometimes neglected.

The next important variable is water and its physical availability. Frequently, in arid areas of the world, water is the limiting factor. However, the capability of the available water supplies (precipitation, surface runoff, and groundwater) for plant production is highly dependent on the efficiency with which the water is used, which in turn is a function of both economic and institutional factors. Besides physical limitations, the questions of economics in supplying water must be answered to ensure that costs are commensurate with planning goals.

Once the *general scope* of an irrigation project has been determined, then more detailed design procedures can follow. The *critical factor* at this point becomes the *infiltration* characteristics of the soil. Unfortunately, infiltration is a complex phenomenon, and the intake function of a particular field will vary during the irrigation season, as well as varying from season to season. There are a number of laboratory and field methods available for determining the intake characteristics of a soil.

Using climatic data, the potential evapotranspiration of the various crops can be calculated. These computations will provide the information regarding water consumption with time, provided that sufficient moisture is made available in the root zone.

The next important step is designing the farm irrigation layout so that sufficient moisture will be available in the root zone when required by the plants. The root zone is capable of storing moisture for future plant use. Again, soil characteristics determine the amount of storage, as well as the capability of the plant to extract the moisture from this "reservoir." At the same time, the leaching requirement for maintaining a salt balance in the root zone must be kept in mind. Consequently, the farm irrigation layout must be capable of supplying not only the plant water requirement, but also the leaching requirement.

The proper design of the farm irrigation layout is *crucial* for (1) uniformly distributing the necessary moisture throughout the field; and (2) minimizing deep percolation losses so as not to aggravate problems in the water removal subsystem (e.g., waterlogging and salinity).

Generally, the development of irrigation projects has not entailed the design of farm irrigation layouts suited to the individual characteristics of each field. Instead, only the general method of irrigation may be adopted (e.g., basin or furrow irrigation). The farmer is usually left to his own means in irrigating his fields, without having the benefit of technical (or economic) assistance. The situation is further aggravated because, along with adopting a general method of irrigation, an average irrigation efficiency for this method is "pulled out of the air." If this so-called average water use efficiency was even close to being correct, it would be most fortunate—let alone taking into account the variability of this "magic" number on any specific field throughout the irrigation season, as well as the variability from field to field. Usually, the application of this average efficiency results in large quantities of deep percolation during the early part of the irrigation season, which in turn contributes to waterlogging of the soils and consequent poor crop yields (or eventual failure of the irrigation enterprise).

Water Delivery Subsystem

The design of the individual farm irrigation layouts should dictate the design of the water delivery subsystem. The irrigation layout design, if properly accomplished, will show the necessary quantities and timing of water deliveries at the farm inlets. The water delivery network must be designed to meet the farm water requirements. Except for alluvial channels conveying large sediment loads, the design of the conveyance works is rather "mechanical."

One of the essential facilities for successfully operating an irrigation conveyance network is adequate and numerous flow measurement devices. To begin with, since each farm has a particular water requirement, then the only means by which the *proper* amount of water can be delivered is by measuring the water at the farm inlet. After all, the farmer cannot be expected to use good water management practices if he doesn't even know the quantity of water being managed. Besides each farm inlet, a flow measurement structure should be provided at all division points in the water delivery subsystem.

The real problem in the water delivery subsystem is the *institutional* framework controlling the operation of this portion of the irrigation system. Generally, the operation of the conveyance facilities has not been related to the requirements for sustaining a long-term productive agriculture. In particular, institutional factors have acted as constraints to improved water management or increased agricultural production.

The primary requirement for sustaining an irrigation system is an institutional framework that is compatible with the design requirements for the water delivery subsystem, which in turn has been dictated by the proper design of the farm irrigation layouts, as well as any constraints imposed by the water removal subsystem. Thus, even if all three components of the irrigation system have been properly designed, the lack of an adequate institutional framework for operating the system in accordance with the design criteria will likely lead to either failure of the system, or at least having agricultural production levels below (or far below) expectations. Many LDC institutions for irrigation systems were developed by outsiders for other than the modernization of agriculture. In some LDC's, the objective was a simple system that would almost operate itself. Some of these institutions were regulatory rather than service-oriented, and little attention was ever given beyond the canal outlet.

Water Removal Subsystem

The principal function of the water removal subsystem is to allow proper drainage below the root zone so that adequate leaching of salts from the root zone will occur. The most satisfactory mechanism for ensuring adequate drainage is proper operation of the water delivery subsystem. By so doing, a drainage problem will probably not occur. This is much better than allowing the problem to occur, then constructing drainage facilities to correct the damage. Unfortunately, the usual solution consists of constructing additional facilities. Frequently, project reports will state "...drainage facilities will be designed after the project has been in operation for a number of years in order more precisely to ascertain drainage requirements." This is the same naive, single-discipline thinking referred to in the previous section, and is the rule rather than the exception. Perhaps the statement is correct that the history of irrigation systems teaches one that man often does not learn from history.

Another important consideration in the water removal subsystem is water quality. If canal seepage and cropland deep percolation losses result in water quality degradation of the underlying groundwater supplies, then the use of these supplies may become impaired. Also, the return flows to the river may limit the usefulness of the river water to downstream users. Numerous examples of this situation can be cited throughout the world. Again, this is the rule rather than the exception.

Administration of Irrigation System

The proper functioning of an irrigation system is highly dependent on an institutional framework that is compatible with the design criteria used in developing the system, as well as providing flexibility in achieving improved water management as the need arises. Satisfying on-farm water management objectives cannot be achieved without controlling water deliveries. Therefore, the administration of the irrigation system requires that satisfactory legal mechanisms exist that control water deliveries.

The failure of new irrigation developments to meet estimated production goals has largely resulted from the lack of followup in providing essential agricultural inputs and services to farmers. Again, the lack of training in farm water management and other farm practices by those responsible for the planning, design, and construction of the irrigation development leads to shortcomings in output. Also, the lack of cooperation between disciplines is certainly a detriment in any country, but is particularly noticeable in the developing countries. Interdisciplinary team research is a noble concept, but it is seldom internalized by researchers and implemented effectively. Perhaps the most expedient solution is to expand the training of those personnel involved in the planning, design, and construction of irrigation enterprises. Such training might then provide feedback into the planning and design of future irrigation developments.

The provision of adequately trained personnel for the operation and maintenance of irrigation systems may be understood, but is sometimes difficult to accomplish. Much focus must be given to training not only the engineers and technicians, but those who work directly and indirectly with the farmer. Although agricultural experiment stations may exist, there is usually a severe limitation in accomplishing on-farm improvements because of insufficient numbers of adequately trained farm-level advisors capable of transferring applied research results directly to the farmers. In many countries, the components of research and advisory services have never been brought together in a carefully planned marriage relationship.

We are all aware of the need for improved delivery systems for technology once it has been made available by research. How best can this transfer of technology in water management be made to the end user, the farmer? Here we need the help of the applied social sciences such as economics, sociology, anthropology, political science, and even social psychology. Some extension system is required that will take the proven findings of research and utilize the knowledge of the social sciences "package" in such a way as to convince the end user to move from the trial to the final adoption stage.

Major Water Quality Problems

Usually, the quality of water coming from watersheds is excellent. At the base of the hills or mountain ranges, large quantities of water are diverted to valley croplands. Much of the diverted water is lost to the atmosphere by evapotranspiration (perhaps one-half to two-thirds of the diverted water), with the remaining water supply being irrigation return flow. This return flow will either be surface runoff, shallow horizontal subsurface flow, or will move vertically through the soil profile until it reaches a perched water table or the groundwater reservoir, where it will remain to be pumped or be transported through the groundwater reservoir until it reaches a river channel.

That portion of the water supply which has been diverted for irrigation but lost by evapotranspiration (consumed) is essentially salt-free. Therefore, the irrigation return flow will contain most of the salts originally in the water supply. The surface irrigation return flow will usually contain only slightly higher salt concentrations than the original water supply, but in some cases the salinity may be increased significantly. Thus, the water percolating through the soil profile contains the majority of salt left behind by the water returned to the atmosphere as vapor through the phenomena of evaporation and transpiration. Consequently, the percolating soil water contains a higher concentration of salts. This is referred to as the "concentrating" effect.

As the water moves through the soil profile, it may pick up additional salts by dissolution. In addition, some salts may be precipitated in the soil, and there will be an exchange between some salt ions in the water and in the soil. The salts picked up by the water in addition to the salts that were in the water applied to the land are termed salt "pickup." The total salt load is the sum of the original mass of salt in the applied water as the result of the concentrating effect plus the salt pickup.

Whether irrigation return flows come from surface runoff or have returned to the system via the soil profile, the water can be expected to undergo a variety of quality changes as a result of varying exposure conditions. Drainage from surface sources consists mainly (there will be some precipitation runoff) of surface runoff from irrigated land. Because of its limited contact and exposure to the soil surface, the following changes in quality might be expected between application and runoff: (1) dissolved solids concentration only slightly increased; (2) addition of variable and fluctuating amounts of pesticides; (3) addition of variable amounts of fertilizer elements; (4) an increase in sediments and other colloidal material; (5) crop residues and other debris floated from the soil surface; and (6) increased bacterial content (Skogerboe and Law, 1971).

Drainage water that has moved through the soil profile will experience different changes in quality from surface runoff. Because of its more intimate contact with the soil and the dynamic soil-plant-water regime, the following changes in quality are predictable: (1) considerable increase in dissolved solids concentration; (2) the distribution of various cations and anions may be quite different; (3) variation in the total salt load depending on whether there has been deposition or leaching; (4) little or no sediment or colloidal material; (5) generally, increased nitrate content unless the applied water is unusually high in nitrates; (6) little or no phosphorus content; (7) general reduction of oxidizable organic substances; and (8) reduction of pathogenic organisms and coliform bacteria. Thus, either type of return flow will affect the receiving water in proportion to respective discharges and the relative quality of the receiving water.

The quality of irrigation water and return flow is determined largely by the amount and nature of the dissolved and suspended materials they contain. In natural waters, the materials are largely dissolved inorganic salts leached from rocks and minerals of the soil contacted by the water. Irrigation, municipal and industrial use, and reuse of water concentrates these salts and adds additional kinds and amounts of pollutants. Many insecticides, fungicides, bactericides, herbicides, nematocides, as well as plant hormones, detergents, salts of heavy metals, and many organic compounds, render water less fit for irrigation and other beneficial uses.

Potential Solutions

Achieving high levels of water use efficiency in order to minimize canal deliveries and prevent or control waterlogging and salinity due to irrigation return flow is both difficult and expensive. Potential solutions and control measures involve physical changes in the system, which can be brought about by constructing sufficient improvements to new or existing systems, or by placing new institutional influences on the system, or by a combination of both. Since irrigation water deliveries and return flow are an integral part of the hydrologic system, control measures for managing the water deliveries and return flows from an irrigated area must be compatible with the objectives for water resource management and development in the total system.

In most instances, the quantity and quality problems in the water removal subsystem are minimized by having highly efficient water delivery and farm subsystems. Minimizing the quantity of surface runoff will assist in alleviating quality problems resulting from sediments, phosphates, and pesticides; whereas minimizing deep percolation losses from irrigated lands will reduce waterlogging and quality problems due to salts, including nitrates.

Water Delivery Subsystem

The importation of high-quality water from adjacent river basins, weather modification to increase precipitation and runoff from the watersheds, bypassing mineralized springs, evaporation reduction from water surfaces, and phreatophyte eradication are some of the available measures for improving the quality of water diverted from a river. Consequently, they play a role in the management of the irrigation system. More feasible approaches may be found in the control of losses from storage and conveyance systems.

Canal Lining

Many unlined irrigation canals traverse long distances between the diversion point and the farm land. Seepage losses may be considerable, resulting in low water-conveyance efficiencies. Canal lining has traditionally been employed to prevent seepage, and the economics of lining have been justified primarily on the basis of the value of the water saved. The possibility that water seeping from canals may greatly increase the total contribution of dissolved solids to receiving waters has only recently been given serious attention, whereas canal lining to prevent waterlogging has been recognized for a long time.

If soils along the canals are high in residual salts, the salt pickup contribution from this source may easily exceed that leached from the irrigated land to maintain a salt balance. The time required to leach these residual salts will depend on the quantity of seepage and the quantity of salts. In addition to the quantity of water saved, the salt from this source could be largely eliminated by canal lining. The value of improved water quality is another benefit to be claimed in the economic justification of canal lining.

Closed Conduits

Evaporation losses from canals commonly amount to a few percent of the diverted water. The installation of a closed conduit (pipeline) conveyance system has the advantage of minimizing both seepage and evaporation losses. Either lined open channels or closed conduits will reduce evapotranspiration losses due to phreatophytes and other noneconomic vegetation along canals. The closed conduit system uses less land and provides for better water control than a canal system. Water quality improvement may very well prove to be the greatest economic justification for closed conduit systems because of minimal seepage losses and considerable flexibility in water control.

Flow Measurement

A key element that must be provided in the water delivery system is flow measurement. The amount of water passing key points in the irrigation delivery system must be known in order to provide water control and attain a high degree of water use efficiency. Many present-day systems employ no flow-measuring devices, and, in some cases, the individual farmer operates his own turnout facilities with no close control of the amount diverted to the farm.

Project Efficiency

Economics play a major role in existing project-irrigation efficiencies, and a close correlation exists between abundance and/or cost of water and project efficiency. For example, where water is scarce of high in cost, the efficiencies are found to be higher. Project management, as well as farm management, involves balancing the immediate cost of water against the higher labor and investment costs required to use it more efficiently (Jensen et al., 1967). The costs of inefficient water use often are not recognized immediately but may be reflected in reduced yields due to nutrient losses or increased salinity, or in extra drainage facilities required later to control rising water tables.

On-Farm Water Management

The most significant improvements in reducing water requirements and controlling waterlogging or salinity will potentially come from improved on-farm water management. This will be particularly true for areas containing large quantities of natural pollutants, such as salts, in the soil profile. In such situations, the key is to minimize the subsurface return flows, thereby minimizing the quantity of pickup. Poor irrigation practices on the farm are the primary cause of overly large water diversions, as well as being the primary source of present return-flow quality problems. Besides improvements at the source, other improvements can be accomplished in the water removal system. Because of the nature of irrigated agriculture, whereby salts must be leached from the root zone, an optimum solution will, in most cases, require improvements in on-farm water management. Numerous technological and institutional concepts could be utilized to accomplish improved water quantity and quality management. Some of the technological possibilities are cited immediately below.

Cultural Practices

When the soils to be irrigated are tight (low infiltration rate and low permeability), and the water supply delivered to the farm is highly saline, cultural practices become extremely significant if crops are to be grown successfully. This situation is aggravated even further in irrigated areas having extremely high summer temperatures. Under these conditions, the management alternatives become: (1) use more salt-tolerant plants (which are usually lower in cash value); (2) use special soil tillage practices (which cost more); (3) leach in the off-season; (4) leach the field one year and plant a crop the next year; (5) prepare the seedbed more carefully; or (6) control the timing and amount of water being applied. Usually, these problems must be faced in the lower regions of a river basin, where the accumulative effects of upstream water quality degradation, along with having finer soils resulting from river deposition, create difficult management conditions.

In general, the deeper water is stored in the soil, the more slowly it will be removed by evapotranspiration. Soil structure, texture, and stratification are the principal properties that control distribution of water storage in the soil. In extreme cases, deep tillage may be required to disrupt slowly permeable layers and permit greater water storage capacity, as well as deeper root penetration. At the same time, excessive or unnecessary tillage can be detrimental to stored soil water, increasing evaporative losses when the crop needs it most. Therefore, cultural practices can play a significant role in overall farm management.

Fertilizers

There is a strong relationship between water use efficiency and fertilizer use efficiency. Applying excessive quantities of water to croplands results in leaching of fertilizer materials below the root zone, where they are unavailable for plant growth. One real potential for improving nitrogen use efficiency over some present management practices would be the use of slow-release fertilizers. There is still a need for improved technology for slow-release fertilizers to match nitrogen release with nitrogen needs by various plants. The use of slow-release fertilizers also has the advantage that by a proper match between nitrogen release and nitrogen needs by plants, only one fertilizer application would be required per season, rather than two, on vegetable crops. When applying fertilizer to crops that are not very salt-tolerant, it becomes necessary to limit the amount of fertilizer being applied. Another solution to this problem would be the application of fertilizer in small amounts with the irrigation water throughout the growing season, essentially spoon-feeding to meet crop requirements. Continual application of nitrogen fertilizer may impair ripening of certain crops.

Water Control

In order to attain high irrigation application efficiencies, positive control of the timing and amount of water being delivered to the farm is required. The irrigator must also be able to control the water supply as it moves across the farm. The water delivery rate must be subject to regulation as well as the quantity applied at any given irrigation. Reducing seepage losses from watercourses, improving water distribution over the field, and reducing unnecessary deep percolation losses are probably the most significant areas for improvement. Related to distribution system losses is water use by noneconomic vegetation in or adjacent to water channels. Such plants extract water not only directly from the supply, but also from the soil under and adjacent to the watercourse. This extraneous vegetation retards flow in the watercourse and increases seepage and evaporative losses, and, in extreme cases, may cause water waste by overflowing or breaking the ditchbank. Reduction of these losses is essential to water control on the farm.

Application Methods

The choice of irrigation method is crucial in dictating water use efficiencies that can be achieved. However, the effect of methods of application on the quality and quantity of return flow has not received detailed study. Conventional surface methods are most commonly used because of their low initial cost, whereas sprinkler methods are used because of their adaptability to a wide range of field and surface conditions and possibilities for reduced labor costs. In most areas there is a real need to "tune up" the existing irrigation systems, thereby attaining the highest practicable irrigation application efficiency that can be achieved with these systems. New and unique approaches to application methods need to be found.

Evapotranspiration Control

Control of evaporation and/or transpiration offers another means of increasing irrigation water use efficiency and improving the quality of irrigation return flows. Such practices as mulching and reduced tillage can be highly advantageous in reducing soil water evaporation. Blevins et al. (1971) demonstrated, with no-tillage studies on corn, a significant decrease in soil water evaporation and greater ability of the soil to store water for use by the crop. By conserving soil moisture, higher corn yields were produced. Surface mulching either with crop residues or artificial barriers has proven effective in reducing water vapor transfer to the atmosphere. Certain surface-active agents also have been shown beneficial in more rapidly establishing a dry barrier at the soil surface and thus reducing evaporative water loss (Law, 1964). Improved irrigation methods offer a great potential for reducing the nonbeneficial evaporative losses of irrigation water applied, thus increasing irrigation efficiencies.

Robins (1967) provided an excellent review of the subject of evapotranspiration control as related to irrigation water requirements. Attainment of any reduction of evapotranspiration, either beneficial or nonbeneficial, would reduce the quantity of irrigation water required for successful crop production.

Optimum irrigation scheduling to extend the irrigation interval and apply water when needed and in the correct amount can exercise beneficial control over evapotranspiration, particularly during periods when crop cover is incomplete.

Irrigation Scheduling

Many studies have shown that in most irrigated areas the amount of water applied and the timing of this water application are quite random. For example, often when the farmer finds that his field is dry, he will irrigate, but the irrigation application may be more than is really needed by the crop. Thus a twofold problem occurs where the plant has already been stressed because of the field being too dry, which means that the yield has already been reduced. The second problem results from more water being applied than was really necessary. In extreme cases, this might even lead to a problem of reduced aeration of the soil.

One of the more interesting areas of water management control presently being explored is that of optimum irrigation scheduling. The purpose of irrigation scheduling is to advise a farmer when to irrigate and how much water should be used (Jensen, 1969; Jensen et al., 1970). Primarily, a farmer relies on visual indications of crop response to decide when to irrigate, or he may have to irrigate on a fixed water rotation system. Irrigation scheduling is geared toward taking soil moisture measurements, along with computing potential consumptive use for the crops being grown, to determine when to irrigate and the quantity of water to be applied.

Irrigation scheduling consists of two primary components, namely, evapotranspiration and available root zone soil moisture. Evapotranspiration is calculated by using climatic data. The other major category of required data pertains to soil characteristics. First, field capacity and wilting point for the particular soils in any field must be determined. More important, infiltration characteristics of the soils must be measured. Only by knowing how soil intake rates change with time during a single irrigation, as well as throughout the irrigation season, can meaningful predictions be made as to: (1) the quantity of water that should be delivered at the farm inlet for each irrigation; and (2) the effect of modifying deep percolation losses. With good climatic data and meaningful soils data, accurate predictions as to the next irrigation date and the quantity of irrigation water to be applied can be made. In order to ensure that the proper quantity of water is applied, a flow measurement structure is absolutely required at the farm inlet.

Efforts look extremely promising, and farmers are claiming a significant benefit from irrigation scheduling. Yields have been increased because water was applied when needed rather than after the crops were stressed. In most cases to date, there has been very little reduction in water use, although it would seem likly that a decrease in water use would occur with time as the farmer gains more knowledge of what is actually occurring in the soil profile. Another benefit to the farmer from this program is that he can anticipate the dates when irrigation is to be accomplished. This allows him to schedule irrigation along with the other duties that must be performed on the farm, and relieves him of the responsibility of deciding exactly when is the best time to irrigate.

The results of studies by Skogerboe et al. (1974) indicate that irrigation scheduling programs have a limited effectiveness for controlling salinity unless irrigation scheduling is accompanied by flow measurement at all the major division points, farm inlets, and field tailwater exits. In addition, it is necessary for canal companies, irrigation districts, or government agencies to assume an expanded role in delivery of the water. The results of these studies show that irrigation scheduling is a necessary, but not sufficient, tool for achieving improved irrigation efficiencies. The real strides in reducing the salt pickup resulting from overirrigation will come from the employment of scientific irrigation scheduling in conjunction with improved on-farm irrigation practices.

Water Removal Subsystem

Surface Runoff

The water removal subsystem consists of removing surface runoff from agricultural lands (if not captured and pumped back on the farm) and receiving deep percolation losses from irrigation. The surface runoff, or tailwater, from one farm may become all or part of the water supply for an adjacent farm, may flow back into the water delivery system at some downstream location, or may be transported back to the river via an open drain, either natural or man-made. Before surface return flows reach the receiving stream, there are essentially three alternatives for alleviating waterlogging and preventing or minimizing the quantity of pollutants discharged into the river. A bypass channel could be constructed to some location where the flows could be discharged without returning to the river. A second alternative would be to store the return flows in shallow storage reservoirs and allow the water to evaporate, leaving behind the pollutants. Seepage must be controlled in bypass channels or storage reservoirs; otherwise, the groundwater may become contaminated. This second alternative has the disadvantage that pollutants are being collected, rather than discharged to the ocean, which may eventually create a real disposal problem.

The third alternative for minimizing the quality degradation in the receiving stream due to surface irrigation return flow would be to treat the return flow. Desalination processes could be used to restore the water supply to a desired quality level, but methods for disposing of brine wastes must be considered.

Drainage and Salinity Control

Waterlogging and salinity pose a serious threat to many irrigated areas. Any expansion upslope from existing irrigated lands becomes a direct threat to the waterlogging of downslope areas (Donnan and Houston, 1967). For example, many of the fertile lands in the San Joaquin Valley of California are now threatened by upslope irrigation development, and some areas in the Yuma Valley of Arizona have been rendered unproductive by irrigation development on the Yuma Mesa. Equally dangerous threats exist from the salt balance problem of these areas. Recirculation of water by pumping or reuse of return flows results in a buildup of salinity. Concomitant with increased salinity are corresponding increases in the leaching requirement and drainage needs. Irrigation development, including impoundment, conveyance, and application upsets the natural hydrologic cycle of an area. Recognition and solution of drainage and salinity problems in such areas requires an intensive application of control measures based on sound scientific knowledge.

Tile Drainage

Tile drainage is a very effective means for removing the less saline waters in the upper portions of the groundwater reservoir, thereby lowering the water table and facilitating the movement of salts from the root zone into the drains, as well as reducing the mass of salts returning to the river. By using tile drainage, salts are allowed to accumulate below the drains. This is particularly true for soils high in natural salts. Tile drainage will not completely remove all of the water moving below the root zone unless the water table is lowered below the natural groundwater outlet. Usually, some water will still move through the groundwater reservoir and return to the surface river, but the quantity of such groundwater return flows can be reduced considerably by tile drainage. The quality degradation to receiving streams from tile drainage outflow can be minimized by treating the outflow. This points out another advantage of tile drainage. Tile drains allow the collection of subsurface return flows into a master drainage system for ease of control and treatment.

Pump Drainage

In Pakistan, tube wells have been effectively used for lowering the water table. The water pumped by tube wells is discharged into watercourses for irrigating croplands. The salinity of groundwater supplies varies widely, with some supplies being too saline for irrigation. Thus, for good-quality groundwater supplies, the tube wells serve the dual purpose of alleviating waterlogging and providing additional irrigation water supplies.

In other situations, pump drainage has been used to remove highly saline groundwater in order to alleviate waterlogging and increase crop production. The primary difficulty becomes the disposal of these highly saline flows, with potential solutions being ponding and evaporation, conveyance and disposal in the ocean, deep well injection, or desalination.

Improving Existing Irrigated Lands

Much of the emphasis in the future will have to be on "improving what we have." In other words, water management will have to be improved, waterlogging and salinity problems alleviated, and crop production increased on existing irrigated lands. Technology alone will not usually bring about the necessary improvements. Instead, a combination of *technological changes* and *institutional modifications* will usually be required *to manage existing irrigated lands* effectively. An approach for defining technological solutions to these problems will be described below.

Inflow-Outflow Analysis

An inflow-outflow analysis is usually the first step in analyzing water management and salinity poblems. A schematic example of irrigated tracts is shown in Fig. 2, in which an inflow-outflow analysis is represented by the difference in the two stream gaging stations, A and B. The well-known hydrologic equation,

$$I - O = \Delta S \tag{1}$$

where I is the inflow, O is the outflow, and ΔS is the change in storage.

Representing the volume of salts passing stream gaging station A over any desired period of time as A_s,

$$A_s - B_s = \Delta s \tag{2}$$

Initially, the analysis is undertaken using annual data over the time length of available data for the two stream gaging stations. (A word of caution — unmeasured subsurface flows underneath the two stream gaging stations could have

Impact of Irrigation on the Quality of Groundwater and River Flows 141

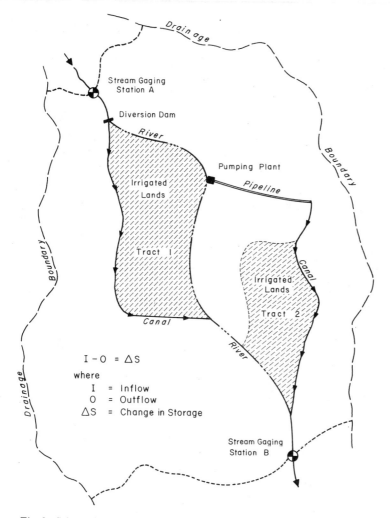

Fig. 2 Schematic example of inflow-outflow analysis.

a significant effect on the analysis.). If ΔS is positive, then salts are being accumulated in the soil profile of the irrigated lands. If $\Delta S = 0$, a salt balance is being maintained. A long-term (e.g., 10, 20, ...50 years) salt balance is necessary for irrigated lands in order to maintain agricultural productivity.

If ΔS is negative in Eq. (2) ($B_s > A_s$), then salt pickup is occurring and the difficulty comes in determining the source of this salt pickup. Also, this would indicate that at least the shallow groundwater flows would be expected to be highly saline. A long-term annual time history of salt pickup will disclose (if sufficient data are available) whether or not the salt pickup rate is increasing, remaining fairly constant, or declining. A declining salt pickup rate indicates that the natural salts in the soil profile (either irrigated lands or watershed lands) are being taken into solution and conveyed into the groundwater reservoir, and then returned to the river. A relatively constant salt pickup rate indicates a large source of salts in comparison with the amount of water percolating through the soil profile and moving through the groundwater reservoir; however, additional analysis (e.g., hydrosalinity modeling) will be required to determine how much of this salt pickup is the result of irrigated agriculture and how much occurs from natural runoff. Also, mineralized springs could be a significant contributor of salts. An increasing salt pickup rate indicates man-made activities (e.g., increasing irrigated acreages or poorer water management on existing irrigated lands).

The schematic in Fig. 2 shows two tracts of irrigated lands. Thus, an inflow-outflow analysis between stream gaging stations A and B does not yield any information regarding the differences between the two tracts. The installation of an additional stream gaging station on the river between the two tracts may allow an inflow-outflow analysis that has only one tract of irrigated land inside the natural surface boundary, but this depends on the geology and its effects on subsurface return flows.

An inflow-outflow analysis is useful primarily for identifying which areas are contributing salts to a river or whether or not an area is maintaining a salt balance. This analysis will not provide the solutions to indicated problems, but will disclose which areas require additional study in order to arrive at necessary technological solutions.

Hydrosalinity Model

Salinity problems from irrigated agriculture are the result of subsurface return flows. Therefore, it becomes highly important to model subsurface, or groundwater, flows. The capability of a *hydrosalinity model* to provide the necessary information for arriving at technological solutions is highly dependent on the accuracy of the groundwater field data and analysis.

A difficulty often encountered while preparing water and salt budgets is the variability in the accuracy and reliability with which the hydrologic and salinity parameters are measured. Usually, the measurement precision varies with the scope of the investigation and the area of the study.

Since the hydrological system is difficult to monitor and predict, it is impractical to expect models to operate without applying some adjustments in order that all components will be in balance. In short, the budgeting procedure is usually the adjustment of the segments in the water and salt flows according to a weighting of the most reliabile data until all parameters represent the closest approximation of the area that can be achieved with the input data being used. The vast and lengthy computation procedure of calculating budgets is facilitated by a mathematical model programmed for a digital computer. A schematic diagram of a general hydrosalinity model is shown in Fig. 3.

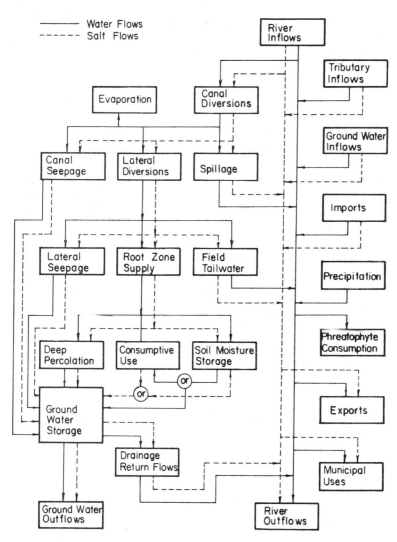

Fig. 3 Schematic generalized hydro-salinity model.

The model was developed in three general sections:

1. all diversions from the canals through small turnouts into the lateral network are distributed onto the farmland after taking into account lateral seepage losses;
2. flow within the root zone including evapotranspiration, tailwater runoff, and deep percolation losses; and
3. groundwater return flows resulting from seepage and deep percolation return to the river system with their salt loads through both surface and subsurface drainage routes.

The details of the computer program are given by Walker (1970).

Cropland Diversions

The irrigation supply is usually diverted by means of diversion dams and then conveyed through the irrigated lands with water being lost by seepage, spilled into wasteways, evaporated, and discharged through turnout structures into laterals.

Natural washes, as well as man-made drains, may be located throughout the irrigated lands to serve as wasteways for canal regulation operation, along with serving as outlets for subsurface return flows. Diversions into the lateral system are also reduced by seepage. Evaporation is insignificant. Of the flows reaching the cropland, only a fraction of the water actually enters the root zone; the remaining flow is field tailwater and becomes surface return flow (where it may be rediverted onto other croplands, or consumed by phreatophytes, or lost by seepage and become subsurface return flow).

Root Zone Flows

The goal of irrigation is to recharge the soil moisture reservoir with sufficient water to meet the growing crops needs until the next irrigation, as well as to maintain an acceptable salt concentration in the root zone. The tendency to overirrigate has produced high water tables (waterlogging) and salinity problems in many areas. The purpose of the root zone submodel is to separate the various flows occurring within the root zone in sufficient detail to quantify the salinity problem.

The important water movements within the root zone are evapotranspiration and deep percolation, with water storage changes also occurring. The separation of these flows by measurement is impractical on a large scale. Consequently, empirical computation methods are employed. The model described here accounts for these basic water and salt flows only by a budgeting process. The

Impact of Irrigation on the Quality of Groundwater and River Flows 145

assumptions made regarding the operation of this model include that the diversions are applied uniformly over each acre of cropland. Phreatophyte vegetation in the area is assumed to extract water only from the groundwater flows or to

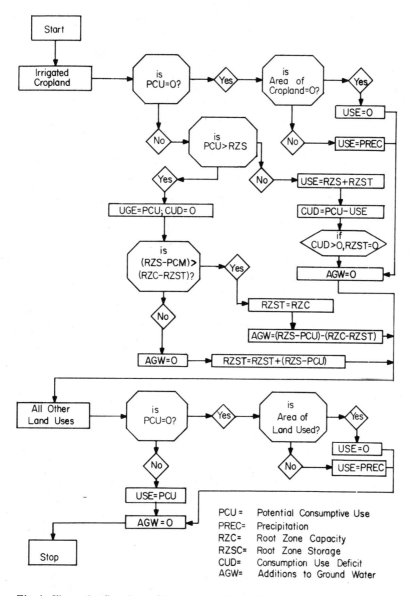

Fig. 4 Illustrative flowchart of the root zone budgeting prodedure.

use only precipitation entering the root zone of these plants. A generalized flowchart of the root zone budgeting procedure is presented in Fig. 4.

Several applicable methods of estimating evapotranspiration can be used in this model. The Blaney-Criddle method provides an acceptable degree of accuracy; however, the Jensen-Haise method is becoming increasingly popular, whereas the Penman method is used whenever the necessary field data are available.

With the evapotranspiration data and field measurements of moisture-holding capacity, texture, infiltration rates, and rooting depths, the budgeting scheme proceeds with the computation of deep percolation losses from the root zone. The calculations are initiated by assuming that the crops use soil moisture at the potential rate until the wilting point is reached. The calculated potential use is then limited to the water added by irrigation and the existing available soil moisture storage. If the supply to the root zone from irrigation is insufficient to meet the crop demands, but the available soil moisture storage is sufficient to make up the difference, then the crop demand is satisfied. The assumption is made that while the soil moisture reservoir is below field capacity, no deep percolation occurs. If the total available moisture in a period is insufficient to meet the total demand, the crops use all water available. A term called *consumptive use deficit* is defined as the difference between the potential and actual uses. Deep percolation losses and leaching occur when the supply is more than enough to meet the crop demands and fill the soil moisture reservoir to field capacity.

The salts in the applied water move with the water into the root zone where they are concentrated by the evapotranspiration process. The behavior of specific ions is complex and has not been considered in this particular model, but will be discussed in the following section, "Predicting Chemical Quality."

Groundwater Model

Most of the water in the soils and shallow groundwater aquifers originates as seepage from canals and laterals, as well as deep percolation from the irrigation of croplands. The groundwater discharges eventually reach the river as surface drainage interception or subsurface return flows. The flows in the surface drainage system can be measured by installing flow-measuring devices at the outflow points. The subsurface return flows are estimated from water table elevation data and the hydraulic gradients in the aquifers. Considerable effort must be made to evaluate properly the necessary parameters in the groundwater computations. For purposes of this model, Darcy's steady-state equation is used,

$$Q = AK \frac{dh}{dk} \qquad (3)$$

in which Q is the discharge, A is the cross-sectional area of flow, K is the hydraulic conductivity, and dh/dk is the hydraulic gradient in the direction of flow.

The groundwater analysis, illustrated in Fig. 5, begins by comparing the values for subsurface return flow obtained from a mass balance of the area to the values obtained by calculation using the field data. It is possible to formulate two estimates of the subsurface return flows. Then, by adjusting the model until both methods yield the same values, a satisfactory alignment between the hydrologic and salinity parameters can be obtained. Because the model focuses

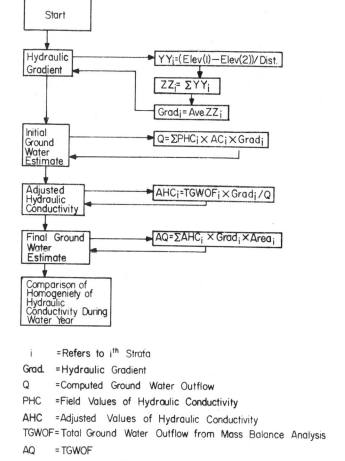

i = Refers to i^{th} Strata
Grad = Hydraulic Gradient
Q = Computed Ground Water Outflow
PHC = Field Values of Hydraulic Conductivity
AHC = Adjusted Values of Hydraulic Conductivity
TGWOF = Total Ground Water Outflow from Mass Balance Analysis
AQ = TGWOF

Fig. 5 Illustratrative flowchart of the groundwater modeling procedure.

attention only on the relative magnitude of hydraulic conductivities, the cross-sectional areas of the strata need only be in proper proportion with respect to depth, and the width can be any convenient value. Then, the values for cross-sectional area can be adjusted with the known hydraulic conductivities. The model adjusts the values of strata hydraulic conductivity until both estimates of the flows are equal. Since this is done on a monthly basis, the model calculates 12 values of hydraulic conductivity for each strata for each year. When adjustments in the model finally result in homogeneous values of hydraulic conductivity, the model represents the "best fit" between monitored and estimated data.

The groundwater modeling procedure can also be descirbed mathematically. The form of Eq. (3) for a number of strata can be written,

$$Q = \sum_{i=1}^{n} A_i K_i' \frac{dh_i}{dx_i} \qquad (4)$$

where A_i is the cross-sectional area of the ith strata, K_i' is the actual measured conductivity of the ith strata, and dh_i/dx_i is the gradient in the flow direction acting on the ith strata. The value obtained from Eq. (4) is then used to adjust the model values of hydraulic conductivity,

$$K_i = \frac{TGWOF}{Q} K_i \qquad (5)$$

where K_i is an adjusted hydraulic conductivity encompassing adjustments for units and strata areas, Q is the value obtained from Eq. 4, and TGWOF is the subsurface return flow estimate from the mass balance analysis.

Generalizing the Model

The mathematical model attempts to simulate the hydrologic conditions of an agricultural system. The concepts are general and can be extended to many areas with some modifications likely required for each area. The program is written in individual but interconnected subroutines that provide a measure of flexibility during operations by separating the calculation phase from either input or output phases. Thus, several of the subroutines become optional if their functions can be replaced by input data, or if certain outputs are not desired.

The main portion of the program is used to read necessary input data and to control the order of water and salt budget calculations. There are certain advantages in separating the input, output, and computational stages of a program, including the following:

1. Input order is not important, as the data are completely available at all stages of computation.
2. Variable sets of data can be utilized in the model when several budgets are desired, or when some form of integration is desired. This is especially useful when an area can be broken down into smaller dependent areas (e.g., tracts 1 and 2 in Fig. 2).
3. The functions of the subroutines are independent of input, thereby making each subroutine a unit that can be implemented in other programs.
4. Corrections and adjustments are easily made without detailed consideration to other segments of the program.

In controlling the computational order of the program, the main program separates the calculation of the water and salt budgets. Consequently, the modeling procedure involves only the water phase of the flow system. This is possible when sufficient field data have been collected. Once the water flow system has been simulated, the individual flows are multiplied by measured salinity concentrations and converted to units of tons per month. At this point in the formation of the budgets, careful attention must be given to the salt flow system since irregularities may be present, thereby necessitating further model adjustments. Thus, when the final budgets have been generated, the salt system, groundwater system, and surface flow system must be reasonably coordinated and additional reliability is assured.

Predicting Chemical Quality

The hydrosalinity model describes the present situation in an agricultural area regarding water and salt flows. However, the only method for predicting the reduction in salts returning to the groundwater and river through implementation of any salinity control measure(s) is by assuming a one-to-one relationship between water (reduction in subsurface return flows) and salt pickup. That is, if the subsurface return flow is reduced by 50%, the salt pickup is also reduced by 50%.

The primary objective is to model the transport of salts through the soils. The first portion of the flow of water and consequent transport of salts is through the root zone, which is usually a zone of partial saturation. A numerical model of the moisture flow and chemical and biological reactions occurring in the root zone has been developed by Dutt et al. (1972) (this reference contains a complete listing of the computer program). This is the basic model that will be used to describe salt transport.

The model consists of three separate programs. The first program describes the soil moisture movement and distribution with time. The second program

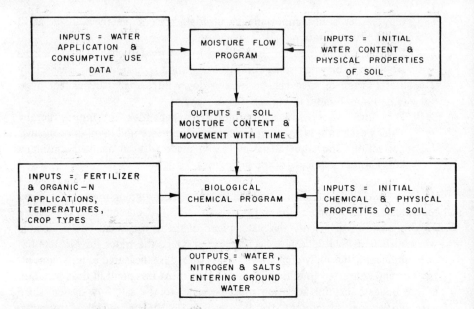

Fig. 6 Generalized block diagram of the model.

interfaces the soil moisture movements with the chemical-biological model. This is needed because the horizons used in the calculations of soil moisture and chemistry differ. The third program computes the chemical and biological activity occurring in the soil profile. Figure 6 is a block diagram of the overall model. A brief description of the moisture flow and chemical-biological models is included to serve as a basis for understanding the data collection requirements.

The flow is one-dimensional and was developed using the Richards equation with a sink term. Schematically, the model is given in Fig. 7. Mathematically, the flow is described using the Richards equation in the form

$$\frac{\partial \theta}{\partial t} = \frac{\partial}{\partial x}\left(D\frac{\partial \theta}{\partial x} - K\right) - S \tag{6}$$

where
- θ = volumetric water content
- t = time
- x = length
- K = hydraulic conductivity
- S = sink term
- D = diffusivity

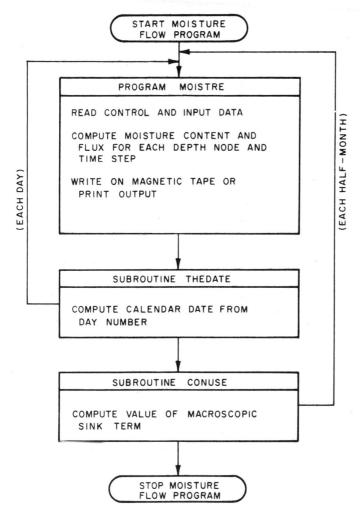

Fig. 7 Generalized block diagram of moisture flow program.

This is the diffusivity form of the equation, which means that only flow in the partially saturated zone of the soil profile can be described. The sink term, S, is computed using the Blaney-Criddle equations for evapotranspiration (other equations could be adapted to the model) with the loss due to evapotranspiration being distributed through the soil profile by assuming a specific root distribution for the crop. The root distribution and coefficients for the Blaney-Criddle equations are supplied by the user. Actual values of evapotranspiration can be used in the sink term when they are known.

Salt transport is described by the following equation in one dimension:

$$\frac{\partial c}{\partial t} = \frac{\partial}{\partial z}\left(D\frac{\partial c}{\partial z}\right) - v\frac{c}{z} \qquad (7)$$

where
 c = solute concentration
 t = time
 D = dispersion coefficient
 z = depth
 v = flux or Darcy velocity

By assuming that the term

$$\frac{\partial}{\partial z}\left(D\frac{\partial c}{\partial z}\right)$$

is negligible compared to

$$v\frac{\partial c}{\partial z}$$

the equation reduces to

$$\frac{\partial c}{\partial t} = -v\frac{\partial c}{\partial z}$$

This assumption implies that transport due to dispersion in partially saturated soils is negligible compared to the convective transport which occurs. This is generally a good assumption.

The model computes the moisture flow (v) and couples the flow with the chemical changes $\frac{\partial c}{\partial z}$ computed in the biological-chemical program to give the salt transport. This technique is the basis for the mixing-cell concept.

The chemical exchange model computes the equilibrium chemistry concentrations for calcium, magnesium, gypsum, sodium, bicarbonates, carbonates, chlorides, and sulfates. The nitrogen chemistry including ammonium, nitrates, and urea-nitrogen uses a kinetic instead of an equilibrium approach. The kinetic approach is needed since microbial activity involved in nitrogen transformation occurs over a period of weeks and days instead of minutes and seconds. The equilibrium chemistry for inorganic salts is a good approximation

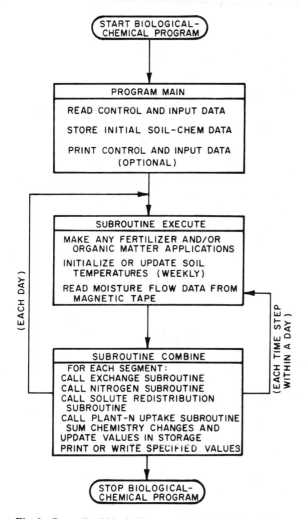

Fig. 8 Generalized block diagram of biological-chemical program.

since the reactions describing their chemistry occur in a matter of minutes or seconds in a flow regime that is changing very slowly. A block diagram of the biological-chemical model is given in Fig. 8.

Once the necessary field data have been collected, equations can be developed to predict the variation in chemical quality (including ionic constituents) of the moisture movement through soil profile, as well as the salt pickup (or salt precipitation) resulting from movement of subsurface irrigation return flows.

These results, when combined with the hydrosalinity model, will allow an evaluation of various salinity control measures on salinity reaching the groundwater and returning to the river.

Cost Effectiveness of Technological Solutions

The results from the hydrosalinity model will provide considerable insight as to which technologies might be most appropriate in order to reduce subsurface return flows (since salinity is a problem associated with subsurface return flows). The goal in reducing subsurface return flows might be: (1) to lower groundwater levels in order to alleviate waterlogging, thereby allowing the leaching of salts in the root zone, which in turn facilitates increased crop production; (2) reducing downstream water quality degradation resulting from salt pickup; (3) improved on-farm water management to increase crop production; or (4) any combination of the first three goals.

Field studies must be undertaken to evaluate the effectiveness of any particular technology, or combination of technologies, in achieving the desired goal(s). An example will be used to illustrate the development of cost-effectiveness functions. If seepage loss measurements are made along many sections of each canal in an irrigated area, there will usually be a considerable variation in seepage loss rates. If the cost of canal lining is compared with seepage loss reduction, then the greater benefits accrue by first lining those sections of canal that have the least cost per unit of seepage loss reduction. Finally, the last sections of canal lining may be quite costly per unit of seepage loss reduction. This concept is illustrated in the upper diagram of Fig. 9. The maximum seepage loss reduction shown on the ordinate in the upper diagram of Fig. 9 represents the total seepage loss from all canals; but, since canal linings still have some seepage losses, there will still be some seepage losses after lining all canals. The lowering of groundwater levels resulting from canal lining is illustrated in the lower diagram of Fig. 9. In order to compute the downstream salt load reduction resulting from various levels of seepage loss reduction, the chemical-biological model described in the previous section must be utilized. This same model can be utilized to determine the reduction in ionic constituents making up the total salt load reduction, thereby allowing the prediction of downstream ionic chemical quality. The analysis could be carried forward another step if data are available regarding the effect of groundwater levels on production of various crops.

The above example illustrates one of the simpler technologies for developing a cost-effectiveness function. Lateral lining becomes somewhat more complex because of the combined effects of seepage loss reduction and providing water control, which may be highly effective in improving irrigation applica-

Impact of Irrigation on the Quality of Groundwater and River Flows 155

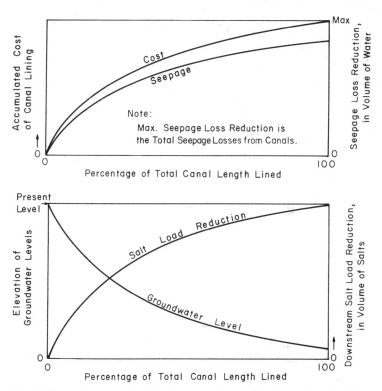

Fig. 9 Illustrative example of development of cost-effectiveness function for canal lining.

tion efficiencies, which in turn reduces deep percolation losses and increases crop production. The addition of flow measuring devices at all junctions in laterals, as well as at farm inlets, can prove highly effective in achieving improved water management. The development of improved on-farm water management technologies can require considerable field data collection. One of the most appropriate on-farm technologies is "tuning up" the present irrigation practices.

Cost-effectiveness functions should be developed for each appropriate technology that can be identified. Then, cost-effectiveness functions should be developed for each combination of appropriate technologies. These cost-effectiveness functions, which relate subsurface return flow reductions to the desired goal(s) resulting from a specified investment, would then be assessed in an optimizational format to arrive at the least cost combination for achieving a desired level of the stated goal(s). Thus, this analysis details the optimal strategy for implementing various levels of individual technological improvements into a comprehensive technology package.

Conclusions

Irrigation system design begins with the farm layouts taking into account climate, soils, and crops. Ideally, the design of farm irrigation layouts should be suited to the individual characteristics of each field. Then, the water delivery subsystem can be designed to provide the necessary quantities and timing of water deliveries at the farm inlets. The seepage losses from the water delivery subsystem and the deep percolation losses from the farm subsystem must be taken into consideration so as to prevent waterlogging and salinity problems in the root zone that would impair the productivity of the cropland. In general, this design process begins at the lowest elevation field in the irrigation system and continues until all of the lands to be irrigated are encompassed, thereby resulting in the design for the total system.

In contrast, the process is reversed when evaluating existing irrigation systems in order to alleviate salinity detriments and improve agricultural productivity. In this situation, we first begin with an inflow-outflow analysis of the total system in order to identify whether or not an area is maintaining a salt balance, or which areas are contributing salts to a river. Next, a hydrosalinity model is developed for the irrigation system. Since salinity problems are the result of subsurface return flows, such a model must be capable of accurately describing groundwater flows, as well as the hydrologic characteristics of the surface irrigation system. The hydrosalinity model describes the present hydrologic situation in an agricultural area. Therefore, in order to evaluate the impact of technological improvements on the hydrosalinity system, the capability must be developed through soil chemistry modeling to predict the impact of such improvements on underlying groundwater reservoirs and downstream river flows. Developing this predictive capability is rather difficult, but once accomplished, allows the generation of cost-effectiveness curves for various technologies and combinations (or packages) of technologies.

Much emphasis will be given in the future to improved irrigation water management to reduce downstream water quality degradation and minimize diversions in order to provide water supplies for new demands, but more important, to alleviate waterlogging and salinity problems in order to increase crop production on existing irrigated lands. In many cases, the key will be improved on-farm water management. However, technology alone will not usually bring about the necessary improvements. A combination of technological changes and institutional modifications will usually be required to effectively alleviate salinity problems resulting from irrigated agriculture.

References

Blevins, R.L., Cook, D., Phillips, S. H., and Phillips, R. E. (1971). Influence of no-tillage on soil moisture. Agron. *93*: 593-596.

Donnan, W. W., and Houston, C. E. (1967). Drainage related to irrigation management. In *Irrigation of Agricultural Lands*. ASA Monograph No. 11, chap. 50, pp. 974-987.

Dutt, G. R., Shaffer, M. J., and Moore, W. J. (1972). Computer simulation model of dynamic bio-physicochemical processes in soils. Technical Bulleting 196, Department of Soils, Water and Engineering, Agricultural Experiment Station, University of Arizona, Tucson.

Jensen, M. E. (1969). Scheduling irrigations with computers. Soil Water Conservation, *24* (5): 193-195.

Jensen, M. E., Robb, C. H., and Franzoy, E. C. (1970). Scheduling irrigations using climate-crop-soil-data. J. Irrigation and Drainage Div. ASCE, *96* (IR1): 25-38.

Jensen, M. E., Swarner, L. R., and Phelan, J. T. (1967). Improving irrigation efficiencies. In *Irrigation of Agricultural Lands*. ASA Monograph No. 11, chap. 61, pp. 1120-1142.

Law, J. P., Jr. (1964). The effect of fatty alcohol and a nonionic surfactant on soil moisture evaporation in a controlled environment. Soil Sci. Soc. Amer., *28* (5): 695-699.

Skogerboe, G. V. (1974). Agricultural systems. In *Human Ecology*. North Holland, Amsterdam, chap. VII, pp. 127-145.

Skogerboe, G. V., and Law, J. P., Jr. (1971). Research needs for irrigation return flow quality control. Report No. 13030-11/71, U.S. Environmental Protection Agency, Washington, D. C.

Skogerboe, G. V., and Walker, W. R. (1973). Changing role of water conveyance systems. ASCE National Water Resources Engineering Meeting, Meeting Preprint 1880, Washington, D. C., January-February.

Skogerboe, G. V., Walker, W. R., Taylor, J. H., and Bennett, R. S. (1974). Evaluation or irrigation scheduling for salinity control in grand valley. Environmental Protection Technology Series, EPA-660/2-74-052. Office of Research and Monitoring, U.S. Environmental Protection Agency, Washington, D. C.

Robins, J. S. (1967). Reducing irrigation requirements. In *Irrigation of Agricultural Lands*. ASA Monograph No. 11, chap. 62, pp. 1143-1158.

Walker, W. R. (1970). Hydrosalinity model of the Grand Valley. M.S. thesis CET-71WRW8. Civil Engineering Department, College of Engineering, Colorado State University, Fort Collins, Colo.

Wiener, A. (1972). *The Role of Water in Development*. McGraw-Hill, New York.

7

Economic Evaluation of Irrigation with Saline Water Within the Framework of a Farm, Methodology and Empirical Findings: A Case Study of Imperial Valley, California

Charles V. Moore

U.S.D.A., E.S.C.S., University of California, Davis, California

Introduction

The Imperial Valley of California is located in the southeast corner of the state, bounded on the south by Mexico and on the east by Arizona. The sole water supply for the Valley is diverted from the Colorado River by the Imperial Irrigation District (IID), which delivers this water by gravity flow to approximately 475,000 acres of land. In 1973, the gross value of crop production in the Valley was $296.7 million, making it one of the most productive areas in the nation.

Resource Base

Irrigation water from the Colorado River has shown a trend toward higher salt levels ever since irrigation development commenced at the turn of this century. Figure 1 shows this trend over the past 30 years and the projections made for the future by the Colorado River Board (1970). These data indicate a current salinity level of about 950 ppm of total dissolved solids or an electrical conductivity (EC) of 1,500 μmho. The Colorado River Board (1970) projects the salinity of the River to rise to 1,210 ppm by the year 1990 if anticipated upstream developments are made and if sources of salts entering the River are not removed or other mitigating investments made.

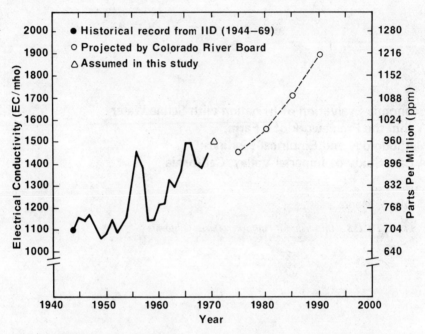

Fig. 1 Historical record and projected trend of water salinity in the Imperial Valley.

The long-term average water supply to the Valley has been more than adequate for intense agricultural production. The 20-year average deliveries per acre for the period 1948-1967 was 4.66 acre feet/year of 14,200 m^3/ha.

The soils of the Imperial Valley are composed of alluvial deposits of fine-textured clays, silts, and fine sands deposited by floodwaters of the Colorado River and coarse-textured wind-transported materials. Both materials occur in strata and lenses of different textures and in various combinations. Crop productivity and drainage practices are influenced by the particular textural combinations encountered in localized areas. Using the local terminology, 71% of the Valley's soils are classified "heavy," that is, clays; 17% "medium," that is, loams and fine sandy loams; and 12% "light," that is, loamy very fine sand and loamy fine sand.

The climate of the Imperial Valley can be characterized as arid, with an average annual rainfall of 2.85 in. (72.4 mm). Most of this scant rainfall occurs in the months of December, January, and February. Hot summers and mild winter temperatures allow year-around farming operations.

Table 1 Number of Farms[a]

Year	1-99	100-219	220-499 (acres)	500-999	1,000+	Total
1950	1,178	362	306	177	108	2,131
1960	603	218	208	155	122	1,306
1974[b]	374	103	112	91	150	830

[a]From Census of Agriculture, U.S. Department of Commerce.
[b]Estimated from preliminary data.

Farm Size Distribution

As elsewhere in California and the United States, the average size of farms in the Imperial Valley has been increasing over time. The data in Table 1 indicate that the number of farms with less than 1,000 acres of land has been decreasing and those over 1,000 acres have been increasing. If we assume that farms of less than 100 acres are part-time or hobby farms, it appears that commercial farms are nearly uniform in distribution with respect to size.

Necessary and Sufficient Conditions for a Long-Term Irrigated Agriculture

Salinization or salts in soils is an extremely important problem in irrigated arid regions of the world. Productive soils may be salinized at different rates, depending on the amount and composition of dissolved salts imported with the irrigation water. Bernstein (1967) considers three factors or conditions affecting water quality determinations—salt tolerance of crops, soil permeability, and drainage.

Because of the physical conditions of climate, soil permeability, drainage (natural or artificial), and the salt tolerance of the crops adaptable to a specific location, irrigation water of a given quality may or may not be usable. These limitations are not absolute, and there is a degree of substitutability among them. For example, artificial tile drainage can be substituted for natural drainage, or the quantity of water can be substituted for quality of water through the use of a higher leaching fraction. Also, crops with a higher salt tolerance can be used to replace salt-sensitive crops as the quality of water deteriorates. In the concept of a long-run steady state, these physical factors can be used to described the limitations to the possibility of a long-term irrigated agriculture. That is, the physical factors determine the necessary conditions for a long-term irrigated agriculture, but they do not specify the sufficient conditions.

The sufficient conditions are specified by the economic parameters that influence the economic viability or irrigated farms. Assume a present-value profit equation in the form

$$\pi_{pv} = \sum_j B_j \left(\sum_i P_{ij} Y_{ij} - \sum_k C_{ikj} X_{ikj} \right) \qquad (1)$$

where $Y_{ij} = f^*(X_{ikj})$ and $k = 1, \ldots, s$, and where B_j is the discount factor, P_i is price of the ith commodity and Y_i is its level of output, and C_k is the unit cost of the X_k^{th} factor input in the ith commodity and the jth year.

By use of the calculus, the optimum use rate for each resource in each time period can be determined. However, the profit equation was constrained so that income in any subplanning period does not become negative. The subplanning period can exceed one year, but cannot be so long that resources with critical zones with respect to, say, salinity in the root zone passes some irreversible level.

The objectives of this chapter are (1) to report the methodology and empirical results of generating a farm firm production function where the return to land and water is a function of both quantity and quality of the water supply; and (2) to estimate a farm firm demand schedule for irrigation water with varying supply and quality levels.

Effects of Supply and Quality of Irrigation Water on Individual Crops*

Assume a production function for an individual irrigated crop of the form

$$y = g(w_q, w_i | K, L, R, \ldots) \qquad (2)$$

where y is yield per acre (hectare), w_q is the water supply variable measured in inches or centimeters, w_i is the quality of the irrigation water supply measured in terms of its electroconductivity, and K, L, and R are capital, labor, and rainfall, respectively.

Following an earlier study (Moore, 1961), we assume that relative plant growth is a function of the mean moisture stress in the root zone of the plant. Then the index of crop growth for one irrigation cycle can be stated as

$$I_{\theta_i} = \frac{\int_0^{\theta_i} g(m)\,dm}{\theta_i \cdot 100} \quad m = (0, \ldots, i) \qquad (3)$$

---

*The author acknowledges that portions of the following section were taken from Moore et al (1974) with the permission of the copyright holder, the American Geophysical Union, publishers of *Water Resources Research*.

A Case Study of Imperial Valley, California

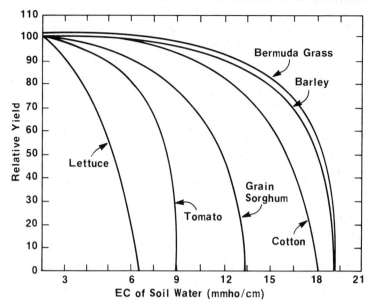

Fig. 2 Salt tolerance of selected crops.

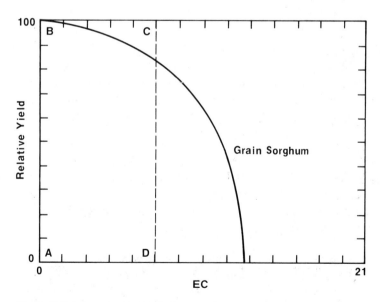

Fig. 3 Salt tolerance growth relation.

where I is the growth rate for one irrigation cycle, θ_i is the soil moisture depletion percent at the time of a subsequent reirrigation, and g(m) is the functional relation between plant growth and soil moisture stress.

The variable w_i in Eq. (2) represents the quality dimension of the irrigation water input. Results of studies by the U.S. Salinity Laboratory at Riverside, California (Bernstein, 1967) indicate that most plants respond negatively to increases in soil salinity.† Representative growth-response curves relating crop yield to the electroconductivity of the soil saturation extract are presented in Fig. 2. The growth index, $I_{\theta j}$, which represents one of these curves for one irrigation cycle, can be defined as the definite integral of a specific growth curve as a fraction of the rectangle ABCD in Fig. 3. This can be expressed as

$$I_{\theta_j} = \frac{\int_0^{\theta} jf(s)ds}{\theta_j \cdot 100} \tag{4}$$

where θ_j is the electroconductivity of the saturation extract at the time of a reirrigation and f(s) is the functional relationship between plant growth and soil salinity.

We can combine Eqs. (3) and (4) into Eq. (5):

$$I_{\theta_{ij}} = \frac{\int_0^{\theta_i} \int_0^{\theta_j} jf(m, s) \, dm \, ds}{(\theta_i)(\theta_j)(100)} \tag{5}$$

Using this theoretical production response relation, 36 possible combinations of irrigation treatment, water quality level, and leaching fraction for each of three soil types and for each crop were budgeted (see Sun, 1972 for additional details of this procedure). Production efficiency frontiers were drawn using these budgets (see Fig. 4A) showing the locus of the budgeted points using the highest physical yield per unit of water as the efficiency criteria. Only those budgeted combinations falling on the efficiency frontier were considered for further analysis.

In the example for cotton in Fig. 4A, yield declines in response to additional units of irrigation water applied. Increased salinity in the irrigation water also has a negative effect on yields. As the electrical conductivity (EC) of the irrigation water increases (quality decreases), yield is reduced but at a decreasing rate.

There is a trade-off or substitutability between water quality and water quantity. This is shown by the isoquants in Fig. 4B. Figure 4B was constructed

---

†See also Chap. 3.

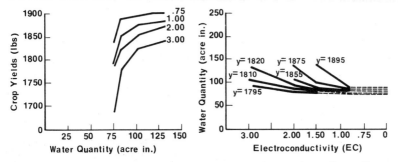

Fig. 4 Efficiency frontiers and production isoquants of cotton in medium soil.

from the data in Fig. 4A and indicates that in order to obtain the same yield of cotton, a larger and larger quantity of water must be applied as the EC of the water increases. Note the increasing distance between isoquants as the EC increases.

Model of the Farm Firm

In order to estimate the effects of different levels of water quality and water supply, a linear programming model was constructed. As was done in developing the efficiency frontiers, cost and return budgets were constructed for each point on the efficiency frontier for each of nine alternative field crops. Three separate models representing three farm sizes were constructed.

Because of the long-term planning horizon under consideration, the objective function for the linear programming model was written to maximize the return to land and water that would be equivalent to maximizing returns for one subplanning period in Eq. (1). This allowed us to take into consideration changes in land values as water quality declined over the planning period.

The objective function was maximized subject to the constraints of a limit to the total quantity of water available based on water rights on the Colorado River, and a seasonal peak water supply based on the physical capacity of the canal system. The supply of land was divided into three soil types. A restriction was placed on the model that lettuce acreage could not exceed 10% of the total irrigated acreage on a farm because of the sensitivity of lettuce prices assumed to a significant increase in winter lettuce production. The acreage of sugar beets that could be grown was limited to 12.5% of the irrigable land for nematode control. Based on historic data, a constraint of 9% of irrigable crop land was placed on cotton acreage because of the cotton allotment program that was in effect at the time of the study. Also because of federal production control

Table 2 Annual Return to Water and Land, Four Water Qualities by Farm Size[a]

Farm size	Water quality (EC)			
	0.75	1.50	2.00	3.00
Small	$ 40,854 (113.8)	$ 35,888 (100.0)	$ 30,687 (85.5)	$ 25,492 (71.0)
Medium	$ 93,309 (112.0)	$ 83,316 (100.0)	$ 72,823 (87.4)	$ 61,448 (73.8)
Large	$317,348 (112.0)	$283,882 (100.0)	$249,068 (87.7)	$210,054 (74.0)

[a]Figures in parentheses indicates index of income with current condition = 100.

programs then in effect, a constraint that at least 30% of the land must be planted to alfalfa or remain fallow was imposed on the model.

Effect of Water Quality on Return to Water and Land

Table 2 summarizes the results of the linear programming model for four levels of water quality and three farm sizes in the Imperial Valley.

Net returns may be expected to decrease by 12 to 15% from the current level if and when the salinity level of the lower Colorado River declines to an EC of 2.0 from the current EC of 1.5. This level (2.0) is projected for the year 2000. In the unlikely event that the salinity level were to increase to an EC of 3.0, a 26 to 29% reduction in farm net returns to land and water could be expected. A level of salinity of this magnitude is far beyond anything anticipated in the future and must be considered a very-low-probability event.

On the other hand, if by dilution or desalination it was possible to reduce the salinity level in the lower Colorado River to an EC of 0.75, that is, half the current level, net returns would increase 12 to 14% above their current level.

Changes in net returns may be explained by changes in total crop acreage, changes in the proportion of high-valued/salt-sensitive crops, adjustments in the leaching fraction, and the irrigation regime that is followed.

Changes in total crop acreage were a major factor influencing returns to water and land was water quality changed. The data in Table 3 indicate that both the amount of double cropping and total crop acreage declines as salinity levels in the lower Colorado River increase and that this pattern holds for all farm sizes.

Changes in the crop mix grown also have important effects on the return to land and water as salinity levels increase. The data in Table 4 indicate that

A Case Study of Imperial Valley, California

Table 3 Estimated Total Crop Acreage and Double Cropping for Four Levels of Water Quality by Farm Size, Imperial Valley

Farm size	Item	Water quality (EC)			
		0.75	1.50	2.00	3.0
Small	Total acreage	387.4	379.0	366.1	359.5
	Crop percent[a]	121.1	118.5	114.4	112.3
Medium	Total acreage	774.8	758.3	732.9	732.3
	Crop percent	121.0	118.3	114.5	114.4
Large	Total acreage	2,612.8	2,516.7	2,435.0	2,433.1
	Crop percent	123.0	118.5	114.6	114.5

[a](Total crop acreage/irrigable acreage) × 100.

Table 4 Crop Acreage for Four Levels of Water Quality by Farm Size, Imperial Valley

Crop	Farm size	Water quality level (EC)			
		0.75	1.50 (acres)	2.00	3.00
Alfalfa	Small	8.2	–	–	–
	Medium	16.5	–	–	–
	Large	96.1	–	–	–
Barley	Small	123.9	123.9	123.9	155.1
	Medium	247.8	247.8	222.4	310.3
	Large	822.2	822.2	740.5	946.1
Cotton	Small	28.8	28.8	28.8	28.8
	Medium	57.5	57.5	57.5	57.5
	Large	191.0	191.0	191.0	191.0
Grain sorghum	Small	123.9	123.9	110.9	135.6
	Medium	247.8	247.8	222.4	284.3
	Large	822.2	822.2	740.5	946.0
Early grain sorghum	Small	31.2	31.2	31.2	–
	Medium	62.5	62.5	87.8	–
	Large	207.5	207.5	289.2	83.7
Lettuce	Small	312.	31.2	31.2	–
	Medium	62.5	62.5	62.5	–
	Large	207.5	207.5	207.5	–
Sugar beets	Small	40.1	40.1	40.1	40.1
	Medium	80.2	80.2	80.2	80.2
	Large	266.3	266.3	266.3	266.3

under our assumptions, alfalfa, which is salt-sensitive at the germination stage, is profitable only at the higher water quality levels (EC = 0.75). The normative solution indicates that land should be left fallow at higher salinity levels. Other than alfalfa, little change occurs in the optimum crop mix until salinity levels reach EC 3.0 (a level far above projections for the lower Colorado River). If salinity levels ever reach this high level, lettuce (a salt-sensitive crop) would no longer be produced and a large decrease in the acreage of early planted grain sorghum would occur.

The remaining crops—barley, cotton, and sugar beets—are tolerant to moderately tolerant to salts, and little or no change is observed in their acreage except the modest increase in barley acreage at the highest salinity level.

In this study, the initial objective function was to maximize the return to land and water. In order to separate out the return going to land and the return going to water, certain assumptions were made: (1) The current average cash rent per acre of land was a valid estimate of the marginal value product (MVP) of the land resource. (2) As water quality declined, the MVP of land would decline

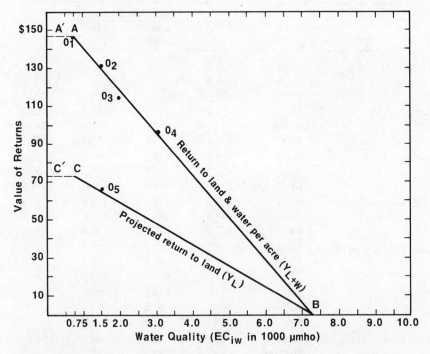

Fig. 5 Projection of return to land and return to water of varying water qualities in the Imperial Valley.

A Case Study of Imperial Valley, California

until it reached zero at a water quality level at which farm income also became zero. Stated another way, we made the assumption of "no crop – no land rent." (3) A linear interpolation between these two points would trace the path of land rents as water quality declined. (4) At water qualities higher than the current level, land rent would increase at the same rate to a point where the EC of the irrigation water was equal to 0.75 and no increases would be expected after that. The results of these assumptions are displayed in Fig. 5.

In Fig. 5, the results of the above assumptions are displayed graphically. The return to land and water (Y_{L+W}) is plotted freehand through the points 0_1, EC = 0.75, 0_2, EC = 1.50, 0_3, EC = 2.0, and 0_4, EC = 3.0. Extrapolation to the horizontal axis indicates a point B at EC = 7.25 where the return to land and water is zero. Separation of the return to land (Y_L) alone is made following these assumptions, where point 0_5 is the current cash rent per acre under existing water quality conditions and is projected to intersect the horizontal axis at EC = 7.25 (the assumption of no water–no land rent).

Technical economies of farm size are present in the Imperial Valley, and if these economies are imputed to the irrigation water input, they become reflected in higher returns to this resource as farm size increases. The data in Table 5 reflect the results of separating out the projected land rent under varying conditions of water quality and the effect of farm size on the return to water based on our assumptions.

Table 5 Average Return to Water for Four Water Quality Levels, by Farm Size, Imperial Valley

Item	Farm size	Water quality (EC)			
		0.75	1.50	2.00	3.00
			(dollars)		
Return to land & water per acre	Small	127.67	112.15	95.90	79.66
	Medium	145.80	130.18	113.79	96.01
	Large	149.34	133.59	117.21	98.85
Estimated land rent		73.50	65.00	59.00	48.00
Return to water per acre	Small	54.17	47.15	36.90	31.66
	Medium	72.30	65.18	54.79	50.85
	Large	75.84	68.59	58.21	50.85
Return per acre foot	Small	9.95	8.66	6.78	5.81
	Medium	13.28	11.97	10.06	8.82
	Large	13.93	12.60	10.69	9.34

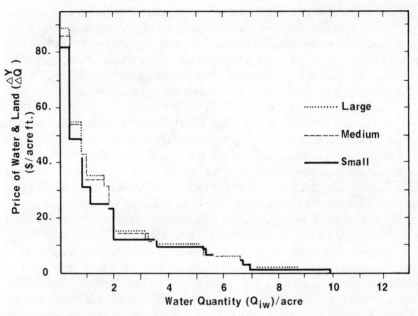

Fig. 6 Price of water and land relative to farm size, with current water quality (EC_{iw} = 1.5).

Demand for Irrigation Water

The linear programming model may be utilized to obtain a static-normative derived demand function for irrigation water. By parametrizing the price of water over the relevant range of prices to the farm, price-quantity relations can be traced. The data shown in Fig. 6 display these relations for the three sizes of farm under consideration in terms of acre feet per acre of irrigable land as opposed to the traditional practice of determining the total quantity of water used. By calculating demand on an acre feet per acre basis, differences in water use by dissimilar-sized farms can be compared directly. These data, estimated for the current water quality level of EC = 1.5, indicate only minor differences between farm sizes in the net return to land and water. This result is not surprising in that the technology and feasible cropping alternatives are virtually the same for all commercial farms in the Imperial Valley. Returns to land and water for the small farm are somewhat below those of the two larger farm sizes, primarily because of the economies of size discussed above.

The effect of water quality on the individual farm demand for irrigation water is equally as dramatic as the price effect. The return to land and water for four levels of water quality for the three prototype farms is shown in Table 6. These data indicate, for an EC of 1.5 (the current situation in the lower

A Case Study of Imperial Valley, California

Table 6 Marginal Return to Land and Water for Three Water Supply Levels by Farm Size and Water Quality Level

Water supply per acre	Farm size	Water quality level (EC)			
		0.75	1.50	2.00	3.00
		(dollars)			
1.61 acre ft	Small	27.83	25.00	23.39	10.87
	Medium	37.69	34.29	32.25	11.39
	Large	39.28	35.88	33.93	11.94
5.45 acre ft	Small	9.00	6.60	8.68	7.54
	Medium	8.99	6.60	9.92	9.36
	Large	9.52	6.60	10.26	9.77
8.0 acre ft	Small	2.06	1.75	1.97	2.06
	Medium	2.71	1.74	1.80	2.06
	Large	2.72	2.08	1.80	2.06

Colorado River), the return for three specific points on the water supply axis of Fig. 6. Second, reading horizontally, at very limited water supply levels the effect on the return to land and water is very pronounced. For example, on the large farm the marginal return to land and water at EC = 0.75 and a water supply of 1.1 acre feet/acre is $39.28/acre and declines to $11.94 when EC increases to 3.0.

The effect of increasing salinity declines as irrigation water supplies become more abundant. At a water supply of 8.0 f/acre (which is greater than the current entitlement of the district), the water quality effect is much smaller. For the large farm, the marginal return to land and water for an EC of 0.75 is $2.72/acre f and declines to about $2.06/acre f with a water quality level of EC = 3.0

Summary and Conclusions

In this chapter, we have attempted to analyze the effect on individual farm changes in the quality and quantity dimension of the water supply using a case study of the Imperial Valley of California.

Utilizing the knowledge available at this time (estimates could be improved as additional information is made available through research in plant soil moisture-salinity relationships), we have developed a production function for irrigated crops where water quantity and water quality are independent variables. As was expected, crop response to additional water applied exhibits diminishing marginal returns. Declining water quality (increasing salinity levels) has a negative effect on yield. It was shown that the negative effects of salinity can be offset by substituting additional units of water. In other words, there exists, at least to a limited extent, a trade-off between water quantity and water quality.

With irrigation technology fixed at the current level of technology, as the water quality in the Imperial Valley decreases from the 1970 level of EC = 1.5 to its projected level of EC = 2.0 by about the year 2000, net returns to land and water in the Valley will be expected to decrease by 12 to 15%. If the quality of the water in the lower Colorado River continues to deteriorate beyond the year 2000 level, to, say, EC = 3.0, farm incomes will be proportionately lower, threatening the economic viability of the Valley's agricultural sector.

Operators of larger-sized farms will be in a better position to survive economically under either conditions of severe limitations on the supply of water available or high salinity levels. This result follows primarily from the economies of large-scale farming that can be imputed to the residual return to land and water.

Fruitful areas for future research include improvement in the estimation of the production function based on empirical field test plots where soil moisture tension, electrical conductivity, and leaching fractions are experimental variables. Second, increased knowledge of alternative irrigation technology, including flood, furrow, sprinkler, and drip irrigation application, will become increasingly important. Third, better methods of estimating field drawdown curves with artificial drainage and the effect of water table levels on plant-soil moisture-salinity relationships need to be developed.

We have attempted to show the economic implication of a production response surface. As knowledge is gained in this area, we shall be able to place increased confidence on the analytical results.

References

Bernstein, L. (1967). *Salt Tolerance of Plants*, U.S. Dept.Agr. Bull. 217, 21 pp.
Colorado River Board of California (1970). *Need for Controlling Salinity of the Colorado River*, Resources Agency, State of California.
Moore, C. V. (1961). A general analytical framework for estimating the production function for crops using irrigation, J. Farm Econ. *43*(4); 876-888.
Moore, C. V., Snyder, J. H., Sun, P. (1974). Effects of Colorado River water quality and supply on irrigated agriculture, Water Resources Res. *10*(2).
Sun, P. C. (1972). An economic analysis of the effects of quality and quantity of irrigation water on agricultural production in Imperial Valley, California. Unpublished Ph.D. thesis, Department of Agricultural Economics, University of California, Davis, Calif.

8

Physical and Economic Evaluation of Irrigation Return Flow and Salinity on a Farm

R. J. Hanks and Jay C. Andersen

Utah State University, Logan, Utah

Introduction

The physical and economic problems in water application, return flow, and reuse present a difficult challenge to users and policy makers. The conflicts that develop have been treated with little proficiency because of lack of information on the basic elements of the problem.

The physical aspects of the problem arise because of natural sources and irrigation return flow, which constitute a large portion of the water in streams and rivers of many parts of the world. In some river basins, such as the Colorado in the United States, some water may be used for irrigation several times before the little that remains enters the ocean or other repository. Since this use involves the evapotranspiration process, which accounts for the major water loss by crops, there is an inevitable buildup of salt concentration in irrigation return flows. In some cases, salt deposits are leached from the soil. Thus, both concentration of the water and loading of the stream may occur. This is seen in the salinity of the Colorado River, which ranges from less than 50 mg/liter (total dissolved solids) in the upper basin mountains to about 850 mg/liter at the Imperial Dam in the lower basin. Even though irrigation return flow may be involved in only part of the salinity concentration, it has been suggested as one of the major areas capable of management. Management of irrigation water to influence downstream salinity has not been considered extensively in the past and, therefore, little is known about the manifold effects of such management.

The economic conflict is generated because the well-being of some users of a river usually conflicts with the well-being of others. There is no solution that makes everyone better off. The only possibility is to arrive at an efficient solution such that no alternative pattern of resource use would make anyone better off without making someone else worse off. The ideal market solution for managing water quality does not exist for at least two reasons. First, prices do not reflect the social value of resources and commodities. The individual decision maker has no incentive, except for his own conscience or goodwill, to take all costs or benefits into account in making a resource allocation decision. Thus, misallocations of resources occur. Second, producers of public goods are unable to collect all revenues from beneficiaries, since users cannot be excluded for nonpayment of the price. Each user may expect to reap benefits from these public goods whether or not he pays the cost. The product market is, therefore, unlikely to suppy optimal amount of goods with collective consumption characteristics.

The salinity problem in large river basins like the Colordao exhibits both of these aspects of market failure. More than half of the salinity concentration in the river is due to natural causes, but if there were no man-made effects, the concentration might not be sufficient to trouble downstream users. Improved levels of water quality have a public good character since such improvements would benefit all users. Private markets, of course, would not succeed in attaining the most socially desirable level of salinity.

The analysis reported here is an evaluation of the physical and economic aspects of management to control salinity discharges from a farm. The study involves two parts, first the development of a physical model to predict the response of the soil water and crop factors to farm irrigation, and second the development of an economic model using the physical model for basic data which predicts the cost effectiveness of irrigation management as related to return flow salinity. The possibilities for avoiding the deterioration of the irrigation return flow are assessed. No conclusions are to be drawn as to who should meet the cost of pollution abatement. The study concerns only the physical nature and the cost effectiveness of possible abatement management schemes.

Model Description

At the beginning of the study, the physical model to supply basic data was carried out somewhat independently of the economic model of cost effectiveness. However, it was soon apparent that much interchange was necessary. The physical model originally produced much information not needed for the economic model and did not supply some needed basic data. Thus, considerable modification of the physical model was necessary. Similarly, the economic model originally devised assumed basic information that was physically unobtainable. The economic model, therefore, had to be adjusted to use the data

Evaluation of Irrigation Return Flow and Salinity

available. The details of the two models are discussed in the following sections separately for purposes of organization and convenience of the reader.

The Physical Model

Effects of salt in the soil on crop production are important to the economic evaluation of irrigation and management practices. Although a large amount of background information is available, recent field work has shown that many situations are much more complicated than can be handled easily by present models. The field situation in this report, for example, was studied by Gupta (1973) and King and Hanks (1973). They found the model modifications of previously published models gave good predictions for water movement but poor predictions for salt movement. Further, where water of different salt concentrations had been added, there was a very small effect on resulting concentrations of the soil solution as a result of different salt concentrations in the added water. It appeared that the salt acted as a large buffer that was influenced only slowly by relatively small salt additions or removals through irrigation and drainage. It thus became evident that the inclusion of complicated physical chemical reactions used by Dutt et al. (1972) were of little practical use because they were not completely accurate and, further, required considerable computer time. Consequently, a simplified salt flow model to simulate the long-time effects of salt buildup by varying the initial salt concentration conditions was devised.

The model is based on the work of Nimah and Hanks (1973a, 1973b) which is concerned with the soil water flow in response to varying irrigation management inputs. The general equation for water flow is given as Eq. (1):

$$\frac{\partial \theta}{\partial t} = \frac{\partial}{\partial z}\left(K\frac{\partial H}{\partial z}\right) + a(z) \tag{1}$$

where θ is the water content, H is the matric potential, K is the hydraulic conductivity, t is time, z is depth, and a(z) is the root extraction term.

The root extraction term is somewhat more complicated because it has plant and soil characteristics in it, as the following equation shows:

$$a(z) = \frac{[H_{root} + 1.05z - h(z,t) - s(z,t)] \cdot RDF(z) \cdot K}{\Delta z \cdot \Delta x} \tag{2}$$

where H_{root} is the water potential at the surface of the root, which is modified to have a different water potential due to gravity and a small friction resistance term of 0.05; h(z, t) is the soil solution matric potential; s(z, t) is the osmotic potential; and RDF(z) is the root density function. Δx is the distance between the plant root and the point in the soil where h(z, t) is realized (we assume equal to 1.0). Depending on the climate and the plant and soil conditions, water may be extracted from the soil without any limitations so that the transpiration is equal to that potential transpiration. However, if the osmotic concentration or the matric potential is sufficiently low, keeping in mind the negative sign, the soil water system will not be able to supply sufficient water to the plant to maintain the transpiration at potential transpiration and then the transpiration falls off. These equations have been discussed in considerably more detail in Nimah and Hanks (1973a) and Childs and Hanks (1975).

The salt flow portion of the model is given as follows:

$$\frac{\partial(C\theta)}{\partial t} = \frac{\partial}{\partial z} D \frac{\partial C}{\partial z} - \frac{d(Cq)}{dz} \tag{3}$$

where C is the salt concentration, D is the combined diffusion and dispersion coefficients, and q is the mass flux of water.

Salt is assumed to move within the soil profile according to mass flow of water and subject to the diffusion restrictions. No consideration is taken for source or sink term where precipitation or solutions of salts could come out of the solid phase of the soil. A numerical approximation of both the water flow and moisture flow part of the model has been written as described by Bresler (1973), Nimah and Hanks (1973) and Childs and Hanks (1975), as well as Hanks et al. (1974). To determine the influence of the salinity on crop yield, another component of the model has to be added. This is done by using the assumptions that have been described by Hanks (1974) and Childs and Hanks (1975), where the relative yield of a crop is related to the relative transpiration. The validity of this assumption for saline conditions has been tested by Childs and Hanks (1975). The data indicate a good linear relationship between relative transpiration and relative yield. Relative yield is here restricted to the dry matter yield and does not include the yield of grain, which might be considerably more complicated.

The estimation of a relative yield is necessary to interface with the economic model discussed later. The variations that are sensed by the model are the result of various initial conditions or boundary conditions that change with time at the top and bottom of the soil. The soil conditions also influence the results as well as the crop conditions, because soil properties influence water uptake and water infiltration in the soil. The plant grown also influences root uptake as well as the boundary conditions of the surface.

As described in detail by Childs and Hanks (1975), it is necessary to determine what the potential evapotranspiration or the potential infiltration rate for the soil would be for any kind of a management system that is imposed. This is done by either measurement of the potential evapotranspiration such as described by Nimah and Hanks (1973b) or by using some method such as described by Jensen (1973) to compute potential evapotranspiration. This model does not require an estimation of the crop coefficient but requires that the potential evapotranspiration be split down into potential transpiration as described by Childs and Hanks (1975).

The basic input data required for the model are given in detail by Nimah and Hanks (1973) and Childs and Hanks (1975), but are summarized as follows: (1) hydraulic conductivity, water content, and matric potential water content data covering the range of water content to be encountered during the period of interest (soil property), (2) air dry soil water contents (soil property), (3) root water potential below which the root will not go where presumably the plant wilts and the actual transpiration will be less than the potential transpiration (plant property), (4) root distribution function for the period of study (plant property), (5) water content and soil solution concentration data at the beginning as a function of depth (initial condition), (6) potential transpiration, potential evapotranspiration rate, and potential irrigation or rainfall as a function of time for the whole period of the run (boundary condition) [potential evapotranspiration assumed equal to that from a free water surface could be calculated by the use of the Penman or other equation as described by Jensen (1966)], (7) osmotic potential of irrigation water (boundary condition), (8) presence or absence of a water table at the bottom of the soil profile (boundary condition). The root density function may be changed as a function of time and depth as the root system grows, as described by Childs and Hanks (1975).

The output data can be selected from among many variables that are computed within the model from a list of the following: (1) cumulative evapotranspiration, transpiration, and evaporation as a function of time, (2) volumetric soil water content and soil pressure head as a function of time and depth, (3) cumulative water flow upward or downward through any boundary within the profile or at the surface, (4) the value of H_{root} as a function of time, or many other other factors. The main item of interest in this computation is the relative transpiration, which is the transpiration computed from the particular management system compared to what the potential transpiration would have been at the same condition if soil water were not limiting.

This analysis does not consider the contribution from natural sources (dissolution) nor the precipitation of salts within the soil profile. We are pursuing studies of these problems now, but they appear to be very complicated although very important.

The Economic Model

The economic model is designed to suggest ways to minimize the income losses imposed by restraints on salt outflow due to irrigation on the farm. It is based on the physical model and a set of cost and return data for the farm. The beginning point is to assume that any amount of salt can be allowed to leave the farm. The model is set to maximize income under this assumption, which has been the policy in the past. The model is then successively constrained to allow the smaller amounts of salt to leave the farm. Of primary concern is the income reduction that accompanies this constraint on resource use. Also of concern are the crops grown, irrigation management practices, and quantity of water applied. As the salt outflow and income incrementally change, the model develops as a by-product the marginal relationship between salt outflow and income. This value can then be compared with alternative ways of reducing salinity in the river or the damages that accrue to downstream users. The implementation of the economic model is in the form of the linear programming model of economic behavior.

The Linear Programming Model of Salt Outflow

The linear programming model used in this study is a profit-maximizing model that has the algebraic form of:

maximize $\quad Z = CX$
subject to $\quad AX \leq B$
$\quad\quad\quad\quad\;\; X \geq 0$

where
Z = net revenue (or profit)
C = the row vector of net revenue per unit of activity
X = the set of activities or production processes
A = matrix of technical coefficients (or production relationships)
B = the column vector of constraints of resource availability

Linear programming and the economic concepts utilized are discussed by Leftwich (1970). The application to the present study is as follows: (1) Select the combination of crops produced, subject to the constraints in certain fixed inputs such as land. The selection is based on the operating costs and the relative prices of the crops. (2) Many of the inputs are not fixed, thus the optimal combination of these variable inputs is selected for the production of the crops produced based on their productivity and the cost of inputs. (3) The level of output per acre is selected based on producing up to but no more than the level where the value of the incremental unit of production equals the cost of the incremental inputs unit of input.

In this study, the various components of the model are defined and constructed as follows:

Alternative processes and activities. Production activities (the x_j) have been developed that are most relevant. These are activities such as growing corn

Evaluation of Irrigation Return Flow and Salinity

silage or oats or alfalfa hay. Each of these can also be treated in alternative ways, such as with different quantities of irrigation applied by sprinkling or flooding. All combinations of these alternative actions were used in this study except that flood irrigation was not used in the lowest three levels of water application. It would be impossible to distribute the small amounts of water uniformly over the season by flooding.

Resource constraints. Limits on resource availability (the b_i) used in this study include the quntities of each of three land classes based on the beginning salinity levels of the soil profile. It was assumed that the farmer had 10 acres with each of three soil characteristics described earlier. Unlimited salt outflow was allowed in the drainage water (which level was subsequently reduced to determine the profitability to the farm operator of letting salt flow into the drains and streams). There were also constraints to force growing of crops in rotation such as to provide for nurse crops for new seedings of alfalfa and limits on corn production for disease control. Net profitability for each unit of production was based on approximate current prices for products and the costs of various farming supplies and operations. Yields were estimated using the 1971 data for the farm as a base with the relative yields predicted in the physical model to give specific values for the rates of water applied as influenced by the initial salt concentrations shown in Tables 8 and 9. The profit function is based on the price of alfalfa, $45 a ton, corn silage, $13 a ton, and oats, $1.60 a bushel. These prices represent approximately the current prices but are adjusted somewhat to a normal long-run relationship to each other.

Situations Studied

There were several situations studied in terms of water management. The data for Vernal, Utah, 1971, as described by Nimah and Hanks (1973b) were taken for the initial conditions, and water was applied in different amounts but at the same frequency as given in the 1971 data. The irrigation water quality used throughout was 6.35 meq/liter, which was equivalent to the present conditions at the Vernal, Utah, farm.

To simulate the effects of soil salinity storage within the root profile, three different levels of soil salinity were studied—20 meq/liter uniform throughout, which is approximately the condition on most of the farm at present; 50 meq/liter uniform throught; and 200 meq/liter uniform throughout.

There were three crops that were studied on the farm. They were corn, oats, and alfalfa. Because of data collected from various sources, the root distribution functions were chosen arbitrarily, as shown in Table 1 for the three crops studied. The corn and oats were modeled as annual crops with different values of crop cover as a function of time during the year. This had an influence on the

Table 1 Relative proportion of Roots at Different Depth Increments at Maturation Assumed for the Calculations

Depth (cm)	Corn	Alfalfa	Oats
2.5-10.5	0.09	0.14	0.18
10.5-25.5	0.20	0.30	0.40
25.5-52.5	0.34	0.33	0.42
52.5-91.5	0.25	0.23	0
92.5-140.0	0.12	0	0
140.0-235.0	0	0	0

potential evapotranspiration distribuiton as described by Childs and Hanks (1975).

Two different irrigation systems were studied. The first was a solid-set sprinkler system with a coefficient of uniformity of 0.88, which is approximately the same as is now in place on the experimental farm being studied. This was compared to a very poor gravity system, which was on the farm before the sprinkler system was applied. The coefficient of uniformity is useful for comparison of the effect of a range of application uniformities.

Results

The Physical Relationships

The results of varying the water added and initial salt concentration on various soil and water properties for corn are shown in Table 2. The data on T/T_p are of primary interest because they seem to be directly related to relative yield. The data of Table 2 show that T/T_p increases as the irrigation applied is increased up to about 46 cm, after which the ration was essentially 1.0 for all initial salt concentration. T/T_p was less than 0.9, however, where irrigation was less than 6, 9, and 26 cm for an initial salt concentration of 20, 50, and 200 meq/liter, respectively.

There was relatively little difference between the T/T_p resulting when the initial salt concentration was 20 or 50 meq/liter, but there was a marked difference when the initial salt concentration was modeled at a level of 200 meq/liter. The 20-meq/liter initial salt concentration was unimportant, so that yield differences were due to water influences only. Note that where the irrigation and rain was less than about 20 cm, there was equal upward flow so that the amount of flow was limited by soil water transmission and plant root extraction. In cases where the initial salt concentration was 200 meq/liter, upward flow was about

Table 2 Comparison of Irrigation Water Applied and Initial Salt Concentration on Relative Transpiration of Corn T/T_p, Evaportranspiration, ET, Drainage, Salt Flow to the Groundwater, and Average Final Salt Concentration[a].

Irrigation and rain (cm)	ET (cm)	T	T/T_p	Drainage (cm)	Salt flow to groundwater (cm)	Initial salt concentration (meq/liter)	Final salt concentration average (meq/liter)
5.6	40.3	35.3	0.81	−14.2	−284	20	62
5.6	38.6	33.5	0.77	−14.2	−710	50	127
5.6	26.2	20.6	0.48	−11.6	−2,320	200	305
15.0	47.7	38.6	0.97	−14.0	−280	20	56
15.0	46.3	37.2	0.93	−13.9	−695	50	116
15.0	34.6	25.1	0.64	−11.4	−2,280	200	296
40.8	50.4	37.6	0.99	−8.7	−174	20	27
40.8	48.3	35.9	0.98	−7.1	−355	50	604
40.8	48.1	35.8	0.97	−6.2	−1,240	200	227
56.4	51.9	37.3	1.00	0.91	19	20	23
56.4	52.2	37.3	1.00	1.0	49	50	50
56.4	56.7	37.3	1.00	1.1	214	200	189
66.7	51.7	37.3	1.00	10.5	210	20	20
66.7	51.6	37.3	1.00	10.6	532	50	42
66.7	51.6	37.3	1.00	10.5	2,160	200	153

[a]Each line represents a computation with some irrigation frequency but different amounts of water applied for climatic conditions of 1971 at Vernal, Utah. A negative sign indicates upward flow of salt and water. Rain was 5.6 cm.

2.5 cm less than for the lower initial salt concentrations. However, drainage (downward flow) was influenced very little by initial salt concentrations.

One feature of the data shown in Table 2 that may be somewhat unique is the large influence of water movement up from the water table (at a depth of 235 cm). The soil properties at the Vernal farm seem to be especially conducive to high water flow in both directions. Other situations with other soils would probably not result in as much upward flow.

The data shown in Table 2 are only a small part of the data collected in attaining these summary values. Each line represents one season where data have been computed at several depth increments and at no greater than 2- to 3-hour increments. Thus, data within the season are also available.

Table 3 show the computations of T/T_p made for alfalfa. The data show more decrease of T/T_p for low irrigation rates than was shown for corn. This was

Table 3 Comparison of Irrigation Water Applied and Initial Salt Concentration on Relative Transpiration of Alfalfa, T/T_p, Evapotranspiration, ET, Drainage, Salt Flow to the Groundwater, and Average Final Salt Concentration[a].

Irrigation and rain (cm)	ET (cm)	T	T/T_p	Drainage (cm)	Salt flow to groundwater (meq)	Initial salt concentration (meq/liter)	Final salt concentration average (meq/liter)
5.6	29.5	25.8	0.52	−9.7	−196	20	43
5.6	28.2	26.6	0.50	−9.4	−472	50	97
5.6	19.8	16.0	0.33	−7.8	−1,561	200	277
15.0	37.6	32.8	0.68	−9.3	−154	20	43
15.0	36.5	31.8	0.66	−9.2	−458	50	94
15.0	28.8	23.7	0.49	−7.6	−1,840	200	268
40.8	51.7	46.7	1.00	−7.4	−148	20	30
40.8	51.3	46.3	1.00	−6.7	−370	50	64
40.8	48.1	43.2	0.93	−5.6	−1,340	200	228
56.4	53.4	48.2	1.00	0.0	0	20	24
56.4	53.9	47.9	1.00	0.4	22	50	52
56.4	53.9	47.9	1.00	0.3	61	200	195
66.7	53.5	48.3	1.00	8.8	178	20	22
66.7	53.1	48.3	1.00	9.3	467	50	44
66.7	53.2	48.3	1.00	9.4	1,882	200	158

[a]Each line represents a computation with the same irrigation frequency but different amounts of water applied for climatic conditions of 1971 at Vernal, Utah. A negative sign indicates upward flow of salt and water. Rain was 5.6 cm.

due to longer season for active water use by alfalfa and for a much greater proportion of transpiration to evapotranspiration for alfalfa than for corn—especially during early season when corn was just planted. Upward water flow was less for alfalfa than corn, probably because of alfalfa's assumed shallow root distribution. This result is probably not representative of other situations, where alfalfa normally roots deeper than corn. The alfalfa root distribution was measured at the site, but the corn data were measured at another location. Like corn, the alfalfa data show little difference between the 20- and 50-meq/liter initial salt concentrations but fairly large differences with 200-meq/liter initial salt concentration. Thus, the T/T_p depression at 20-meq/liter initial salt concentration is due to inadequate irrigation. The differences in T/T_p at any one irrigation level, for initial salt concentrations between 20 and 200 meq/liter, were due strictly to

Evaluation of Irrigation Return Flow and Salinity

Table 4 Comparison of Irrigation Water Applied and Initial Salt Concentration on Relative Transpiration for Oats, T/T_p, Evapotranspiration, ET, Drainage, Salt Flow to the Groundwater, and Average Final Salt Concentration[a].

Irrigation and rain (cm)	ET (cm)	T	T/T_p	Drainage (cm)	Salt flow to groundwater (meq)	Initial salt concentration (meq/liter)	Final salt concentration average (meq/liter)
5.6	18.3	13.3	0.29	−3.8	−74	20	33
5.6	18.0	12.9	0.28	−3.8	−191	50	78
5.6	14.3	8.2	0.18	−3.6	−718	200	248
15.0	27.1	20.2	0.46	−3.8	−76	20	33
15.0	26.7	19.4	0.44	−3.8	−189	50	76
15.0	22.9	13.3	0.32	−3.5	−700	200	242
40.8	46.0	35.2	0.89	−2.5	−50	20	26
40.8	45.7	35.1	0.88	−2.4	−120	50	58
40.8	42.3	31.5	0.80	−1.2	−240	200	208
56.4	53.6	38.5	0.97	1.3	26	20	24
56.4	53.4	38.8	0.98	1.3	66	50	52
56.4	51.4	37.0	0.93	2.5	490	200	185
66.7	52.5	38.6	0.99	10.0	198	20	20
66.7	52.5	38.6	0.99	10.0	495	50	43
66.7	52.5	38.6	0.99	0.9	1,975	200	157

[a]Each line represents a computation with the same irrigation frequency but different amounts of water applied for the climatic conditions of 1971 at Vernal, Utah. A negative sign indicates upward flow of salt and water. Rain was 5.6 cm.

a salt effect—where 15 cm of irrigation and rain was applied, T/T_p was 0.68 to 0.49 resulted from the high initial salt concentration.

Table 4 shows the computed data for oats when irrigation water was managed in a manner similar to corn and alfalfa. The values of T/T_p were smaller for oats for a given irrigation regime than for corn or alfalfa. This was due mainly to a more shallow root depth, but was also partly due to a difference in the relative relation of T_p to ET_p. Because of the shallow root zone, upward flow was less than 4 cm. This caused the ratio, T/T_p, to be less than 0.9 (for 20 meq/liter initial salt concentration) when irrigation and rain was less than about 52 cm. As was the case for alfalfa and corn, the T/T_p results with 50-meq/liter initial salt concentration were only slightly different than for 20 meq/liter, whereas the results for 50 meq/liter were considerably larger than with an initial salt concentration of 200 meq/liter.

There is a feature of the computation that is especially noticeable in Table 2 for corn. The computer program allows for the possibility that, if evaporation is less than potential evaporation, the difference, E_p-E, can be used in transpiration. Thus, potential transpiration is not a constant in Table 2 but increases as the irrigation and rain applied decreases. For a rain of 5.6 cm, T_p for corn was 40.3 and for irrigation and rain of 56.4 cm, T_p was 37.3 cm. Hanks et al. (1971) have demonstrated that this energy "trading" occurs, but it may be that the model computation overcorrects for it.

Figure 1 shows a 10-year computation during which irrigation and rain were about one-half ET. The data indicate no decrease in the T/T_p ratio until the seventh year, after which it fell rapidly, leveling off at the tenth year. Figure 1 shows the average salt concentration building up to about 260 meq/liter at the tenth year. When T/T_p decreases, the transpiration also decreases. After the tenth year of cropping, ET had decreased by 15 cm, which was only 9 cm above the water added. The difference between the water added and ET came from soil water storage and flow upward from the water table. Note that the particular results computed for a simulated run of year, shown in Fig. 1, are highly dependent on the particular situation. If a crop with more shallow roots had been used, an entirely different situation would have resulted.

One of the purposes of the computation shown in Fig. 1 was to see how these results compared with the data of Table 2 where different initial salt concentrations were used to simulate salt buildup. For the same irrigation schedule, the data of Table 2 indicate a T/T_p ratio of 0.90 for an initial salt concentra-

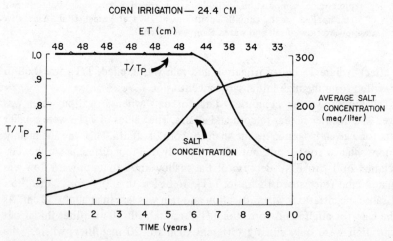

Fig. 1 Relative transpiration and average salt concentration for corn with deep roots irrigated at a rate of 24.4 cm/year as influenced by year.

tion of 200 meq/liter and ending up with an average of 296 meq/liter. The data of Fig. 1 indicate essentially the same ratio of T/T_p, although the salt concentration at the end of the year is not as high as that shown in Table 2. Thus, using a uniform salt concentration profile as the initial condition gives the same result as using the profile existing at the end of the previous crop years. In fact, the uniform profile is probably more accurate, since the upward and downward diffusion and mass flow due to evaporation and drainage tends to equalize the salt in the profile over the winter.

The single point values, relating water added to the T/T_p, are somewhat unrealistic in a real field situation because water is not distributed uniformly. Even in the best system there are part of the field that receive more water than others. To account for this, engineers have defined a uniformity coefficient as follows:

$$Cu = 1 - \frac{D}{M} \tag{4}$$

where M is the average irrigation rate and D is the average deviation (sign ignored) about the average rate. It should be noted that if Cu = 1.0, water application would be completely uniform. To add this factor to the computations, it was necessary to assume a distribution pattern and the extent of coverage that might apply for some mean water application rate. From the distribution pattern, a new value of T/T_p results from integrating T/T_p over the water distribution pattern. This also provides salt outflow information. These data were calculated assuming a uniformity of 1.0 for all of the data presented up to this point. Considering nonuniform coverage, the relationship of T/T_p to average water added by irrigation can then be constructed. These data are shown in Tables 5, 6, and 7 for the three crops for three different uniformities. The amount of salt outflow is also shown. These tables show essentially the same ratio of T/T_p for all uniformities provided the water application is insufficient to allow any drainage (and thus salt outflow). However, once the irrigation rate is high enough to result in some drainage, the ratio of T/T_p is 1.0, 0.98, and 0.90 for a Cu of 1.0, 0.88, and 0.42, respectively (20 meq/liter initial salt distribution). This ratio variation results from poor uniformity due to irrigation greater than ET — above which point there is no further increase of T/T_p as the water added increases. The same result is also shown in Table 8, where the average water application is greater than ET. For this situation, some part of the field received water at less than ET, resulting in T/T_p less than 1.0.

These results point out another situation of great practical importance involving some present concepts of low leaching ratios. If water distribution is nonuniform and the average leaching ratio is low, than there will be part of the field that is not leached at all and salts will accumulate. This could be a serious problem when the same uniformity distribution pattern prevails year after year.

Table 5 Relative Yield of Corn, Equal to T/T pot, as Influenced by Two Different Values of Cu, Water Applied, and Initial Salt Concentration[a].

Irrigation and rain (cm)	Initial salt (meq/liter)	T/T pot Cu = 0.88	Salt outflow Cu = 0.88 (meq/cm^2)	T/T pot Cu = 0.42	Salt outflow Cu = 0.42 (meq/cm^2)
15.0	20	0.94	0	0.93	0
15.0	50	0.92	0	0.91	0
15.0	200	0.64	0	0.64	0
40.8	20	1.0	0	0.98	89.
40.8	50	0.99	0	0.97	216.
40.8	200	0.96	0	0.88	892.
56.4	20	1.0	60.	0.99	357.
56.4	50	0.99	158.	0.98	821.
56.4	200	1.0	644.	0.91	3,563.
66.7	20	1.0	239.	0.99	703.
66.7	50	1.0	581.	0.98	1,575.
66.7	200	1.0	2,398	0.92	7,099.

[a]Deep roots assumed.

Table 6 Relative Yield of Alfalfa, Equal to T/T pot, as Influenced by Two Different Values of Cu, Water Applied, and Initial Salt Concentration[a].

Irrigation and rain (cm)	Initial salt (meq/liter)	T/T pot Cu = 0.88	Salt outflow Cu = 0.88 (meq/cm^2)	T/T pot Cu = 0.42	Salt outflow Cu = 0.42 (meq/cm^2)
15.0	20	0.68	0	0.68	0
15.0	50	0.66	0	0.65	0
15.0	200	0.49	0	0.49	0
40.8	20	0.98	0	0.90	86.
40.8	50	0.97	0	0.89	212.
40.8	200	0.91	0	0.81	804.
56.4	20	1.0	44.	0.92	449.
56.4	50	1.0	124.	0.92	996.
56.4	200	0.99	512.	0.86	3,492.
66.7	20	1.0	232.	0.94	1,007.
66.7	50	1.0	571.	0.93	2,128.
66.7	200	1.0	2,170.	0.89	7,158.

[a]Medium-depth roots assumed.

Table 7 Relative Yield of Oats, Equal to T/T pot, as Influenced by Two Different Values of Cu, Water Applied, and Initial Salt Concentration[a].

Irrigation and rain (cm)	Initial salt (meq/liter)	T/T pot Cu = 0,88	Salt outflow (meq/cm^2)	T/T pot Cu = 0.42	Salt outflow (meq/cm^2)
5.6	20	–	–	–	–
5.6	50	–	–	–	–
5.6	200	–	–	–	–
15.0	20	0.45	0	0.45	0
15.0	50	0.43	0	0.43	0
15.0	200	0.31	0	0.32	0
40.8	20	0.87	0	0.79	84.
40.8	50	0.87	0	0.78	209.
40.8	200	0.78	17.	0.71	818.
56.4	20	0.97	63.	0.84	365.
56.4	50	0.97	157.	0.84	918.
56.4	200	0.93	738.	0.79	3,161.
66.7	20	0.99	225.	0.87	780.
66.7	50	0.99	563.	0.87	1,967.
66.7	200	0.98	2,178.	0.82	6,492.

[a]Shallow roots assumed.

The Economic Alternatives

The physical relationships discussed above are the basic data for the economic analysis. From the physical data, the relevant information on growing corn silage, oats, or alfalfa hay was accumulated. Decision options that included water application by sprinkler or by flooding at rates (from irrigation and rain) of 10.3, 15.0, 22.0, 40.8, 56.4, and 66.7 cm for each of the crops were utilized.

Limits on resource availability (the B_i) or right-hand-side values used in the linear programming study include the quantities of each of three land classes based on the beginning salinity levels in the soil profile. It was assumed that the farm under study had 10 acres with each of three soil characteristics (20, 50, and 200 meq/liter) as described earlier. Also, an unlimited quantity of salt outflow was allowed in the drainage water (which level was sequentially reduced to determine the loss in profitability to the farm from letting salt flow into the drains and streams). There were also constraints to force growing of crops in rotation

Table 8 Predicted Yield of Crops Under Sprinkler Irrigation by Initial Salt Content of Soil, by Water Application Rates[a].

Initial salt content of soil (meq/liter)	Water level (irrigation plus rain) (cm)	Crop yield		
		Alfalfa (medium roots) (tons)	Oats (shallow roots) (bushels)	Corn silage (deep roots) (tons)
20	10.3	3.3	34.0	20.5
	15.0	3.7	44.2	21.6
	22.0	4.4	55.7	22.8
	40.8	5.3	80.1	22.8
	56.4	5.5	89.0	22.8
	66.7	5.5	91.3	22.8
50				
	10.3	3.2	32.8	19.7
	15.0	3.6	39.8	21.1
	22.0	4.3	54.4	22.6
	40.8	5.3	79.8	22.8
	56.4	5.5	89.2	22.8
	66.7	5.5	91.4	22.8
200				
	10.3	2.2	22.2	12.9
	15.0	2.7	28.8	14.7
	22.0	3.5	43.3	17.9
	40.8	4.9	71.9	22.8
	56.4	5.4	85.3	22.8
	66.7	5.5	90.1	23.0

[a] Based on Tables 5, 6, and 7, above, and assuming a coefficient of uniformity of application (CU) = 0.88.

such as to provide for nurse crops for new seedings of alfalfa and limits on corn production for disease control.

The net profit values for each unit of production were based on approximate current prices for products and the costs of various operations. Yields were estimated using the 1971 data for the farm as a base and the relative yields predicted in the physical model to give specific values for the rates of water applied as influenced by the initial salt concentration in the soil as shown in Tables 8 and 9.

The profit function is based on a price for alfalfa, $45/ton; for silage, $13/ton; and for oats, $1.60/bushel. These represent approximately the current prices, but are adjusted somewhat to a normal long-run relationship to each other. The cost of raising crops was computed as shown in Table 10.

Table 9 Predicted Yield of Crops Under Flood Irrigation by Initial Salt Content of Soil, by Water Application Rates[a].

Initial salt content of soil (meq/liter)	Water level (irrigation plus rain) (cm)	Crop yield		
		Alfalfa (medium roots) (tons)	Oats (shallow roots) (bushels)	Corn silage (deep roots) (tons)
20	40.8	4.9	72.4	22.6
	56.4	5.0	77.5	22.7
	66.7	5.1	80.0	22.7
50	40.8	4.9	71.9	22.4
	56.4	5.0	77.1	22.5
	66.7	5.1	79.7	22.6
200	40.8	4.5	65.4	20.2
	56.4	4.7	72.3	20.9
	66.7	4.9	75.7	21.3

[a]Based on Tables 5, 6, and 7, above, and assuming a coefficient of uniformity of application (CU) = 0.42.

Results of the Economic Analysis

Two main sets of results were desired in order to draw conclusion. These were the set of production activities that would maximize farm profit at each level of salt outflow and the loss in income from not allowing an incremental ton of salt to flow out. The latter may also be characterized as its mirror image, the value to the farm of allowing an additional ton of salt outflow.

A number of different situations were modeled to determine the effects of differences in assumed depth of roots for alfalfa and corn silage, effects of irrigation method and rate of application, and restrictions on the crop combinations.

Situation 1: Unrestricted Corn in the Rotation, Corn Roots Deep, Alfalfa Roots Shallow, Sprinkle or Flood Irrigation

Without any constraint on corn in the rotation, the production activities in the optimal production pattern included nothing other than corn. In Fig. 2, the most profitable production activities are summarized. Note that the tons of salt outflow for the entire 30 acres are on the scale at the bottom of the figure. The set of crops that is optimal is plotted for the 10-acre units by soil type (where initial soil salt is at the high, medium, or low level) for each level of salt outflow. For instance, at a level of 60 tons of salt outflow, the model indicates

Table 10 Cost Components of Crop Production by Crop and by Method of Water Application.

Crop	Fixed cost ($/acre)	Growing cost ($/acre)	Irrigation costs				Harvest cost $
			Water level (Irrigation plus rain) (cm)	Sprinkler construction cost ($/acre)	Energy cost ($/acre)	Flood ($/acre)	
Alfalfa hay	13.65	27.09		24.22		9.63	7.50/ton
			10.3		1.22		
			22.0		3.30		
			40.8		6.59		
			56.4		8.91		
			66.7		10.71		
Oats	13.65	58.11		24.22		9.55	0.16/bu
			10.3		1.22		
			15.0		1.65		
			22.0		3.30		
			40.8		6.59		
			56.4		8.91		
			66.7		10.71		
Corn silage	13.65	70.39		24.22			
			10.3		1.22		
			15.0		1.65		
			22.0		3.30		
			40.8		6.59		
			56.8		8.91		
			66.7		10.71		

that for the low salt condition, the entire 10 acres should be in corn irrigated at the fifth level (next to highest) by flooding. For the medium salt condition, there should be about 4 acres of corn at the fourth level of water application. On the saltiest land, there should be 10 acres of corn irrigated at the fifth level by sprinkling.

In meeting the requirement for low salt outflow, sprinkler systems and low rates of water application were required in the model. As the allowable salt outflow was increased, the irrigation was increased and the method of irrigation changed to flooding. Net profit increased by about $900 (or $30 per acre) as the salt outflow constraint was relaxed. Almost all of this profit increase occured in the first 20-ton increment. Only about $100 of additional profit (Fig. 3) for the 30-acre block of land could be attained beyond this first 20-ton increment. In a practical management situation, all 30 acres would be irrigated by sprinkling or by flooding, rather than a combination of systems.

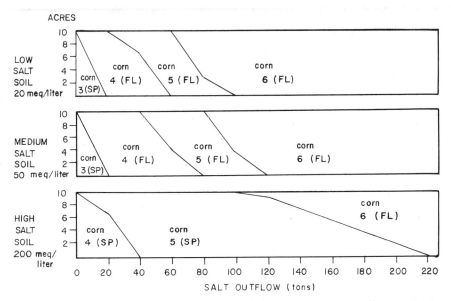

Fig. 2 Optimal cropping and irrigating pattern for high, medium, and low initial soil salt conditions where corn roots are deep and alfalfa shallow and either flooding or sprinkling is allowed as an irrigation method.

There are two main reasons for obtaining these results. First, it was assumed that corn was a deep-rooting plant so that this crop was profitable at low levels of irrigation, since in the physical model the corn obtained considerable water from deep soil moisture or underground supply. In a static one-year analysis with a light application of water, there would be no outflow of salts, but there would be an accumulation in the soil profile. Second, corn was the most profitable crop assuming that yields can be maintained.

In Fig. 4, the value to the farm of an additional ton of salt outflow as a function of salt outflow is shown. Note that the cost to the farm of reducing the outflow of salt (or value for letting an additional ton flow out) is very low compared to any possible cost of removal by desalination.

Situation 2: Corn Restricted to One-Half of the Acreage, Corn Roots Shallow, Alfalfa Roots Deep, Sprinkle or Flood Irrigation

This situation was tested for several reasons. Corn could probably not be grown exclusively for several years because of varied needs for livestock feed, disease and fertility problems on the land, and grower preference for multiple crops. Also, the depth of corn roots may be somewhat shallower than the perennial alfalfa crop. The data that indicated corn was deep-rooted and alfalfa somewhat shallower were from separate experimental plots and may not be appropriate for the area of this study.

Under these assumptions, the cropping patterns over the range of salt outflow are shown in Fig. 5. Alfalfa would be profitable, and the required nurse

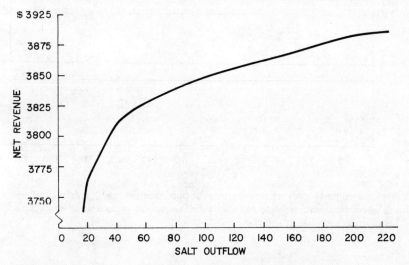

Fig. 3 Net revenue by amount of salt outflow for the 30 acres for situation 1.

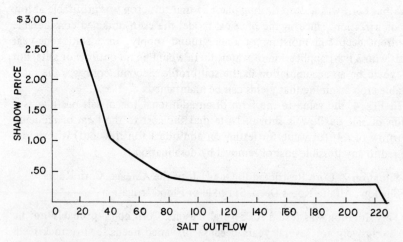

Fig. 4 Shadow price or value of an additional ton of salt outflow for the 30 acres for situation 1.

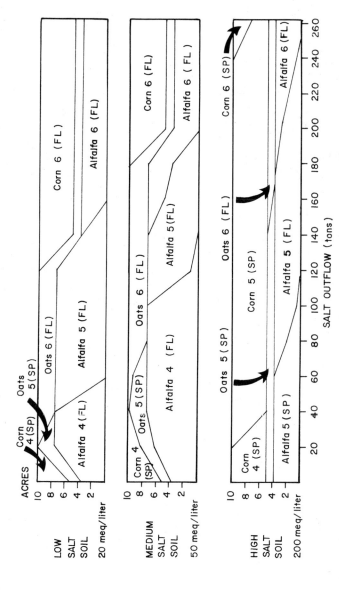

Fig. 5 Optimal cropping and irrigating pattern for high, medium, and low initial soil salt conditions where corn roots are shallow and alfalfa deep and either flooding or sprinkling is allowed as an irrigation method.

crop would accompany low salt outflows since alfalfa roots are assumed deep where more soil moisture or groundwater can be obtained, and heavy water application is not required for reasonably good yields. Low levels of irrigation are again optimal at low levels of salt outflow. Higher levels of water application are most profitable for high salt outflow. Note that compared to the previous situation in which corn was unrestricted and the corn roots were deeper than alfalfa, a higher total salt outflow is more profitable than if there are no restrictions on these factors. This higher level of salt outflow is caused by the requirement for a mix of crops and by shallow corn roots, which do not tap the underground water supply. As before, the most restrictive constraints in salt outflow are the most costly to the farm plan. Very high levels of additional salt outflow add little to the profit (Figs. 6 and 7).

Situation 3: Corn Restricted to One-Half of the Acreage, Corn Roots Shallow, Alfalfa Roots Deep, Flood Irrigation Only

Under this assumption (flooding only), a relatively small amount of corn would be produced except at high levels of water application and for high levels of salt outflow (Fig. 8). This result is due to alfalfa being able to obtain water from underground sources so that fairly good yields can be obtained without high levels of salt output resulting from the leaching due to heavy water application.

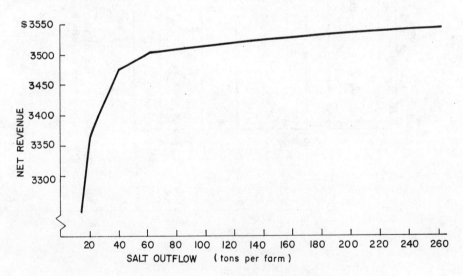

Fig. 6 Net revenue by amount of salt outflow for the 30 acres as shown in Fig. 8.

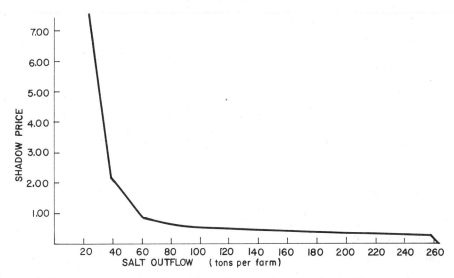

Fig. 7 Shadow price or value of an additional ton of salt outflow for the 30 acres as shown in Fig. 8.

Note that the water application levels on alfalfa are largest on the low salt soil and then lower successively to the high salt soil remain idle at low levels of permissible salt outflow. The system cannot meet the tight constraint on salt if all land is used, since flood irrigation is possible only at the three highest levels of water application.

The highest level of salt output is much higher, nearly 100 tons, than with the previous situations in which sprinkling is one of the options. The highest penalties for restricting salt output, as usual, are where the salt constraint is most restrictive as shown in Figs. 9 and 10. But once the constraint is relaxed to more than 3 tons/acre, the value is less than $1/ton.

Comparison and Evaluation of Situations

In comparing the different situations studied, it is clear that the crop that has the assumed deep roots is generally more profitable. As mentioned, this results from extraction of water from underground sources, alleviating the demand for the heavy applications of water and the salt leaching that accompanies heavy watering. This net upward flow leads to salt accumulation with time, so these one-year results do not apply for a period of years where net leaching does not occur. In other situations in which groundwater would not be available, such a result would not be expected. Without constraints on salt outflow, it appears that flood irrigation is most profitable to the farm. The advan-

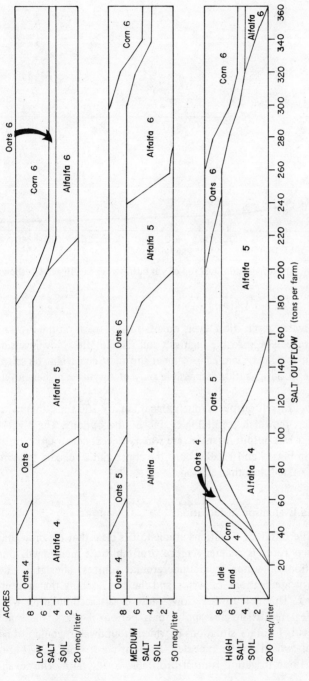

Fig. 8 Optimal cropping and irrigating pattern for high, medium, and low initial soil salt conditions where corn roots are shallow and alfalfa deep and where flood irrigation only is allowed.

Evaluation of Irrigation Return Flow and Salinity

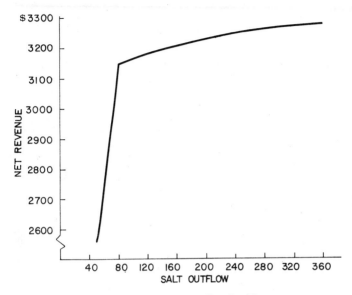

Fig. 9 Net revenue by amount of salt outflow for 30 acres.

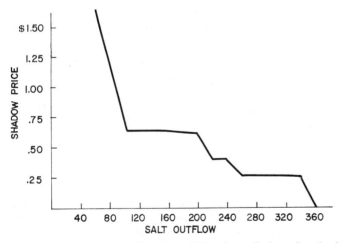

Fig. 10 Shadow price or value of an additional ton of salt outflow for 30 acres.

tages of better yields and the lower water use cost were not sufficient to make sprinkling generally profitable. It was found that net profit at the maximum was about $8/acre less ($250 for the 30 acres) if the irrigation system was constrained to sprinkling. If the farm was constrained to 1 ton salt output per acre, sprinkling would be more profitable by a few hundred dollars. At 2 tons/acre, sprinkling would be more profitable than flooding by about $300 ($10/acre). This difference depends on leaving some land idle under flooding to meet the restriction in addition to the yield advantages and lower water costs resulting from sprinkling.

In evaluating the shadow prices of salt output (value to the farm of an additional ton of salt output), it is clear that the first ton or two of salt per acre under any assumptions are most critical. It is not known just how much salt is presently coming from cultivation of lands of this type, but the amount may be somewhat higher than 1 or 2 tons. Therefore, it may well be possible under any set of management objectives to reduce salt outflow considerably with minimum cost (usually less than $1/ton). This value surely is much less than other cost estimates of salt reduction in the Colorado River. The Bureau of Reclamation currently estimates other control measure at $9 to $30/ton of salt (U.S. Bureau of Reclamation, 1974).

Policy Implications

One of the most controversial issues we face is the one of who should pay the costs for water pollution control. Is it the upstream users who have been utilizing their resources in the same way for decades? Or should it be the downstream users who might bribe those upstream to cease using those practices that harm those downstream? A third alternative is for public assumption of the burden since the incidence of costs and benefits is very difficult to identify. Such judgements are for the democratic processes of government and the courts to make.

This research indicates that whoever assumes responsibility may minimize costs using various agricultural management practices like the ones modeled in this study. It is further apparent that a zero discharge standard is very costly compared to reduction of salt output to moderately low levels. In an economic sense, the optimal policy strategy would be to provide for agricultural management practices that reduce salt outflow to a level such that the incremental cost of reduction is equal to the alternative costs of salt removal and to the damages accruing from an amount of salt received by downstream water users. This assumes that an agreement can be reached on what these costs and damages are.

In a practical sense, it may not be possible to measure or otherwise determine salt outflow from individual farms and to provide the enforcement to assure reductions in salt outflow. The most workable program might then be a policy of subsidized or enforced improvement in water management such as can be provided by sprinkling systems, ditch lining, field leveling, and other practices. Of these, sprinkling seems to be very promising. The government, quite obviously, should carefully investigate practices such as sprinkling before vast investments are made in desalination plants to achieve equivalent reductions in salt levels in the Colorado and other rivers.

References

Bresler, E. (1973). Simultaneous transport of solute and water under transient unsaturated flow conditions. Water Resources Res. 9(4): 975–986.

Childs, S. W., and Hanks, R. J. (1975). Model of soil salinity effects on crop growth. Soil Sci. Soc. Amer. Proc. 39: 617–622.

Dutt, G. R., Shaffer, M. J., Moore, W. J. (1972). Computer simulation model of dynamic bio-physiochemical processes in soils. Arizona Agr. Exp. Sta. Tech. Bull. 196.

Gupta, S. C. (1973). Salt flow in soils as influenced by water flow, root extraction and exchange. Ph.D. dissertation, Utah State University.

Hanks, R. J. (1974). Model for predicting plant growth as influenced by evapotranspiration and soil water. Agron. 65: 660–665.

Hanks, R. J., Allen, L. H., and Gardner, H. R. (1971). Advection and evapotranspiration of wide-row sorghum in the Central Great Plains. Agron. 63: 520–527.

Hanks, R. J., Andersen, J. C., King, L. G., Childs, S. W., and Cannon, J. R. (1974). An evaluation of farm irrigation practices as a means to control the water quality of return flow. Utah Agr. Exp. Sta. Res. Rept. 19.

Jensen, M. E. (ed.) (1973). *Consumptive Use of Water and Irrigation Water Requirements.* American Society of Civil Engineers, New York.

King, L. G., and Hanks, R. J. (1973). Irrigation management for control of quality of irrigation return flow. U.S. Environmental Protection Agency, EPA–R2–73–265, Washington, D. C.

Leftwich R. H. (1970). *The Price System and Resource Allocation.* Dryden Press, Hinsdale, Ill., chap. 18.

Nimah, M. N., and Hanks, R. J. (1973a). Model for estimating soil water and atmospheric interrelations: I. Description and sensitivity. Soil Sci. Soc. Amer. Proc. 37: 522-527.

Nimah, M. N., and Hanks, R. J. (1973b). Model for estimating soil water and atmospheric interrelations: II. Field test of the model. Soil Sci. Soc. Amer. Proc. 37: 528-553.

U.S. Bureau of Reclamation (1974). Colorado River Water Quality Improvement Program: Status Report. January.

9

Economic Impacts of Regional Saline Irrigation Return Flow Management Programs

Robert A. Young and Kenneth L. Leathers

Colorado State University
Fort Collins, Colorado

Introduction

This chapter examines the economic impacts of alternative on-farm water management policies for controlling salinity in irrigation return flows. The study focuses on the Grand Valley in west-central Colorado, an area thought to be representative of irrigation return flow problems in the Upper Colorado River Basin. Direct economic impacts, defined as increased costs or reduced income to farmers, are estimated with linear programming models of representative farms. A regional interindustry model is used to trace indirect economic impacts on related economic sectors of the local economy.

Background and Assumptions

Salinity (dissolved solids) in water supplies is thought to cause significant economic damages to agricultural, municipal, and industrial water users in the Lower Colorado River Basin (U. S. Environmental Protection Agency, 1971). Dissolved solids in the Colorado River arise from both natural causes (salt springs, surface runoff) and manmade causes (agriculture and industry). Total salt contributions from irrigation in the Colorado River basin have been estimated to account for 38% of the total damages that accrue to downstream water uses (U. S. Environmental Protection Agency, 1971, Table 5). The saline irrigation return flow problem in the Upper Basin is unusual in that substantial amounts of salts are "picked up" from ancient marine deposits beneath the irrigated lands. The more familiar processes of nitrate leaching and concentration of dissolved solids via evapotranspiration account for only a small part of the total return flow salinity.

The Grand Valley, the area selected for case study, is located in west-central Colorado at the confluence of the Gunnison and Colorado Rivers. The Valley lies about 4,400 ft above sea level, and the normal growing season averages about 190 days. Approximately 56,000 acres of land is presently cultivated. Major crops grown include corn, alfalfa, sugar beets, small grains, and permanent pasture. Slightly less than 15% of the irrigated acreage is planted to pome and deciduous orchards and other speciality crops. With an annual rainfall seldom exceeding 10 in, irrigation is necessary to maintain a viable commercial agriculture in the Valley.

The primary cause of return flow salinity is thought to be extremely saline aquifers (containing dissolved solids concentrations as high as 10,000 ppm) overlying a marine-deposited formation called Mancos Shale. Lenses of salts contained in the shale are dissolved by water entering and coming into chemical equilibrium with the shale formation before returning to the river channel. Water enters the aquifers by seepage from delivery canals, laterals, and drains and from deep percolation from fields as a result of excessive application of irrigation water. The present analysis is based on an estimate of 460,000 tons (per average year) of salt pickup attributable to irrigated agriculture, or over 8 tons per irrigated acre. An additional 140,000 tons is thought to be contributed by natural sources such as surface runoff from surrounding desert in the study reach (W. T. Franklin, personal communication). However, a detailed analysis of nature-source controls was beyond the scope of study.

A comment on the uncertainty surrounding estimates of the specific mechanisms and sources of salt pickup is in order at this point. The various academic and federal agency teams studying the problem have proposed conflicting estimates of the proportion of salt pickup in the study area as a result of natural versus man-made causes, and within the man-made category, the proportions due to seepage from, respectively, (1) the main canal system, (2) the on-farm distribution systems, and (3) the fields themselves. Since the appropriate control program differs among these sources, it is important that these relationships be well understood. The estimates used in our analysis represent a compromise among the competing hypotheses and might not be endorsed by any of the research groups.

The multiplicity of sources of salinity, the variety of potential means for their abatement, and the fact that the receptors of damage from salinity pollution are located in different political jurisdictions than the principal sources of pollution creates a unique and complex regionwide water quality management problem.

Concepts and Procedures

A complete economic analysis of a proposed pollution control policy should attempt to measure (1) the benefits of pollution control (in terms of willingness of receptors to pay to avoid the damages, (2) the costs of reducing or removing waste discharge, (3) the costs of monitoring and enforcement of regulations, and (4) where relevant, indirect or secondary costs and benefits. A feasible policy is one by which the incremental benefits exceed the incremental costs. Further, the implications of proposed cost sharing or pricing arrangements should be established.

Several policy instruments by which public agencies might regulate salinity discharge can be identified. These include direct efffluent standards, taxes on effluent, public subsidy of private-sector abatement activities, marketable effluent rights, direct public investment in abatement facilities, and indirect controls on use of resources complementary to the production process from which pollution emanates (i.e. water and land) (Peskin and Seskin, 1975; Kneese and Schultze, 1975).

In the Grand Valley, salinity is produced by several hundred independent commercial farms, and by many more noncommercial "farms" (small acreages with water rights whose owners largely derive income from nonfarm sources). In this case, the monitoring and enforcement costs of most of the direct approaches (standards, taxes, and effluent permits) appear likely to become exceedingly large. (As noted above, measurements of the total amount of salinity emanating from the irrigated crop sector are subject to dispute, yet this task is surely simpler than identifying amounts due to individual firms.) The costs of public investment in direct structural control methods have been reported elsewhere (Skogerboe and Walker, 1972; Skogerboe et al., 1974; Utah State University, 1975). Hence, we report here on studies which assume that the policy instruments would be indirect ones that regulate water or land use in irrigated crop production.

Control of water supply seems to be an obvious place to begin. The prevailing water pricing procedures imply a negligible marginal price to the user, and there is little concern for limiting diversions to the amount specified by the water right. Farmers now divert much in excess of even the most generous estimates of evapotranspiration and leaching requirement. These excess diversions lead, in turn, to relatively large deep percolation or drainage losses. Hence, it is hypothesized that on-farm deep percolation losses (and associated salt pickup) could be reduced by more efficient irrigation practises.

We have studied several possible practices by which farmers could influence the amount of deep percolation. First, irrigators may modify traditional irrigation practices and vary the rate of water applied per unit area in the crop season. Previous research by agricultural engineers has revealed that soil infiltration rates in the study area are relatively high in the early part of the irrigation season, but drop to low levels as the season progresses (Skogerboe et al., 1973). Hence, a large proportion of seasonal deep percolation losses are believed to occur in the first two irrigations. Practices that minimize percolation losses early in the season, at some cost to the farmer, will reduce salt pickup. Since crops typically vary as to deep percolation losses, even with similar irrigation practices, a second method of reducing deep percolation can be achieved by reducing the acreage of the crops that are high contributors in favor of those that are less of a problem. A third approach is to control seepage in the farm delivery system by, for example, substituting aluminum pipe for the present unlined ditches. Each of these alternatives would lead to increased costs or decreased revenues to affected farmers. A more drastic option considers the partial or complete retirement of selected irrigated lands by a purchase or lease program.

Linear programming models of five representative farm situations (each characterizing a different farm size category) provide the basis for deriving estimates of the economic costs of on-farm salinity controls. The model is based on the assumption that farmer resource allocation behavior can be appropriately represented in the activity analysis format. This approach hypothesizes an optimum to be selected among a set of discrete production processes (activities) (Day, 1963). Data for the models were collected by personal interviews with about 100 farmers, nearly 30% of commercial crop farmers in the study area (Leathers, 1976). The models form a Valley-wide characterization of farm sizes, resource levels, cropping patterns, and irrigation technologies. The linear programming model utilized in this study is a conventional short-run land and water allocation model, with constraints on cropland, water, and acreages of specified crops. The objective function is net return (defined as gross crop sales minus operating costs). We believe that the more efficient irrigation practices could be adopted without adversely affecting crop yield, so no yield-reduction cost is included.

Each crop production activity incorporates a coefficient representing annual deep percolation per acre. Our principal results assume that salt is picked up at the rate of 5 tons of dissolved solids per acre foot of deep percolating water. This rate represents the average for the Valley and reflects a compromise among conflicting estimates. However, since this parameter is regarded as the least reliable relationship in the analysis, we also indicate the effect of alternative values.

The model is solved to find the net income-maximizing situation for each of a number of increasingly severe constraints on water supply. Direct costs are computed in terms of net income losses due to hypothesized imposition of water supply would seek modified irrigation practices that would reduce water needs. Some of such practices would reduce deep percolation losses.

Modifications in typical current irrigation practices designed to reduce salt pickup are analyzed in the model by introducing production processes with varying water supplies, application rates, timing, and methods. These alternative processes are based on the following considerations.

The typical soils in the Grand Valley exhibit unusual infiltration characteristics, in that soil intake rates fall dramatically as the growing season progresses. Early in the year, when soils are loose and open following preplant tillage operations, the soils permit rapid infiltration of water. A relatively large amount of water is lost to deep percolation in the early season when water is run for long periods in order to achieve adequate wetting in the lower end of the fields. Later in the year, the soil tightens up and the proportion of deep percolation falls, while the proportion of tailwater increases. It is proposed that a relatively simple adjustment of irrigation practices, namely, reducing the amount of time that water flows in each furrow, could still achieve uniform wetting while effectively reducing deep percolation (Kruse, 1975). This adjustment would be required only during the first two irrigations of the season, since lower infiltration rates and higher evapotranspiration losses during the hotter summer period lead to conditions where significant deep percolation would not be experienced even with longer sets.

Irrigation is one of the most time-consuming tasks on Grand Valley crop farms. Many of the smaller farms, which account for a significant portion of total irrigated acreages, are managed by operators who maintain part-time or full-time jobs off the farm. They accomplish farm work after normal working hours and on weekends. Accordingly, irrigation settings are usually changed at 24- or 48-hr intervals. Because irrigations are worked around other farm tasks, larger farms typically follow the same practice. As noted above, low cost for incremental water deliveries (about $2.00/acre ft) and nonenforcement of water rights provide a situation in which there is little incentive for using water efficiently.

The additional cost of reducing the length of an irrigation set (reducing the amount of water applied and the time of water intake) involves no real resource costs, in that the tasks involved in changing sets must be performed in any case. However, an "inconvenience cost" is incurred, since, for row crops, the operator would have to set up and monitor his water twice each day at 12-hr intervals rather than once per day as is typical at present. There is no particular empirical basis on which to assign an accounting price to this loss of producer utility. Our crude approximation consisted of using a charge for the relevant irrigations amounting to twice the normal irrigation wage to reflect the utility loss of the "overtime" labor requirement. (This figure — twice the prevailing irrigation wage — is also close to the average off-farm wage in the area, and might be construed as an opportunity cost for those employed off the farm.) The costs of adoption more efficient irrigation practices are, to emphasize, not real resource costs, and may be an overestimate of the actual inconvenience to the farmer.

Results and Discussion

We summarize here results of our analysis of two potential regional policy alternatives: (1) constraining water deliveries (so as to encourage farm water management practices that reduce deep percolation and seepage losses), and (2) land requirements (which would completely remove saline irrigation return flows).

Constrained Water Supplies

Aggregated results of that portions of our analysis dealing with actual field percolation are shown in Table 1. Each of five farm size models were solved for a number of alternative constraints on irrigation water supply. The first row portrays our estimates of existing conditions, whereas the subsequent rows represent the predicted effects of successive 5% reductions in the available water supply. Case 1 represents our estimate of the most likely quantity of salt picked up in drainage water (5 tons/acre ft), whereas cases 2 and 3 illustrate the sensitivity of the results to alternative assumptions regarding that parameter.

Substantial reductions in this particular portion of salt discharge appear to be possible at relatively low cost under the case 1 assumptions. The first 20% reduction of water supplies is absorbed inexpensively by more efficient irrigation techniques, yet estimated salt discharge is reduced by 96,000 tons over this range. About 62% of the estimated on-farm initial salt discharge appear to be avoidable at costs to farmers of less than $2.00/ton. However, once water use is reduced to minimum evaporative and leaching requirements, savings take place only by crop substitution and actual cessation of crop production, at much greater cost. Similar conclusions are shown for cases 2 and 3.

A second category of percolation losses associated with on-farm irrigation water management practices stems from the farm water conveyance system. On-farm conveyance losses are estimated to account for about the same amount of salt as the field losses just discussed, 150,000 tons/year.

Losses from the farm conveyance system (farm laterals and head ditches) could also be influenced by constraining water deliveries (or encouraged by a subsidy program). Aluminum pipe to replace laterals and gated aluminum pipe for head ditches represent a means by which 90% of present on-farm delivery seepage can be avoided. We estimate the cost of a 90% reduction in this portion of salt pickup can be achieved at an average cost of about $7.00/ton. The procedure and results supporting this conclusion are described in Table 2. Possible additional benefits that may accrue to farmers as a result of adopting gated pipe, for example, increased yields through better control of irrigation water, and/or irrigation labor savings, have not been included here, so these costs may be somewhat overestimated.

Table 1 Estimated Annual Direct Income and Salt-Reduction Effects of Limiting Irrigation Water Delivered to Farms in the Grand Valley

Irrigation water constraint (1,000 acre ft)	Net crop income ($1,000)	Case 1[a]		Case 2[a]		Case 3[a]	
		Salt discharge in return flows (1,000 tons)	Incremental income loss per unit salt ($/ton)	Salt discharge in return flows (1,000 tons)	Incremental income loss per unit salt ($/ton)	salt discharge in return flows (1,000 tons)	Incremental income loss per unit salt ($/ton)
275[b]	7,915[c]	143[c]		207		89	
261	7,882	132	2.06	185	1.50	79	3.30
247	7,846	112	1.80	157	1.28	67	3.00
234	7,803	87	1.72	122	1.23	52	2.87
220	7,744	52	1.69	73	1.20	31	2.81
206	7,343	30	18.23	42	13.83	18	30.85

[a]Cases 1, 2, and 3 assume 5, 7, and 3 tons of salt per acre foot of drainage water, respectively.
[b]Water measured as deliveries to farm head gates for 56,000 acres of cropland.
[c]Assumed present conditions.

Table 2 Estimated Annual Costs for Reducing On-Farm Conveyance System Seepage with Aluminum Pipe in the Grand Valley

	Item	Per cropped acre	For 56,000 acres
A.	Materials requirements		
	Unlined laterals and ditches replaced[a] by aluminum pipe and structures:	51.2 ft	543 miles
	8 in pipe (laterals)	17.05 ft	180.8 miles
	6 in gated pipe (head ditches)	34.15 ft	362 miles
	Concrete turnouts (one per 12 acres)		4,667
B.	Ownership and operating costs		
	New cost:[b]		
	8 in solid pipe ($1.98/ft)	$33.76	
	6 in gated pipe ($1.62/ft)	53.32	
	Turnouts ($104.51 installed)	8.71	
	Initial investment	$95.79	$5,364,240
	Annualized costs:[c]		
	Depreciation	$ 8.62	
	Interest, insurance, and taxes	4.78	
	Repairs and maintenance	3.83	
	Total	$17.23	$ 964,880
C.	Control effectiveness		
	Salt pickup avoided with 90% efficiency	2.41 tons	135,000 tons
	Average annual cost per ton	$ 7.15	

[a]These estimates are "net" of approximately 33 ft of unlined laterals per cropped acre (350 miles). Although situated within the confines of farms, these laterals are owned and maintained by the various irrigation districts.

[b]Pipe cost estimates (in 1974 prices) were obtained from local dealers. 100% replacement of all unlined laterals, and 50% replacement of all irrigation head ditches owned by farmers is assumed (see footnote [a]). The number of concrete turnouts (from laterals to farm head ditches) reflect an average field size of 12 acres in the Valley.

[c]Calculated using conventional farm budgeting techinques assuming a useful life of 10 years and an interest rate of 6%.

Land Retirement

A more drastic approach to control of saline return flows in an irrigated area would be permanent withdrawal of water supplies. Because of the aridity of the study area, this would imply that the irrigated lands would be permanently retired from crop production. This approach warrants consideration since it is possibly the only policy instrument that would ensure compliance with the goal of zero pollutant discharge by 1985 as set forth in the 1972 U.S. water quality control legislation.

We examine two possiblities. The first option (which we label option I) represents a partial retirement scheme. Selected areas of irrigated land that are markedly less productive (i.e., saline soils and/or soils with serious natural drainage problems that hinder plant growth) are considered for retirement. Chief among the soil groups that fit this description are those silty clays and clay loams derived residually from Mancos Shale (Knobel et al., 1955). Although these soils typically exhibit poor yields as compared with the remainder of the area, conventional irrigation practices are normally followed, resulting in deep percolation and seepage losses equivalent to the area averages. Approximately 8,600 acres in the Grand Valley (as well as 10,200 acres in the neighboring Uncompaghre Valley) fall into this category. Together they represent 15% of the area's irrigated lands and 8% of the area's crop output. Since these areas of relatively unproductive soils are not contiguous in large blocks, retirement of such lands would control deep percolation from fields, farm laterals, and head ditches, but would not account for main distribution system losses.

A different strategy (option II) considers the effect of retiring an entire irrigation district. All canals and laterals controlled as an integrated unit and the acreage (both poor and productive) they service would be withdrawn from production. With this option, land retirement implies the inclusion of canal and lateral seepage losses to salt pickup, with would be excluded under the first option. The government Highline Canal, a Bureau of Reclamation project operated by the Grand Valley Water Users Association and serving approximately 20,500 acres of irrigated crops, was chosen to illustrate the impacts of this option. It is assumed that this district is representative of the Valley as a whole in terms of both productivity and salt pickup, so that results could be generalized to a full retirement program.

Accordingly, we compute the costs of retirement on the basis of two assumed rates of annual salt reduction per acre: 5.4 tons under option I, and 8.2 tons under option II. Further, we consider each land retirement option under each of two assumptions about accounting stance and resource mobility. This procedure permits the establishment of reasonable upper and lower bounds on the social costs of the program.

A national accounting stance and an assumption of instant and costless mobility of nonland resources provide a lower-bound estimate on land retirement costs. This perspective implies that all labor and mobile capital resources employed in irrigated crop production (plus those affected in forward- or backward-linked sectors) can immediately be reemployed elsewhere in the national economy without loss of productivity. In such event, the income losses from a land retirement scheme could be measured by the foregone net income from the land and immobile improvements (mainly the irrigation distribution system). The estimated residual allocated to land resources will vary with farm size and soil productivity, among other factors. Our procedure for determining the residual return to land and immobile improvements for the lower-bound estimate was based on traditional crop share rents in the study area. The landowner typically receives one-third of the gross crop sales, but is responsible for a similar proportion of fertilizer expenses as well as for all real estate taxes and water charges. Gross sales on the low-productivity lands (option I) were estimated at $177/acre (1975 price levels), whereas the estimated average of all lands was $349/acre. Computing the landowner's share (one-third) and deducting fertilizer, tax, and water charges yield annual net returns foregone of $39 for option I lands and $90 for option II. Using these rates of return, our lower-bound cost estimates are $7.22/ton of salts removed annually under option I ($39/5.4 tons), and $10.98/ton ($90/8.2 tons) under option II.

An upper-bound estimate of land retirement cost is obtained by computing the direct and indirect income effects via the use of an interindustry model of the Grand Valley Trade Area (GVTA). A regional accounting stance is assumed and capital, labor, and entrepreneurial resources in both the agricultural production sectors and the related processing sectors are assumed to have only a limited opportunity for reemployment in productive activity elsewhere in the region. This limited resource mobility is reflected in an assumption that 30% of such displaced persons and capital resources will be reemployed within a reasonably short time. [This approach follows that of Kelso et al. (1973)].

We estimated the direct and indirect impacts of the two options by detailed analysis of the community (small-region) economic structure of the Grand Valley Trade Area. A revised (scaled-down) input-output model of the upper mainstem subregion of the Colorado River Basin, reported by Udis and others (1973), was used to characterize economic interactions and levels of income and production for the three-county area of Delta, Montrose, and Mesa, which together define the west-central Colorado GVTA. A total of seven agricultural production sectors and 12 nonagricultural industries form the endogenous processing portion of this model. Local governments and households were also considered endogenous. State and federal governments, inventories, capital investment, and import-export accounts comprise the final demands or exogenous sectors.

A number of modifications in conventional interindustry analysis were necessary to satisfy the purpose of our study. First, given a regional accounting stance, only community impacts are considered, "as if" nonlocal impacts that result from interactions with the GVTA did not matter. Second, the household sector is defined as aftertax return to locally invested capital, resident wages, and profits. Third, the effects of forward linkages in estimating the direct and indirect impacts are omitted. Significant amounts of feed grains are imported by local firms, and the fed livestock and processing sectors do not rely wholly on locally grown produce. The sectoral income multipliers used are adjusted for backward linkages only, and relate to per-dollar changes in output as conceptualized by Martin and Carter (1962), rather than changes in output-per-dollar changes in final demand. The scaled-down multipliers, which ranged from 90 to 99% of their unadjusted values, measure the change in regional wages, profits, and returns to local capital-per-dollar change in crop revenue.

The forage and feed crops sector includes corn grain and silage, permanent pasture, and alfalfa; sugar beets and small grains are handled in the food and field crop sector; the orchard and vegetable enterprises are represented in the fruit and speciality crop sector of the model. Under the first option, since orchard and other high-value crops are rarely grown on poorer soils, the fruit and speciality crop sector is not considered.

The adverse effects of land retirement, measured in terms of the annual dollar value of nonrecoverable community income loss, are $110.23/acre for option I and $214.77/acre for option II. The control cost for option , found by dividing $110 by 5.4 tons of salts avoided annually per acre, is $20.37/ton, and for option II ($215/8.2), it is $26.22/ton. Even though a recoverability of 30% is assumed and forward linkages are ignored, we judge this procedure to yield a rather generous upper-bound estimate of annual retirement costs. [See Leathers (1976) for a more detailed description of the assumptions and procedures of this analysis.[1]]

Summary and Conclusion

We have analyzed the economic impacts of several alternative methods proposed for reducing saline return flows from irrigated lands in the Grand Valley of western Colorado. Our analysis suggests that adjustments in on-farm water management practices are a relatively inexpensive method of dealing with the problem. Other structural or land retirement measures appear to be considerably more expensive. Although conceptual disagreements concerning measures of downstream benefits (damages avoided) are not resolved, it is our judgement

---

[1] One additional procedure for reducing salt loading in the Grand Valley is under consideration by the U.S. Department of the Interior (1974). This plan calls for the lining of all main canals and laterals to prevent seepage. Although the report cited is known to be under revision, the revised estimates are not available to us. The earlier published estimates indicated an investment requirement of $59 million to line approximately 715 miles of canals and laterals. This investment is expected to reduce salt pickup by about 200,000 tons/year. Amortizing the investment over 30 years at 6% yields an annual cost of $4.29 million, equivalent to $21.43/ton of salt removed.

that such benefits are not large enough to warrant undertaking of any of the proposals other than adjustments in on-farm water management. We are dubious about even this alternative since our analysis fails to consider administrative and enforcement costs. Further, the burden of the program falls almost entirely on irrigation water users. We close with a few remarks on these points.

For purposes of our study, we have assumed that farmers could in some way be costlessly persuaded to adopt more efficient irrigation practices. To assure uniform adoption of such practices would no doubt require a considerable regulatory and monitoring effort, which would be both expensive and unpleasant for all concerned. Given the uncertainties about both sources of salinity and downstream damages, we hesitate to recommend such programs strongly.

We also wish to comment on the incidence (that is, which groups bear the costs and benefits) of water-use efficiency measures to reduce salinity. The type of water or land-use controls we have hypothesized have their impacts directly on the wateruser's pocketbook. This is in direct contrast to the structural measures (canal lining, desalination, etc.), which are slated to be financed from federal appropriations. Our calculations indicate that up to about one-half of the estimated Grand Valley salt discharge could be avoided by more efficient water use on the farms themselves. However, the reduced incomes necessary to accomplish this would total about $1 million annually. Although this amount would not destroy the agricultural economy of the Valley, it represents a significant part of the return to their land and managerial resources.

It seems, therefore, appropriate to consider some arrangement to share these costs. Where the market system is being bypassed to serve some presumably overriding public interest, it is desirable to include in the public programs some incentives to held assure an efficient allocation of resources. In this case, we would suggest that at least a half share of the burden of improving water quality be borne by the water-users groups in the Lower Basin. If such a plan were adopted we would be more certain of that region's real need for reduced salinity.

Acknowledgments

We are grateful to Dr. W. T. Franklin, Department of Agronomy, Colorado State University, for assistance in formulating the model of salt pickup employed in the analysis. We are also indebted to Dr. Wynn R. Walker, Department of Agricultural Engineering, and Dr. Gordon Kruse and Dr. Sterling Olsen of the Agricultural Research Service, U.S.D.A. (Fort Collins, Colorado) for providing data and comments on this research. They do not necessarily agree with the analysis nor with conclusions and recommendations.

Funding for the research summarized here was provided over a several-year period by the Colorado State University Experiment Station, the U.S. Environmental Protection Agency, the U.S. Office of Water Research and Technology, and Resources for the Future, Inc. None of these organizations is responsible for the conclusions expressed here.

References

Day, L. M. (1963). Use of representative firms in studies of interregional competition and production response, J. Farm Econ, 45(5).

Kelso, M. M., Martin, W. E., and Mack, L. E. (1973). *Water Supplies and Economic Growth in an Arid Environment: An Arizona Case Study,* University of Arizona Press, Tucson, Ariz.

Kneese, A. V., and Schultze, C. W. (1975) *Pollution, Prices, and Public Policy*, Brookings Institute, Washington, D.C.

Knobel, E. W., Dansdill, R. K., and Richardson, M. L. (1955). Soil survey of the grand Junction Area, Colorado. Series 1940, No. 19, U.S. Dep. of Agriculture, Soil Conservation Service and Colorado Agricultural Experiment Station.

Kruse, E. G. (1975). Alleviation of salt load in irrigation water return flow of the upper Colorado River Basin, FY75 Annual Progress Report of Research conducted by the ARS, USDA, and U.S. Salinity Laboratory for U.S. Bureau of Reclamation.

Leathers, K. L. (1976). The economics of controlling saline irrigation return flows in Colorado. Unpublished Ph.D. dissertation, Department of Economics, Colorado State University, Fort Collins, Colo.

Martin, W. E., and Carter, H. O. (1962). A California interindustry analysis emphasizing agriculture, Giannini Foundation Research Report No. 250, University of California, Berkeley, Calif.

Peskin, H. M., and Seskin, E. P. (eds.) (1975). *Cost Benefit Analysis and Water Pollution Policy.* The Urban Institue, Washington, D.C.

Skogerboe, G. V., Walker, W. R., Taylor, J. H., and Bennett, R. S. (1974). Evaluation of irrigation scheduling for salinity control in Grand Valley, EPA-660-2-74-052 Series, Office of Research and Development, U.S. Environmental Protection Agency, Washington, D.C.

Skogerboe, G. V., Walker, W. R., Bennett, R. S., Ayars, J. E., and Taylor, J. H. (1974). Evaluation of drainage control for salinity control in Grand Valley, EPA 660-2-74-084 Series, Office of Research and Monitoring, U.S. Environmental Protection Agency, Washington, D.C.

Skogerboe, G. V., and Walker, W. R. (1972). Evaluation of canal lining for salinity control in Grand Valley. Report 13030 DOA/1-72. Office of Research and Monitoring, U.S. Environmental Protection Agency, Washington, D.C.

Udis, B., Howe, C. W., and Kreider, J. (1973). The interrelationship of economic development and environmental quality in the Upper Colorado River Basin: An interindustry Analysis. Report to U.S. Department of Commerce, Economic Development Administration, University of Colorado, Boulder, Colo.

Utah State University, Water Research Laboratory (1975). Colorado River regional assessment study, parts 1–4, prepared for the National Commission on Water Quality, Logan, Utah, June 1975 (review draft).

U.S. Department of the Interior, Bureau of Reclamation (1974). Colorado River water quality improvement program: Status report. Denver, January.

U.S. Environmental Protection Agency, Region 8 and 9 (1971). The mineral quality problem in the Colorado River Basin, Summary report and appendices, San Francisco.

10

The Measurement of Regional Economic Effects of Changes in Irrigation Water Salinity Within a River Basin Framework: The Case of the Colorado River

Charles W. Howe

*University of Colorado,
Boulder, Colorado*

Jeffrey T. Young

*Marshall University,
Huntington, West Virginia*

Introduction and Analytical Framework

The economic development of river basins typically results in increasing upstream consumptive uses and in the degradation of water quality. Although uses benefit upstream riparians, they can result in damage to downstream riparians. This is particularly likely in semiarid regions where water scarcity is great and where river flows are nearly totally consumed.

The extent of these upstream-downstream impacts depends on the other land and climatic resources of the upstream and downstream regions and the history of development along the river. As an example, the Lower Basin of the Colorado River develped economically much before significant development was begun in the Upper Basin. Even though a treaty of long standing (1928) legally divides the waters between the Upper and Lower Basins, excess waters have been permitted historically to flow into the Lower Basin where they have been used for irrigation and municipal purposes. As development accelerates in the Upper Basin, water deliveries to the Lower Basin will be reduced, causing reductions in current beneficial uses. Quality also will fall as salinity concentrations are increased by higher salt loadings in irrigation and municipal return flows and by increased consumptive use.

This chapter presents a simple framework for consideration of the management of the waters of the Colorado River. The chapter then presents the results of utilizing existing regional economic models to evaluate the water quantity and quality trade-offs between Upper and Lower Basins. The analysis does not result in a full optimality analysis of management alternatives for the entire basin, but it does generate interesting and useful information which it is hoped will become part of continuing optimization studies of the Basin. Figure 1 shows the region being analyzed, the Colorado River Basin consisting of six subbasins: the Green, Upper Main Stem, and San Juan Basins, which constitute the so-called Upper Basin; and the Lower Main Stem, Little Colorado, and Gila, which constitute the Lower Basin. The major damages from increasing salinity occur in the Lower Main Stem Subbasin and in the Imperial and Coachella Valleys of California to the west of that Basin, whereas the opportunities for reducing salinity loading that are analyzed here lie in the Upper Main Stem Subbasin.

The conceptual framework for the present study consists of the following components:

1. Upstream agricultural, industrial, and municipal diversions that consume water and add to salt loadings.
2. A set of management actions and investments that can be taken to reduce both consumptive use of water and salinity loadings in the Upper Basin. These steps include improved on-farm water management, canal and ditch lining, changing cropping patterns, and the control of important natural point sources. The same steps may involve increased or decreased consumptive uses of water.
3. Downstream water users, agricultural, industrial, and municipal, who are damaged by salinity. Certain actions can be taken to mitigate these damages, including changed irrigation practices, better drainage, changed cropping patterns, and water treatment.

For the upstream area that contributes most of the salt, it is necessary to derive a supply or cost schedule for salt load reductions, whereas for the downstream users it is necessary to derive a schedule of marginal benefits that would accrue from reduced salinity. The study concentrates on the agricultural uses of water in both Upper and Lower Basins, since agriculture accounts for about 90% of the consumptive uses of water in the Basin and about 37% of the salinity concentration. The economically efficient set of salinity-reducing steps can then be selected by comparing the marginal cost and marginal benefit schedules.

The second section presents empirical representations of the agricultural production function in the form of yield responses to changing salinity. The data used are from the Lower Colorado River Basin. The third section presents a model of the regional income-generating mechanism so that the regionwide income impacts of reduced crop output as a result of increased salinity can be evaluated. The fourth section then evaluates the effects of agricultural acreage reduction, one major step that could be taken to reduce salinity and increase

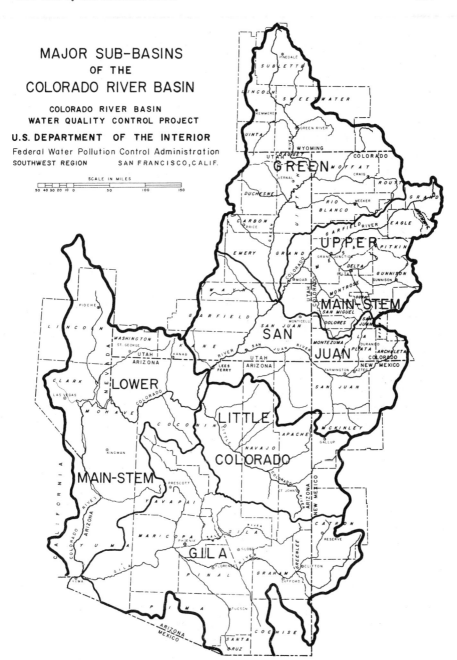

Fig. 1 Major subbasins of the Colorado River Basin.

water availability to the Lower Basin. The last section ties together downstream benefits and upstream costs so that economically efficient steps for mitigating salinity can be selected.

Crop Loss Analysis

This section determines the direct economic impact of salinity increases on agriculture using a simple profit-maximization model. The results are aggregated into sectors conformable to the regional input-output table to calculate direct and indirect economic impacts in the Lower Basin economy.

The Basic Data

The data inputs into the direct agricultural analysis consist of physical yield data that give total crop output as a function of salinity, number of annual irrigations, and soil type; crop production guidelines that give best-practice crop budgets; annual production reports that give crop prices; sprinkler cost estimates; a monthly breakdown for annual irrigation; and price indexes.

Table 1 presents the detailed analysis of Lower Colorado River water quality as it exists at present (900 ppm total dissolved solids, TDS) and as projected through use of regression functions that related each component to TDS.

The physical yield data were provided by Robinson (1974, 1975) and Jackson (1975). By finding a statistical relationship between electrical conductivity (EC) of the soil extract, irrigation frequency and method, and total dis-

Table 1 Conductivity (mmho/cm) and Ion Concentrations (meq/liter) as a function of Total Dissolved Solids (ppm), Lower Colorado River[a]

TDS[b]	EC[c]	Na	Mg	Ca	SO$_4$	Cl	HCO$_3$
900	1,339	6.64	2.94	4.86	7.42	4.04	3.04
1,000	1,484	7.49	3.27	5.19	8.05	4.73	3.11
1,100	1,629	8.35	3.60	5.53	8.68	5.46	3.18
1,200	1,773	9.23	3.93	5.86	9.29	6.22	3.25
1,300	1,917	10.12	4.26	6.17	9.89	7.01	3.31
1,400	2,061	11.02	4.60	6.48	10.48	7.83	3.37

[a]From Robinson, 1975, Table 1, p. 9.
[b]TDS = total dissolved solids.
[c]EC = electrical conductivity.

solved solids in the irrigation water, it was possible to determine a relationship between TDS in the irrigation water and relative yield for each crop, using the California Committee of Consultants' yield declination curves. The relative yield fractions were then multiplied by average pure water yields and estimated acreage to determine output on each soil type as a function of TDS in the irrigation water and the frequency of irrigation, holding the total quantity applied constant.

"Guidelines to Production Costs and Practices," compiled by the Agricultural Extension Service of the University of California (September 1973) provides the crop budgets used in the profit calculation. Similar budgets were used for all study areas in Arizona and California (Agricultural Extension Service, 1968, 1972; Hathorne, 1974). In cases where the same crop appeared in different irrigation districts, the budgets were assumed to be the same. Crop prices were taken from the annual crop reports published for each irrigation district, adjusted for changes in the agricultural price level using the U.S. Department of Agriculture's "Prices Received by Farmers" price index (U.S. Department of Agriculture, 1968-1974 inclusive). Similarly, the "Prices Paid" index was used to adjust cost figures (U.S. Department of Agriculture 1968-1974 inclusive). All price figures are expressed in 1974 dollars. The cost of using sprinkler irrigation was provided by Robinson and was estimated at $93.00/acre per year for full-season sprinkler irrigation.

Optimal Choice of Irrigation Regime

Five irrigation regimes are considered, each giving a different level of output at different levels of TDS. The irrigation regime that maximizes total profit is assumed to be the one chosen. The initial condition was assumed to be a TDS level of 900 ppm with 16 irrigations annually. Incremental profits for moving to annual frequencies of 22, 29, and 35 furrow irrigations or to 35 sprinkler applications are calculated by comparing incremental cost to incremental revenue.

As salinity increases, some crops on some types of soil will begin to show negative profits. In the short term, the farm may continue to grow the crop if it can cover all variable costs and a portion of fixed cost. In the long run, however, all costs are variable and must be covered. To reflect this long- and short-run difference, two cases were run: case I, incorporating the long-run profitability criterion with acreage being taken out of cultivation as profit becomes zero or negative; and case II, reflecting the short-run profit criterion. In this case, acreage was left under cultivation as long as *variable* costs were covered. Each case generated estimates of output for each crop on each soil type at each level of salinity. An example of the computer output for this part of the analysis is presented in Table 2.

Table 2 Sample Output of Profit-Maximizing Program, Imperial Valley, Case 1, Alfalfa (1974 dollars)

Soil type 2 Irrigation increment	900 ppm	1,000 ppm	1,100 ppm	Total dissolved solids 1,200 ppm	1,300 ppm	1,400 ppm
			Marginal Profit per Acre			
16-22 Annual irrigations	− 9.69	5.57	19.73	19.73	19.73	20.82
22-29 Annual irrigations	−11.30	−11.30	− 6.94	− 0.40	0.69	− 0.40
29-35 Annual irrigations	− 9.69	− 9.96	− 9.96	− 5.33	8.84	16.46
			Total Profit per Acre			
16 Irrigations	136.64	121.39	102.86	91.97	76.71	64.73
22 Irrigations	126.95	126.95	122.60	111.70	96.45	85.55
29 Irrigations	115.65	115.65	115.65	111.30	97.30	85.15
35 Irrigations	105.97	105.97	105.97	105.97	105.97	101.61
35 Irrigations by sprinkler	69.22	69.22	69.22	69.22	69.22	69.22
		Total Output at Maximum Profit in 10^3 Dollars				
	17,192	17,192	16,991	16,490	17,192	16,991
Irrigation frequency for maximum profit	16	22	22	22	35	35
		Marginal Output Loss in 10^3 Dollars				
	0.00	0.00	200	501	−702	200

The Case of the Colorado River

Table 3 Aggregate Output by District, Case 1 (thousands of 1974 dollars)

	900 ppm	1,000 ppm	1,100 ppm	1,200 ppm	1,300 ppm	1,400 ppm
Coachella	129,426	129,739	129,399	128,919	127,315	127,135
Gila/Yuma	87,018	89,572	86,849	86,335	82,166	81,893
Imperial Valley	243,900	254,865	233,091	230,534	228,814	204,513
Indian Reservation	49,718	50,553	50,110	49,775	47,675	46,387
Palo Verde	57,277	57,054	53,936	53,484	50,976	50,706

Table 4 Aggregate Output by District, Case 2 (thousands of 1974 dollars)

	900 ppm	1,000 ppm	1,100 ppm	1,200 ppm	1,300 ppm	1,400 ppm
Coachella	129,426	130,291	129,952	129,727	128,624	128,076
Gila/Yuma	87,018	89,572	88,834	88,254	83,984	83,630
Imperial Valley	243,900	254,865	250,759	246,566	231,521	227,920
Indian Reservation	49,718	50,701	50,252	49,913	47,675	47,592
Palo Verde	57,277	59,054	58,534	67,874	55,148	53,101

Table 5 Changes in Crop Output, California, Case 1 (thousands of 1974 dollars and percentages)

I–O Sector	Crops	900–1,000 ppm		1,000–1,100 ppm		1,100–1,200 ppm		1,200–1,300 ppm		1,300–1,400 ppm		Total change	
4	Barley, wheat sorghum, corn		426	−	445	−	326	−	427	−	392	−	1,164
5	Cotton	+1%	0	1%	−36	1%	−137	1%	−104	1%	−121	3%	−398
6	Asparagus, onions lettuce, tomatoes, watermelon, cantaloupe	<1%	−10,067	<1%	−17,263	<1%	−674	<1%	−3,241	<1%	−1,452	<1%	−12,563
7	Dates, grapes	+8%	153	13%	−77	1%	0	3%	−77	1%	77	10%	78
8	Grapefruit, lemons, limes, oranges, tangerines	<1%	42	1%	−197		−431	−1,453		0		−2,039	
		<1%		1%		<1%		1%		0	<1%		2%
9	Alfalfa	<1%	355	11%	−7,213	3%	−1,749	1%	−703		−22,360	50%	−31,658
											41		346
10	Sugar Beets		0		0		−173		173		−346	<1%	
	Total Value		11,054		−25,231		−3,490		−5,832		−24,748		−48,247

Table 6 Changes in Crop Output, California, Case 2 (thousands of 1974 dollars)

I–O	Crops	900–1,000	1,000–1,100	1,100–1,200	1,200–1,300	1,300–1,400	Total
4	Barley, wheat corn, sorghum	426	– 388	– 349	– 298	– 459	– 1,068
	Cotton	0	– 36	– 137	– 104	– 121	– 398
6	Asparagus, onions, lettuce, tomatoes, watermelon, cantaloupe	10,067	–2,752	–2,109	–16,316	–1,452	–12,562
7	Dates, grapes	153	– 77	0	– 77	– 77	– 78
8	Grapefruit, lemons, limes, oranges tangerines	595	– 197	– 171	– 992	– 363	– 1,128
9	Alfalfa	2,366	–1,514	–2,155	–1,263	–3,379	– 5,945
10	Sugar beets	0	0	– 173	173	– 346	– 346
	Total value	13,067	–4,964	–5,094	–18,877	–6,197	–21,525

Table 7 Summary of Changes in Crop Output: California and Arizona (thousands of 1974 dollars)

	900-1,000 ppm	1,000-1,100 ppm	1,100-1,200 ppm	1,200-1,300 ppm	1,300-1,400 ppm	Total change
Case 1 (long-run profitability)						
California	11,054	−25,231	−3,490	−5,832	−24,748	−48,247
Arizona	3,389	−3,166	−849	−6,380	−1,562	−8,568
Total	14,443	−28,397	−4,339	−12,212	−26,310	−56,815
Case 2 (short-run profitability)						
California	13,607	−4,946	−5,094	−18,877	−6,197	−21,525
Arizona	3,537	−1,186	−919	−6,507	−438	−5,513
Total	17,144	−6,150	−6,013	−25,384	−6,635	−27,038

It is frequently the case that output rises rather than falling when salinity increases from 900 ppm to 1,000 ppm. This is partly the result of assuming that 900 ppm and 16 annual irrigations are the initial positions. In many cases this is not the profit-maximizing position. When salinity increases to 1,000 ppm, the profit-maximization criterion comes into play and better irrigation practice is used. In many cases, this effect more than counterbalances the salinity increase, causing total output (but not profitability) to rise.

The results of the crop loss analysis are illustrated in Tables 3 through 6. These show the total estimated output in a variety of ways. Table 7 summarizes the changes in crop output.

Indirect Impact Analysis

The basic task of the indirect impact analysis was to find existing models that would permit the estimation of the regionwide effects of crop losses. Input-output (I–O) models have been the prevalent tool of regional analysis, and the existence of numerous state and multistate models made this approach attractive. On the other hand, the appropriateness of I–O analysis for the salinity problem, involving as it does changes in the input-yield relations and in farm management practices, could very well be questioned.

The technique of input-output analysis has been presented in so many excellent sources that it will not be repeated here (see Miernyk, 1965; or Baumol, 1972). Suffice it to say that the basic model is static and assumes a fixed technology. Thus investment demands, public and private, that may be generated by adaptations of the regional economy to changing salinity conditions are either ignored or taken into account as ad hoc additions to final demands. Changes in input-yield relations that could be expected to change the technical coefficients were found to be insignificant in their impacts on regional income calculations.

The two-state input-output-trade model utilized in the present study was taken from Ireri and Carter, *California-Arizona Economic Interdependence and Water Transfer Problems* (1970). The California State model had originally been constructed by Martin and Carter (1962) for 26 endogenous sectors (emphasizing agriculture) on the basis of 1954 data, but was updated to 1958 by Ireri and Carter. The Arizona model was constructed by Tijoriwala et al., from 1958 data on a sector basis comparable to the California model.

The definitions of the commodities or industries included in each endogenous sector are given as follows:

1. Meat animals and products — beef, hogs, sheep and lambs, wool and mohair

2. Poultry and eggs — chickens, eggs, broilers, turkeys and turkey eggs, other poultry and eggs, and hatcheries
3. Farm dairy products — milk, cream and dairy animals sold for meat
4. Food and feed grains — wheat, rye, rice, corn, barley, oats, sorghum, corn, and sorghum silage
5. Cotton — cotton lint and cottonseed
6. Vegetables — Irish potatoes, sweet potatoes, melons, dry beans and peas, strawberries, and all other vegetables
7. Fruit (excluding citrus) and tree nuts — apples, apricots, cherries, nectarines, peaches, pears, persimmons, plums, prunes, pomegranates, avocados, dates, figs, olives, grapes, gree nuts, and bush berries
8. Citrus fruits — oranges, tangerines, lemons, grapefruit, limes, and satsumas
9. Forage crops — hay and pasture
10. Miscellaneous agriculture — legume and grass seed, vegetable seeds, greenhouse and nursery products, on-farm forest products, sugar beets, oil crops, miscellaneous crops, horses and mules, honey and beeswax, agricultural services, and hunting and fishing

The nonagricultural sectors are (11) grain mill products, (12) meat and poultry processing, (13) dairy products, (14) canning, preserving, and freezing, (15) miscellaneous agricultural processing, (16) chemicals and fertilizers, (17) petroleum, (18) fabricated metals and machinery, (19) aircraft — aircraft and parts, (20) primary metals, (21) other manufacturing, (22) mining, (23) utilities, (24) selected services, (25) trade and transportation, (26) unallocated services.

In the endogenous section of the model, each sector had both an output row and an input column. There were five rows and ten columns in the exogenous portion of the California and Arizona models, the exogenous sectors being (27) scrap and by-products, (28) new construction, (29) maintenance construction, (30) new construction, (31) state and local government, (32) federal government, (33, 34) inventory change, (35) gross private capital formation, (36) households.

Two major changes have been made in the Ireri and Carter model for purposes of the present analysis: (1) The transactions table was updated to reflect 1974 prices; (2) the household sector was "endogenized," that is, brought into the active transactions part of the I–O matrix. The updating to 1974 prices naturally constitutes only a partial updating of the model. Endogenizing the household sector required certain approximations, but was necessary to capture the Keynesian-type expenditure multiplier impacts stemming from decreases in (farm) household incomes related to salinity.

The Handling of Forward Linkages

Significant quantities of the outputs of the agricultural sectors impacted directly by salinity go to other sectors as inputs rather than being delivered to final demand. A glance at the transactions table would show, for example, that the food and feed grains sector delivers large quantities of its output to meat animals and products, poultry and eggs, farm dairy products, and consumes substantial portions of its own output. The technical coefficients matrix indicates a fixed relationship between each of these intermediate inputs and the output of the using sector — one of the basic assumptions incorporated in I—O models. Insofar as this relationship holds, a reduction in availability of an input will cause a multiple reduction in the output of the using industry. For example, if a_{12} = 0.5 and sector 1 reduces its flow of output to sector 2 by $1, sector 2 must reduce its output by $2. The forward linkage *multiplier* would be $1/a_{12}$ = 1/0.5 = 2 for this particular industry pair.

When a supplying industry suffers an output loss, it must choose which customers are to take the cut. If the "final demand customers" — in our case exports out of the California-Arizona region — take the cut, there is no forward linkage effect. But if an endogenous sector takes the cut, there may be a forward linkage effect. This depends on the options the endogenous using sector has for finding substitutes for the diminished input supply. These options may include finding alternative sources of the identical input, substituting different materials, or increasing the efficiency of use of the input (a step usually requiring more of other inputs). If none of these steps is available, production operations will be phased out in proportion to the input reduction.

The impact of reduced input supplies depends greatly on the time frame — on the rate of change. If the cutoff occurs suddenly, there will be short-term difficulties in finding new supplies and the impact may be quite distruptive. On the other hand, if the shortage develops slowly and if it is anticipated, the customer plants have a much greater opportunity to devise solutions.

Identifying from the I—O table the forward linkages that are likely to be important is difficult in theory but simple in practice! A very small a_{ij} coefficient could be interpreted to mean that the input from sector i to sector j was quite unimportant. We know that the demands for inputs that constitute only a small part of total cost are highly inelastic. Thus the using industry will find substitutes somewhere, somehow. For this really to follow, however, it would have to be known that other sources did exist at "reasonable" prices — that the old source was not the sole source. Since these conditions are difficult to ascertain for faraway regions, the actual criterion tended to be a "relatively large direct input coefficient (from 0.02 to 0.20)," combined with an impressionistic evaluation of the nature of the product. Another practical consideration was the base-period distribution of the supplying sector's output: How much in absolute terms went to the various intermediate uses and how much to final demand? Only those sectors receiving approximately 5% or more of a supplying sector's output were considered for possible forward linkages.

The results of applying these considerations to the California-Arizona I-O table are shown in Table 8. For example, if California Sector 7 (fruit, excluding citrus, and tree nuts) suffers a reduction in output of $100, it is postulated (on the basis of base-period data) that California Sector 14 (canning, preserving, and

Table 8 One-Stage Forward Linkages: California-Arizona

Originating sector	Forward-linked sectors	Distribution of outputs %	Forward multipliers ($1/a_{ij}$)	Reduction in deliveries to final demand per $ in original sector
C4	C1	0.18	16.78	3.02
	C2	0.16	12.34	1.97
	C3	0.12	21.19	2.54
	C4	0.54	5.02	2.71
C6	C14	0.31	9.45	2.93
	FD	0.69	–	–
C7	C14	0.66	6.05	3.99
	C15	0.11	46.49	4.45
	FD	0.23	–	–
C9	C1	0.50	4.82	2.41
	C3	0.48	4.23	2.03
	FD	0.02	–	–
C10	C5	0.14	7.93	1.11
	C6	0.11	17.54	1.93
	C14	0.20	28.57	5.72
	C15	0.17	38.76	6.59
	FD	0.38	–	–
A4	A1	0.13	43.67	5.67
	A2	0.06	5.67	0.34
	A3	0.12	8.00	0.96
	A11	0.26	3.50	0.91
	C11	0.04	400.00	15.92
	FD	0.39	–	–
A9	A1	0.72	6.11	4.40
	A3	0.14	5.71	0.80
	A10	0.06	14.49	0.87
	C1	0.04	344.83	13.79
	C3	0.04	303.03	12.12

The Case of the Colorado River

freezing) will suffer a reduction in inputs of $66, California Sector 15 (Miscellaneous Agricultural Processing) a reduction of $11, and deliveries to final demand (FD) will be reduced by $23. The forward multiplier for Sector 15 looks pretty large and might well be judged unreasonably large. The final result of the $100 loss of output in Sector C7 then is postulated to be reduced deliveries to final demand of $399 by Sector C14, $445 by Sector C15, and $23 by Sector C7 itself. These figures are used with the appropriate $(I-O)^{-1}$ when the "one-stage forward linkage" cases are analyzed.

Indirect Impacts

The analysis has covered four combinations of situations: case 1, the long-run full-costs profitability criterion; case 2, the short-run profitability criterion; the now forward linkage case; and the forward linkage case. Obviously, the case 2-forward linkage case is most likely to represent the very short-run, quick phase-out results, whereas the case 1-no forward linkage combination is more likely to represent the actual impact of gradual salinity increase. Given the long life of much farm capital equipment, case 2 may actually be relevant over long periods of time.

The results of the analysis for two cases are shown in Tables 11 and 12, which show for each change of salinity the associate sectoral changes in total gross output (TGO). Sector 27 represents the "government and household" sector defined earlier, so the *TGO changes for Sector 27 represent the changes in payments to households and government – our approximation to the changes in regional income*. The following observations can be made regarding Tables 9 and 10: (1) The effects of salinity on the sectoral TGOs and on regional income (Section 27) are not proportional to the salinity increments and are quite irregular; (2) the multiplier effects are much stronger in the more developed California economy. The overall results in terms of regional income are presented in Table 11. The wide range of income losses among the four cases (531/64 = 8.3) shows how important it is to know which case fits the problem area.

Regional income multipliers can be derived from these figures by comparing the *total loss in regional income* (direct and indirect) to the *direct loss of income* in agriculture and the foward linked sectors, if applicable. The results are summarized in Table 12.

Given the very gradual nature of salinity increases, the most plausible case can be argued to be "no forward linkages: case 1" or perhaps "no forward linkages: case 2" because of the long life of capital equipment.

The nature of regional income losses versus national income (national economic efficiency) losses should be emphasized at this point. Losses incurred by a region may well be made up in other parts of the nation, especially in the case of agriculture. It is well known that, except for unusual periods such as 1973-1974, markets for agricultural commodities are limited. Federal programs sometimes limit production. The development of Western irrigated agriculture in the United States has had the effect of displacing agricultural production from other regions, especially the South and Southeast (Howe and Easter, 1971, especially Chap. 6). The effect of increasing salinity in Western irrigation water supplies may be to reverse some of this trend, to increase the extent of viable agriculture in those other areas. To the extent that this happens, the Southwestern income losses will be offset by income gains in other regions, leaving only interregional distributional effects.

Table 9 Changes in TGO, California, Case 1, No Forward Linkages (thousands of 1974 dollars)

Sector	900-1,100 ppm	1,100-1,200 ppm	1,200-1,300 ppm	1,300-1,400 ppm	Total change
1	0	−100	0	−100	−200
2	−100	−100	−200	−300	−700
3	−200	−200	−100	−600	−1,100
4	−100	−3,400	−600	−600	−4,700
5	0	−200	−100	−100	−400
6	−7,400	−800	−3,500	−2,000	−13,700
7	−100	−100	−200	−400	−800
8	400	−500	−400	−100	−600
9	−4,900	−800	−800	−21,500	−28,000
10	−1,100	−300	−100	−1,600	−3,100
11	−100	−300	−300	−700	−1,400
12	−100	−100	−100	−200	−500
13	−300	−500	−300	−1,300	−1,300
14	−300	−400	−600	−900	−2,200
15	−1,000	−1,100	−1,100	−3,200	−6,400
16	−1,100	−500	−600	−1,900	−4,100
17	−2,200	−1,400	−1,400	−5,300	−10,300
18	−1,900	−700	−1,300	−4,700	−8,600
19	−400	−400	−300	−1,100	−2,200
20	−400	−200	−200	−1,000	−1,800
21	−2,100	−1,900	−1,700	−5,700	−11,400
22	−200	−100	−100	−300	−700
23	−1,400	−1,400	−1,200	−4,200	−8,200
24	−1,200	−1,300	−1,100	−3,700	−7,300
25	−4,800	−5,000	−4,400	−15,100	−29,300
26	−4,900	−5,100	−4,300	−15,400	−29,700
27	−17,100	−21,500	−17,300	−61,200	−117,100

The Case of the Colorado River

Table 10 Changes in TGO, Arizona, Case 1, No Forward linkages (thousands of 1974 dollars)

Sector	900-1,100 ppm	1,100-1,200 ppm	1,200-1,300 ppm	1,300-1,400 ppm	Total change
1	0	0	−100	−100	−200
2	0	0	0	0	0
3	−100	0	−100	0	−200
4	−200	−100	−100	−400	−800
5	0	0	0	0	0
6	700	−800	−6,300	−300	−6,700
7	0	0	0	0	0
8	−200	0	−100	0	−300
9	−1,700	−100	0	−1,000	−2,800
10	−200	0	−100	0	−300
11	0	−100	−100	0	−200
12	−100	0	−200	−100	−400
13	−100	0	−200	−100	−400
14	0	0	−100	0	−100
15	0	−100	−200	−100	−400
16	0	−100	−100	0	−200
17	0	0	0	0	0
18	−100	0	−100	−100	−300
19	100	−100	0	0	0
20	0	0	0	−100	−100
21	−100	0	−500	−100	−700
22	0	0	0	−100	−100
23	−1,400	−100	−800	−200	−2,500
24	−100	−100	−500	−100	−800
25	−500	−400	−2,200	−600	−3,700
26	−800	−600	−2,900	−800	−5,100
27	−1,900	−2,400	−11,700	−3,100	19,100

Table 11 Projected Annual Regional Income Losses from an Increase in Colorado River Salinity from 900 to 1,400 ppm under Alternative Cases (millions of 1975 dollars)

	California	Arizona	Total
Case 1 (long-run adjustment):			
No forward linkages	117	19	136
One-stage forward linkages	492	39	531
Case 2 (short-run adjustment):			
No forward linkages	54	10	64
One-stage forward linkages	201	15	216

Table 12 Regional Income Multipliers (dollars reduction in payments to households and government per dollar direct income loss[a])

	1,000 ppm	1,100 ppm	1,200 ppm	1,300 ppm	1,400 ppm
		No Forward Linkages: Case 1			
Californis	3.55	2.59	3.34	2.96	2.47
Arizona	2.83	2.17	2.67	1.85	1.98
		No Forward Linkages: Case 2			
California	3.44	3.43	3.42	2.50	3.08
Arizona	2.72	2.58	2.66	1.87	2.70
		One-Stage Forward Linkages: Cases 1			
California	7.72	9.98	7.64	8.48	9.78
Arizona	4.90	5.74	5.64	1.81	8.53
		One-Stage Forward Linkages: Case 2			
California	8.08	9.43	10.20	7.82	10.44
Arizona	2.29	2.90	2.87	2.14	5.75

[a]Under 1974 prices, litte acreage reduction takes place. Most agricultural output losses thus approximately equal direct reductions in farm income. This is not true for forward-linked industries.

Table 13 presents the cost estimates in terms of dollars of *regional* income loss per part per million of TDS. Again, the gradual nature of anticipated salinity increases makes "case 1: no forward linkages" the most plausible case.

Direct and Indirect Economic and Hydrologic Impacts of Agricultural Acreage Reduction in the Upper Colorado River

This section calculates the regional economic and hydrologic impacts of a hypothetical program of phasing out economically marginal irrigated land in the Upper Basin thought to contribute heavily to the salt load of the River. The water quality problems of the Colorado River have been studied and modeled extensively. Among the more significant reports dealing with the origins and management of salinity are Hyatt et al. (1970), U.S. Department of the Interior (1971), U.S. Environmental Protection Agency (1971), U.S. Department of the Interior, Bureau of Reclamation (1972), Colorado River-Great Basin Consortium of Water Centers and Institutes (1973), and U.S. Department of the Interior, Bureau of Reclamation (1974). The U.S. Environmental Protection Agency (1971) has given the sources of salt concentration at Hoover Dam as shown in Table 14.

Table 13 Estimated Reductions in Regional Income (thousands of 1974 dollars and $/ppm)

	900-1,100	1,100-1,200	1,200-1,300	1,300-1,400
Case 1, NFL:[a]				
California	17,100	21,500	17,300	61,200
Arizona	1,900	2,400	11,700	3,100
Total	19,000	23,900	29,000	64,300
$/ppm[b]	80,000	239,000	290,000	643,000
Case 2, NFL:[a]				
California	−29,875	17,403	47,294	19,110
Arizona	−6,580	3,092	12,231	1,183
Total	−36,455	20,495	59,525	20,293
$/ppm[b]	−182,275	204,950	595,250	202,930
Case 1, FL:[a]				
California	151,407	49,166	49,466	242,259
Arizona	9,612	5,168	11,367	13,323
Total	161,019	54,334	60,833	255,582
$/ppm[b]	805,095	543,340	608,330	2,555,820
Case 2, FL:[a]				
California	−63,219	51,802	147,808	64,743
Arizona	−4,667	3,327	13,961	2,518
Total	−67,886	55,129	161,769	67,261
$/ppm[b]	−339,430	551,290	1,617,690	672,610

[a]NFL = no forward linkage and FL = forward linkage to processing industries.
[b]This row is in *dollars* per ppm.

Table 14 Effects of Factors on Salt Concentrations at Hoover Dam (1942-1961 records)

Factor	Contribution to concentration (mg/liter)	Percent of total concentration
Natural diffuse sources	275	39
Natural point sources	59	8
Irrigation (salt additions)	178	26
Irrigation (consumptive use)	75	11
Municipal and industrial	10	1
Exports of water	20	3
Evaporation and phreatophytes	80	12
Totals	697	100

The U.S. Bureau of Reclamation had developed a program to deal with important point sources and is in the process of optimizing a program of control covering all sources and reflecting the damages being imposed (U.S. Department of the Interior, 1972; and Maletic, 1974). It is clear from the foregoing table that Upper Basin irrigation contributes substantially to the salt concentration problem, both through its extensive consumptive use of relatively high-quality water and through the salt loadings of its return flows. The U.S. Environmental Protection Agency estimated that the salt pickup from irrigated acreage was approximately as follows: average above Hoover Dam, 2 tons/acre/year; Grand Valley, Colorado, 4 to 8 tons/acre/year; below Hoover Dam, 0.5 to 1.0 tons/acre/year. More recent Bureau of Reclamation data indicate more than 10 tons/acre/year in the Grand Valley (Leathers and Young, 1975). When the large externalities are observed and when the marginal economic condition of a significant number of acres of irrigated land in the Upper Basin is noted, a selective reduction in irrigated acreage naturally suggests itself as a potentially efficient way of reducing both salinity loadings and concentration.

On-farm management practices can also be expected to form part of an optimum program of salinity control. More careful application of water and the lining of ditches, especially in the areas where return flows pick up such great quantities of salt, are likely to be helpful, as are changing furrow length, recycling drain water from tailwater pits, and more extensive use of sprinklers. The results of studies by Leathers and Young at Colorado State University will be quoted in the next section.

The following acreages constitute the most likely candidates for the phase-out[1]:

Crop	Grand Valley acres	Uncompaghre Basin acres
Corn	1,200	1,500
Other grains	2,000	2,300
Alfalfa	2,300	2,700
Pasture	3,300	3,700
Total	8,800	10,200

These acres average (in 1974 dollars) a total output value of about $150/acre/year and yield net incomes to the farmer somewhere in the $30 to $50/acre/year range.

---

[1] Minor amounts of dryland farming might be able to continue.

The Case of the Colorado River

The main tools of analysis used to estimate the impacts of this acreage phase-out have been the integrated input-output and hydrologic models developed by Udis et al. (1973) for the Green, Upper Main Stem, and San Juan River Basins. An earlier application of these models was reported by Howe and Orr (1974a, 1974b). The input-output model of the Upper Main Stem Basin uses 1970 prices, but ad hoc adjustments have been used to raise results approximately to 1974 prices. Whereas the hydrologic models were calibrated using hydrologic records of as great a length as possible, the historical record used in driving the models was 1962-1969, inclusive.

A baseline projection to 1980 of the area is shown in Table 15. Also shown are the payments to households (wages, salaries, rents, dividends, and interest) as percentages of the gross value of output. Payments to households were used as an approximation to regional income. Corresponding to the 1980 economic projections, Table 16 exhibits the projected subbasin surface outflows in acrefeet, based on the 1962-1969 hydrologic record. Three cases have been treated in the following analyses:

Case I: a direct reduction of 8,800 acres of corn, other grains, alfalfa, and pasture in the Grand Valley

Case II: a direct reduction of 10,200 acres of these crops in the Uncompaghre Basin

Case III the combined reductions of cases I and II

These reductions have been valued at $100 (in 1970 dollars) gross output per acre and have been treated as reductions in the "range livestock" sector of the input-output model since that sector is defined to contain pasture, alfalfa, and those grains grown on ranches for winter feed. It has been assumed that the resultant direct reductions in the value of output would take the form of reduced exports from the seven county region. Results of the analysis are given in Table 17.

Summary of Costs of and Benefits from Salinity Reduction in the Colorado River Basin

The preceding sections have estimated regional income losses for the Lower Colorado River Basin (Arizona-California) stemming from agricultural damages imposed by salinity. Regional income losses to the Upper Colorado River Basin that might follow the phasing out of certain economically marginal acreages were also estimated. This section combines those data with data from other sources to construct marginal cost and benefit schedules for varying quantities of salt load reduction. These schedules permit the identification of currently justifiable projects of programs for salinity control.

Table 15 Projected 1980 Output Levels for the Affected Basins (thousands of 1970 dollars)

Sector	Uncompaghre Basin	North Fork of Gunnison	Gunnison[a] Main Stem	Colorado[b] Main Stem	Total 7[c] county area	Percent HP in TGO[d]
1. Range livestock	6,854	1,926	5,794	26,938	41,512	48
2. Feeder livestock	1,545	260	955	1,705	4,465	2
3. Dairy	724	237	639	1,902	3,502	31
4. Food/field	2,734	762	898	3,274	7,668	52
5. Truck crops	417	159	178	429	1,183	33
6. Fruit	357	560	735	2,458	4,110	32
7. Forestry	193	205	213	1,539	2,150	53
8. Other agriculture	411	089	213	898	1,611	28
9. Coal	0	11,218	0	664	11,887	53
10. Oil and gas	0	0	0	8,174	8,174	(NA)
11. Uranium	0	0	0	160,042	160,042	(NA)
12. Zinc and lead	1,108	0	0	3,343	4,451	(NA)
13. Other mining	1,713	125	248	5,646	7,732	60
14. Food/kindred	2,767	529	8,303	17,794	29,393	19
15. Lumber/wood	1,848	396	1,531	1,624	5,399	26
16. Printing/publishing	345	345	173	3,392	4,255	40
17. Fabricated metals	100	0	800	2,900	3,800	30
18. Stone, clay, grass	360	0	480	3,480	4,320	24

19. Other manufacturing	1,540	880	2,860	32,254	34,054	19
20. Wholesale trade	2,075	907	2,081	27,959	33,022	23
21. Service stations	923	199	569	3,146	4,837	68
22. Other retail	8,286	1,810	5,172	37,529	52,797	58
23. Eating/drinking	3,411	460	1,290	11,472	16,633	31
24. Agricultural services	814	455	700	2,427	4,396	40
25. Lodging	761	109	257	9,280	10,407	40
26. Other services	3,489	416	1,551	20,176	25,632	47
27. Transportation	4,747	955	5,172	22,392	33,266	38
28. Electrical energy	2,168	161	375	8,082	10,786	26
29. Other utilities	3,079	862	1,712	15,766	21,419	42
30. Contr. construction	6,177	1,705	3,337	59,801	71,020	27
31. Rental/finance	10,213	1,994	5,680	42,542	60,429	73
Total					684,352	

aGunnison River from North Fork confluence to but not including Grand Junction.
bColorado River from Glenwood Springs, including basins of the Dolores and San Miguel Rivers.
cMesa, Delta, Gunnison, Montrose, Ouray, Hinsdale, and San Miguel Counties.
dPercentage of payments to households in total value of each sector's output.

Table 16 Projected 1980 Surface Outflows Based on Projected 1980 Economic Conditions and 1962-1969 Hydrology (thousands of acre feet)

	Uncompaghre Basin	North Fork of Gunnison	Gunnison[a] Main Stem	Colorado[b] Main Stem
January	4.5	2.6	54.9	167.9
February	5.8	4.5	49.6	164.2
March	7.4	8.6	66.1	203.6
April	20.6	22.4	128.4	347.2
May	38.5	84.0	354.2	881.9
June	27.9	72.4	299.5	881.8
July	18.4	24.2	131.7	385.7
August	11.0	6.7	75.1	207.4
September	20.7	7.4	80.7	175.5
October	27.5	9.4	88.0	254.7
November	15.2	8.0	84.5	238.6
December	8.8	3.8	75.9	206.1
Ave. mon.	17.2	21.2	124.1	342.9
Ann. total	206.3	254.3	1,388.7	4,114.6

[a]Gunnison River from North Fork confluence to but not including Grand Junction.
[b]Colorado River from Glenwood Springs, including basins of the Dolores and San Miguel Rivers.

Table 17 Summary of Upper Basin Impacts of Agricultural Acreage Phase-Out

	8800 acres in Grand Valley	10,200 acres in Uncompaghre	Total
Direct loss in value of output (1970 dollars)	$ 880,000	$1,020,000	$1,900,000
Direct loss (1974 dollars)	1,320 000	1,530,000	2,850,000
Direct + indirect loss of regional income (1970 dollars)	954,000	1,104,000	2,058,000
Direct + indirect loss (1974 dollars)	1,431,000	1,656,000	3,087,000
Regional income loss per acre (1974 dollars)	163	163	163
Total annual reduction in consumptive uses of water (acre feet)	41,800	16,000	30,800
Regional income loss per acre foot of water saved	64	70	67
Reduced salt loads[a]	88,000	102,000	190,000

[a]Using 10 tons/acre/year

The Case of the Colorado River

Data are available on the following projects or programs for salinity control:

1. Paradox Valley of the Dolores River (Bureau of Reclamation project). It is estimated that 180,000 tons/year can be eliminated by undertaking a groundwater pumping and evaporation scheme that would reduce water contact with huge salt domes. The project will cost approximately $16 million. If we assumed indefinite life for the project facilities and a 10% interest rate, the annual cost of $1.6 million implies a cost per ton removed of $8.90. An increase in consumptive water use will also occur. We have assumed this loss to be 10,000 acre/year.
2. Grand Valley (part of Upper Main Stem Subbasin) canal lining scheme (Bureau of Reclamation). It is estimated that canal lining and better irrigation scheduling can reduce the salt load by 200,000 tons/year. The cost is estimated to be $59 million, implying a cost of $30/ton removed. Canal losses of about 40,000 acre ft/year will be avoided and we assume that 20,000 acre ft of this represents actual saving of water that would not have returned to streams for further use.
3. Improved on-farm irrigation practices. Leathers and Young (1975) have estimated that about 110,000 tons/year could be avoided at costs averaging $2/ton. These practices would also save 45,000 acre ft of consumptive use.
4. Modified cropping patterns I. Leathers and Young have estimated that salt loadings of 21,500 tons/year could be avoided in the Grand Valley at costs of about $8/ton plus 4,300 acre ft of added consumptive use.
5. Modified cropping patterns II. Leathers and Young estimate that further modifications could avoid an added 12,500 tons/year at costs of about $19/ton and 2,200 acre ft of added consumptive water use.
6. Howe and Young (see Section 4) estimate that phasing out 8,800 acres of cropland in the Grand Valley would involve a regional income loss of $163/acre/year, but saving 10 tons of salt/acre and reducing consumptive use by 14,800 acre ft/year.
7. Howe and Young (see Section 4) also estimate that phasing out 10,200 acres in the Uncompaghre Valley (part of the Upper Main Stem Subbasin) would also involve a regional income loss of $163/acre, saving an assumed 10 tons of salt/acre/year and reducing consumptive use by 16,000 acre ft/year.

In these data, we have the essential ingredients of a supply schedule for salt removals, that is, a marginal cost schedule for different quantities removed. There are, however, two difficulties in proceeding to construct that schedule.

The first is that a joint product is being produced by most of these activities: salt reduction and a reduction in consumptive use of water (activities 1, 4, and 5 actually increase consumptive use of water). The waters of the Colorado River are fully utilized at the present time, so that water has a positive scarcity value. An acre foot freed from one use will be used beneficially further downstream. An added acre foot of water consumed deprives downstream parties of its use. Howe and Easter (1971, Table 30, p. 93) estimated this opportunity cost to be about $10/acre ft under reasonable assumptions. That figure currently needs to be increased to $15 to allow for inflation.

Thus, in costing out salt reductions, it is possible to subtract from (add to) the cost per ton removed the value of water simultaneously released from (added to) consumptive use and valued at $15/acre ft.

The second problem is that the quantities of salt associated with the various projects or programs listed above are not strictly additive. For example, if canal lining is undertaken, the salt reductions available through improved on-farm irrigation practice may be reduced. It is probably the case that the full benefits from such irrigation improvements could not be realized after modification of cropping patterns. Finally, it is clear that acreage reductions in the Grand Valley will reduce the areas to which improved irrigation practice and cropping patterns can be applied.

To deal with this problem, we make the following assumption: The potential salt savings from activities 2, 3, 4, and 5 (as listed earlier and all being in the Grand Valley) will be reduced in proportion to any acreage phased out in the Grand Valley. Since there are 8,800 candidate acres in the Grand Valley out of a total (non orchard) irrigated acreage of 57,000 acres, this would represent a 15% reduction, or a reduction to 85% of the levels achievable without acreage phase out. Thus if we construct a cost schedule that contains Grand Valley acreage phase out, we must reduce the potential salt savings from activites 2, 3, 4, and 5 by 15%.

The Paradox Valley point-source project and the phase-out of acreage in the Uncompaghre Valley are independent of the other activities.

Table 18 gives the results of these calculations, with the various activities listed in ascending order of cost per ton of salt removed.

Benefits from salinity reduction take the form of damages avoided. The basic data on losses related to agriculture were presented in Table 13. The small portion of M&I use (less than 5% of total withdrawals) undoubtedly involves some damage to residential, public, commercial, and industrial equipment. These damages may not be negligible (see Wesner, 1974; or Tihansky, 1974). Such damages would have to be added to the benefit schedule, but the small amounts withdrawn for M&I uses would make M&I benefits (damages saved) per ton of salt removed quite small. For present analyses, it seems reasonable, therefore, to confine our attention to the salinity intervals 900-1,100 ppm and 1,100-1,200 ppm.

The relationship between a change of 1 ton (2,000 lb) in salt loading in the

The Case of the Colorado River

Table 18 Supply Schedule for Salt Reductions

Activity	Independent salt-saving potential (tons/year)	Salt-saving potential c/ G.V. acre ret. (tons/year)	Project cost per ton of salt removed	Water saved (a.f.)	Value of water saved per ton of salt saved	Net cost per ton of salt saved
On-farm practices	110,000	93,500	$2.00	44,700	$6.10	$-4.10[a]
Paradox Valley	180,000	153,000	8.90	-10,000	-0.80	9.70
Modified Crops I	21,500	18,300	8.00	-4,300	-3.00	11.00
Grand Valley acreage ret.	–	88,000	16.30	14,800	2.50	13.80
Uncompaghre Valley acreage ret.	102,000	102,000	16.30	16,000	2.30	14.00
Modified Crops II	12,500	10,600	19.00	-2,200	-2.60	21.60
Grand Valley Canals	200,000	170,000	30.00	20,000	1.50	28.50
Total	626,000	635,400	–	–	–	–

[a]The negative sign indicates the high desirability of undertaking these activities. Project costs may be somewhat understated. See Leathers and Young (1975).

Table 19 Annual Benefits[a] per Ton of Salt Load Reduction: Alternative Cases and Salinity Levels (1974 dollars/ton)

| Salinity levels | No forward linkage | | Forward linkage | |
	Case 1	Case 2	Case 1	Case 2
900-1,100 ppm	8	0[b]	81	0[b]
1,100-1,200 ppm	24	20	54	55

[a]Damages avoided.
[b]Whereas Table 15 shows negative damages for these cases, we interpret this as no significant damage.

Upper Basin and the TDS concentration at Parker and Imperial Dams is approximately that 10,000 tons equals 1 ppm.[2] This permits the conversion of the loss data of Table 15 into (1974) dollars per ton of TDS. These data are presented in Table 19.

---

[2]This is an average value taken from Bureau of Reclamation studies and given to the present author by John T. Maletic.

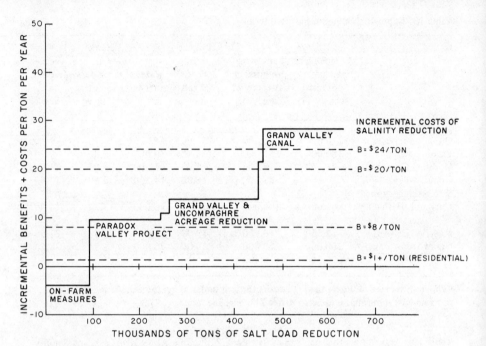

Fig. 2 Incremental costs and benefits from salinity reduction.

Figure 2 provides a visual summary of the data on supply of and benefits from salinity reduction taken from Tables 18 and 19.

Acknowledgements

This work was funded as a matching grant under Public Law 88–379 (as amended) as part of Office of Water Research and Technology Project No. B-107-Utah.

References

Agricultural Extension Service (1968). Coachella Valley, sample costs of production; lemons, orange, Riverside, Calif.

Agricultural Extention Service (1972). Coachella Valley; sample cost of production; grapefruit, dates, and grapes, Calif.

Agricultural Extension Service, University of California (1973). Guidelines to production costs and practices, Imperial County crops, Calif.

Baumol, W. J. (1972). *Economic Theory and Operations Analysis,* Prentice-Hall Englewood Cliffs, N.J.

Colorado River-Great Basin Consortium of Water Centers and Institutes (1973). *Salinity Management Options for the Colorado River*, Center for Water Resources Research, Utah State University, Logan, Utah.

Hathorn, S. Jr., et al. (1974). Yuma County 1974, field crop budgets (preliminary), Cooperative Extension Service, University of Arizona, Tucson, Ariz.

Howe, C. W., and Easter, K. W. (1971). *Interbasin Transfers of Water: Economic Issues and Impacts*, Johns Hopkins Press, Baltimore,

Howe, C. W., and Orr, D. V. (1974a). Effects of agricultural acreage reduction on water availability and salinity in the Upper Colorado River Basin, Water Resources Res. *10*(5).

Howe, C. W., and Orr, D. V. (1974b). Economic incentives for salinity reduction and water conservation in the Colorado River Basin, in J. E. Flack and C. W. Howe, *Salinity in Water Resources*, Merriman Publishing Company, Boulder, Colo.

Hyatt, M.L., Riley, J.P., McKee, M.L., and Israelsen, E.K. (1970). *Computer Simulation of the Hydrologic-Salinity Flow System Within the Upper Colorado River Basin,* Utah Water Research Laboratory, Utah State University, Logah Utah (PRWG54−1).

Ireri, D., and Carter, H. O. (1970). *California-Arizona Economic Interdependence and Water Transfer Models*, Giannini Foundation Research Report No. 313, California Agricultural Experiment Station.

Jackson, E. (1975a). Salinity management options for the Colorado River: Agricultural consequences, Arizona Agricultural Experiment Station, Yuma Branch Station, University of Arizona.

Jackson, E. (1975b). Salinity management options for the Colorado River: Phase 1, Damage estimates and control program impacts − Central Arizona Project Area (mimeo). Agricultural Experiment Station, Yuma Branch Station, Yuma, Arizona, April 24, 1975.

Leathers, K. L., and Young, R. A. (1975). Economic evaluation of nonstructural measures to control saline irrigation return flows, paper given at the Western Agricultural Economics Association meetings, Reno, Nevada, July.

Maletic, J. T. (1974). Current approaches and alternatives to salinity management in the Colorado River Basin, in J. E. Flack and C. W. Howe (eds.), *Salinity in Water Resources*, Merriman Publishing Company, Boulder, Colo.

Martin, W. E., and Carter, H. O. (1962). *A California Interindustry Analysis Emphasizing Agriculture, Part I: The Input-Output Models and Results,* Giannini Foundation Research Report No. 250, California Agricultural Experiment Station.

Miernyk, W. H. (1965). *The Elements of Input-Output Analysis*, Random House, New York,

Robinson, F. E. (1975). Salinity management options for the Colorado River: Agricultural consequences, California Agricultural Experiment Station, University of California—Davis and El Centro, Calif.

Robinson, F. E. (1974). Salinity management options for the Colorado River, Phase 1: Damage estimates and control program impacts, El Centro, California Agricultural Experiment Station,

Tihansky, D. P. (1974). Economic damages from residential use of mineralized water supply, Water Resources Res. *10*(2),

Tijoriwala, A. G., Martin, W. E., and Bower, L. G. (1968). *Structure of the Arizons Economy: Output Interrelationships and their Effects on Water and Labor Requirements. Part I: The Input-Output Model and Its Interpretation; Part II: Statistical Supplement.* Technical Bulletin 180, Agricultural Experiment Station, University of Arizona, Tucson,

Udis, B. Howe, C. W., and Kreider, J. F. (1973). *The Interrelationship of Economic Development and Environmental Quality in the Upper Colorado River Basin: An Interindustry Analysis*, National Technical Information Service, Accesion Number COM—73—11970.

U.S. Department of the Interion (1971). *Quality of Water: Colorado River Basin,* Progress Report No. 5, January.

U.S. Department of Agriculture, Agricultural prices, January 1968-December 1974 inclusive.

U.S. Department of the Interion, Bureau of Reclamation (1972). *Colorado River Quality Improvement Program.*

U.S. Department of the Interior, Bureau of Reclamation. Crop production report, Coachella Division, 1968-1972 inclusive.

U.S. Department of the Interior, Bureau of Reclamation (1974). Economic impacts of changes in salinity levels of the Colorado River, Denver, Colorado.

U.S. Environmental Protection Agency (Regions VIII and IX) (1971). *The Mineral Quality Problem in the Colorado River Basin:* Summary Report.

Wesner, G. M. (1974) The importance of salinity in urban water management, in J. E. Flack and C. W. Howe (eds.) *Salinity in Water Resources*, Merriman Publishing Company, Boulder, Colo.

11

Economic Cost and Trade-Offs in Improving Water Quality and Nonpoint Pollution Through Agriculture: An Interregional Approach

Earl O. Heady, Kenneth J. Nicol, and James C. Wade

Iowa State University,
Ames, Iowa

Introduction

From the standpoint of physical contribution, agriculture is the major source of nonpoint pollution in the United States. It has grown increasingly so with the agricultural sector's technical and economic development. With development, the real cost of capital has declined relative to labor and land. Consequently, farms have mechanized highly, which in turn has caused high fixed costs. High fixed costs encourage large units, to spread fixed costs and allow lower per-unit costs. This tendency, however, also encourages specialization in one or a very few products, since machinery and equipment now are so highly oriented to a particular product. The specialization has nearly encouraged monoculture in the corn belt of the United States, or at least a larger number of farms that specialize only in crops such as corn and soybeans. This tendency also prevails elsewhere in crops, just as it has with large-scale livestock breeding and feeding units. Thus, over a vast space, water runoff from agriculture has been intensified with a greater transport of silt, animal wastes, and residual chemicals and pesticides as a consequence of the economic development of agriculture. The low real cost of capital items such as chemicals and pesticides also has caused them to be substituted for legumes and forages for maintaining soil productivity and livestock feed sources. Their greater use has augmented increased runoff and sedimentation in extended nationwide nonpoint pollution of rivers and water bodies from agriculture.

CARD Studies

The Center for Agricultural and Rural Development (CARD) of Iowa State University has developed an extensive set of models dealing with land and water use and allocation in relation to nonpoint pollution control and water or environmental quality improvement. These models, covering the entire nation but also relating to and producing detailed results for individual producing areas and land resource groups, include all resources and commodities of American agriculture. One of their major orientations is in evaluating different policies and alternative futures in control of nonpoint pollution and stream quality improvement through agriculture relative to trade-offs in farm commodity prices and income, consumer food costs, the nation's food supply and export capacity, and related variables that impact with the manner land and water are managed, water runoff or sedimentation is controlled, and nonpoint pollution generally is restrained to different levels. Our models emphasize, on the environmental side, control of sediment delivery or soil loss, reduced inputs and transport of chemical fertilizers and pesticides, and reduced animal waste transport into streams. But they measure, in the process of solution, the effect of these controls on national and regional resource and production patterns, resource values, farm commodity prices, farm income, consumer food costs, the nation's export capacity, and the distribution of the costs and benefits of pollution control. They are direct models for evaluating the optimal allocation of water and land resources in the United States. We are in the early stage of these models, financed by RANN of NSF, but already have some important solutions and application and will have many more in the near future.[1] A variety of models is being developed. These include a family of large-scale linear programming models, which sometimes have dimensions of around 10,000 equations and 150,000 variables. These programming models include as many as 223 producing areas, 9 land groups per area (thus actually a total of 2,007 regions), 51 water supply regions, and 35 major commodity market regions. We also have models of somewhat smaller dimensions and with lower solution costs. Supplementing the linear programming models emphasized, we have upcoming econometric recursive simulation and quadratic programming models with parallel characteristics. Currently, however, our emphasis is on medium-sized but rather easily managed linear programming models, which contain submodels or sectors for water, land, crops, livestock (every commodity of U.S. agriculture), several environmental dimensions, a transportation network, a commodity demand sector, and an endogenously generated resource pricing sector. Because of space limitations, we summarize only one of these models in this chapter. Many other model solutions and emphases will be available in the future. We first explain a soil loss control

---

[1] The chapter has been written in the fall of 1975.

model (but it also is one of water and land allocation) that includes 223 producing areas, 9 land classes per area (and thus 2,007 actual local regions), 51 water supply regions, and 25 commodity market regions. (We later summarize a sediment delivery model concerned with net discharge into rivers and a parallel chemical delivery model.) The first two of these regional delineations are summarized in Figs. 1 and 2 respectively.

Model Capabilities

The models have the capacity to evaluate simultaneously variables and outcomes in terms of (1) national markets, prices, incomes, and employment; and (b) production patterns, resource use, and economic structure of rather small resource regions, under the posed imposition of alternative policies or futures. We believe that models with these characterisitcs and capabilities are extremely important for the future as national, state, and local entities evaluate and consider implementation of environmental, land use, water, and other resource and technological restraints related to problems emerging under the nation's advanced state of economic development. Otherwise, the programs and policies imposed by states, municipalities, and regional planning bodies on water and land use will encounter unexpected economic effects, causing them to be nullified because they give inequitable distributions of the costs and benefits of the goals attained. For example, initial solutions of our models suggest that individual states that impose restrained patterns of land use, water runoff, sediment delivery, and technologies will find that, through market impacts, producers and resource owners of other states and locations will realize economic gains while those of the imposing state will bear the costs in lower incomes and reduced resource prices. Even for certain quality controls imposed at the national level, relative returns can be positive in some regions and negative in other regions.

Although our current models emphasize land use, water technology, and nonpoint pollution or environmental quality as they relate to agriculture, with modifications to accommodate sedimentation by highway and urban construction they eventually can be adapted to incorporate a major share of society's aggregate land use planning problems at the national level. They can, given various national objective functions and with planning restrained to conform with local goals in land use and technology, trace the impacts back to individual land and water resource regions. Or, conversely, the models also can trace the impacts of plans imposed independently at region or area levels to other regions and to the aggregate national level. Optimal patterns of land, water, and technology (fertilizer, insecticides, livestock scale and dispersion, etc.) use can be evaluated whether they relate to discrete regulations or pricing mechanisms imposed by the public. The models under construction also allow examination of the

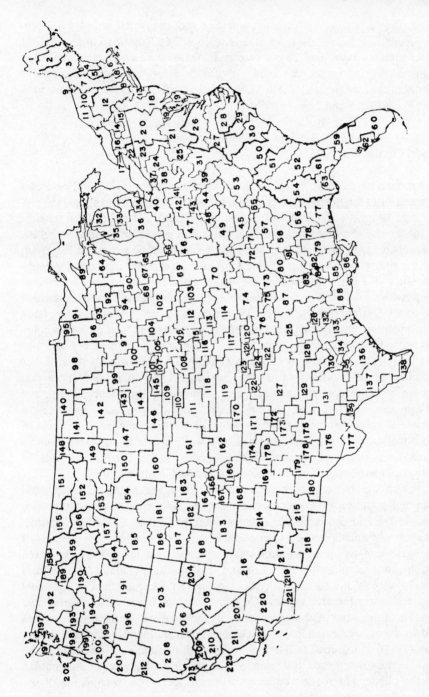

Fig. 1 The 223 producing areas.

Improving Water Quality and Nonpoint Pollution

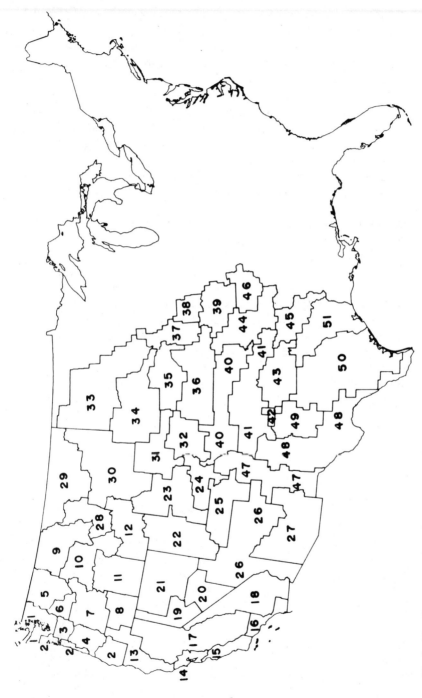

Fig. 2 The 51 water supply regions.

effect of different environmental and land use programs on employment and income generation in rural areas and the potential reflection of these impacts through various agricultural structures as represented by different farm size and employment patterns. We believe that models which specify these possibilities and outcomes at both the national and local levels will prove extremely important in future years and decades as the nation and states come to explore and require particular environmental quality and water and land use controls. The models, which incorporate a transporation submodel for commodities and water and product transfer activities, thus cause and allow interdependency among every commodity, in every producing area, and for every water supply in the nation. They allow selection of optimal resource use patterns and environmental quality impacts for the nation in future time periods. They also reflect comparative advantage in the allocation of land and water to competing alternatives as represented in relative yields, general technologies, environmentallly restrained technologies, production costs, transport costs, and imposed environmental restraints. They allow substitution of land at one location for water or land at another location a thousand miles away (or visa versa). Similarly, they allow and analyze these substitutions when environmental restraints are applied to restrain the technologies used in any one resource region. Finally, they allow evaluation of various policy alternatives in use of land and water resources, and environmental quality controls in interaction with commercial agricultural policies, export goals, and domestic demands in both regional and national markets. These capabilities of environmental and resource models are necessary if evaluations of national possiblities are to be complete in relation to their impacts and equity in the many rural communities that are unique resource areas of the nation.

Nature of Models

The model summarized emphasizes optimal water and land use patterns, agricultural water allocation, agricultural technology, and soil and water conservation methods under environmental restrained soil loss. The objective function is Eq. (1), which minimizes the cost of producing and transporting the various crop and livestock commodities among producing and land resource regions of origin, regions of processing, and regions of consumption to meet domestic and export demands. The costs of water consumption and transfer also are included in Eq. (1). The programming prices and costs cover all factor costs (except land rents, which are reflected in shadow prices) and thus allow simulation of a long-run market equilibrium for each commodity with a national allocation reflecting the comparative advantage of each of the 223 producing regions, 1,891 land regions, and 51 water supply regions—subject ot environmental restraints and the level and the level and location of consumer demands. The objective function (OF) is minimize, where

$$OF = \sum_i \left[\sum_k \left(\sum_m X_{ikm} UC_{ikm} + \sum_n Y_{ikn} UC_{ikn} + \sum_m Z_{ikm} \right) + \sum_p L_{ip} UC_{ip} + DPP_i UC_i + IPP_i UC_i + DWH_i UC_i + IWH_i UC_i + FLG_i UC_i + FP_i UC_i \right]$$
$$+ \sum_w \left(WB_w UC_w + WD_w UC_w + WT_w UC_w \right) + \sum_t \sum_c T_{tc} UC_{tc} \quad (1)$$

The variables, parameters, and other terms are defined in a following section.

Variables and Equations of the Model

Each of the 2,007 land resource groups has alternative crop management systems producing commodities with associated yields and soil loss subject to soil types, average weather prevailing, and conservation tillage practices employed. Data were developed in conjunction with the Soil Conservation Service of the U.S. Department of Agriculture to represent soil loss per acre under various mechanical practices and rotations or land use systems for each land resource group. Soil loss control alternatives were evaluated through a universal soil loss equation (2). The equation used for each crop, each crop management, and for each land resource group is the form

$$SL = K \cdot L \cdot S \cdot R \cdot C \cdot P \quad (2)$$

where
SL = the per acre gross soil loss
K = the erodibility factor associated with the soil type
L = the computed value of relating slope length to soil loss control
S = derived from a non linear function relating slope gradient to level of soil loss
R = an index of erodibility for the rainfall of the area accounting for varying levels of intensity, duration, and measured rainfall
C = an adjustment factor giving an index of the relative ability of alternative cropping patterns to reduce soil loss
P = an adjustment factor to account for the potential soil loss reduction from adopting conservation practices

Each variable in the model represents an alternative crop management system that incorporates a given rotation, crop tillage method, and conservation practice for an individual crop and land resource group. The rotation and tillage methods combine to give the unique C value and the conservation practice determines the P factor. The K, L, and S factors are dependent on the soil characteristics and the regional rainfall patterns determine the R factor.

Per-acre crop costs and crop yields had to be estimated for the alternative crop management systems on each land group. The cost data include depreciation and expenditures on machinery, labor, pesticides, nonnitrogen fertilizers (nitrogen is balanced endogenous to the model), and all other production items for both crops and livestock. Unit costs reflect different efficiencies of farming as land is used with straight rows, contours, strip cropping, terraces, and different rotations. The costs also reflect the higher pesticide requirements and lower machinery and labor requirements for crops grown under a reduced-tillage cultivation pattern. The costs sum to an aggregate that depends on the particular cropping management system and field technology and, when combined with the outputs from the system, reflect the comparative advantage of each crop and livestock system on each land class in each region.

The outputs from the system reflect yields of each crop and the associated quantity of gross soil loss. The interaction within the system also is reflected in a nitrogen balance subsector where the nigrogen flows in the model are examined. The entire cost and yield section of the model is interlocked with alternative technologies, levels of resource input, and alternative input uses to meet domestic and export demands. As an example of other interrelationships in the model, consider the nitrogen-fertilizer-crop yield section. Nitrogen available in an individual region is an independent variable in the crop yield equation, but the source of the nitrogen may vary. It can be supplied from chemical fertilizers, livestock wastes, or nitrogen fixation by legumes. For legumes the amount of nitrogen produced is dependent on their yield, which in turn is dependent on the early spring nitrogen availability (through fertilizer) and the nonnitrogen fertilizer availability over the growing season. The livestock wastes available are dependent on the type and quantity of feed available for livestock and the concentration of the animals in the region. Also affecting the yields of the crop is the land class on which it is grown and the conservation and tillage practice associated with the cropping management system.

Both dryland and irrigated crop variables are included for producing regions in the 17 Western states that grow irrigated crops, and the model allows selection among dryland or irrigated farming for each region. A range of livestock ration (variables) is allowed in all producing regions, since the least-cost mix can be drawn from various grain, forage, and pasture crops grown in the region or imported (where allowed) from others. The model includes variables representing various cropping systems and technologies affecting soil loss, livestock production, commodity transportation, water transfers, consumer demand fulfillment, and alternative export levels.

Each of the 223 producing regions has land restraints of the nature indicated in Eqs. 3 through 9. Each region has a soil loss restraint [as in Eq. (10)]. We analyzed four levels of per-acre soil loss restraints: 3 ton, 5 ton, 10 ton, and no limit. Most of our later illustrations are for the 5-ton level—although we do present some summary figures for other levels. A nitrogen balance equation [as in Eq. (11)], and a pasture restraint [as in Eq. (23)] are included in each region. Each water supply region has a water restraint [as in Eq. (13)], where the variables and parameters are defined subsequently.

Each of the 30 consuming regions has net demand equations for all of the relevant crop and livestock activities as illustrated by Eq. (14). Regional consumer demand quantities were determined exogenously from geographic and national projections of population, economic activity, per-capita incomes, and international exports through the region for 2000. All technology and demand coeffiecients and the solution are for 2000. National demands were defined for cotton and sugar beets as indicated in Eq. (15). Poultry products, sheep, and other livestock were regulated at the consuming-region level. International trade was regulated at the regional levels as indicated in Eqs. (16) and (17).

Commodities included in the endogenous analysis are soil loss, nitrogen, water, corn, sorghum, wheat, barley, oats, soybeans, cotton, sugar beets, tam hay, wild hay, improved pasture, unimproved and woodland pasture, cropland pasture, public grazing lands, forest lands grazed, all dairy products, pork, and beef. Also accounted for prior to solution of the model are crops including fruits, nuts, vegetables, rice, flax, and others, and the livestock broilers, turkeys, eggs, sheep, and other livestock.

Dryland cropland restraint, each region by land class:

$$\sum_m X_{ikm} a_{ikm} \leq LD_{ik} \tag{3}$$

Irrigated cropland restraint, each region by land class:

$$\sum_n Y_{ikn} a_{ikn} + \sum_m Z_{ikm} a_{ikm} \leq LR_{ik} \tag{4}$$

Dryland wild hay restraint, each region:

$$DWH_i a_i \leq ADWH_i \tag{5}$$

Irrigated wild hay restraint, each region:

$$IWH_i \, a_i \leq AIWH_i \tag{6}$$

Dryland permanent pasture restraint, each region:

$$DPP_i \, a_i \leq ADPP_i \tag{7}$$

Irrigated permanent pasture restraint, each region:

$$IPP_i \, a_i \leq AIPP_i \tag{8}$$

Forest land grazed restraint, each region:

$$FLG_i \, a_i \leq AFLG_i \tag{9}$$

Soil loss restraint, each region, each land class, each activity:

$$SL_{ikm+n} \leq ASL_{ikm+n} \tag{10}$$

Nitrogen balance restraint, by region:

$$FP_i + \sum_p b_{ip} L_{ip} + EL_{ic} b_{ic} - EC_i f_{ih}$$

$$- \sum_k \left(\sum_m X_{ikm} f_{ikm} + \sum_n Y_{ikn} f_{ikn} + \sum_m Z_{ikm} f_{ikm} \right)$$

$$- DPP_i f_i - IPP_i f_i - DWH_i f_i - IWH_i f_i - FLG_i f_i = 0 \tag{11}$$

Pasture use restraint, each region:

$$\sum_k \left(\sum_m X_{ikm} r_{ikm} + \sum_n Y_{ikn} r_{ikn} + \sum_m Z_{ikm} r_{ikm} \right)$$

$$+ DPP_i r_i + IPP_i r_i + FLG_i r_i - \sum_p L_{ip} q_{ip} - EL_i q_i \geq 0 \tag{12}$$

Water use restraint, by water region:

$$WB_w \pm WT_w \pm WI_w - WO_w - WX_w - WE_w + WD_w$$

$$- \sum_{i \in w} IWH_i d_i - \sum_{i \in w} IPP_i d_i$$

$$- \sum_k \sum_{i \in w} \left(\sum_m X_{ikm} d_{ikm} + \sum_n Y_{ikn} d_{ikn} + \sum_m Z_{ikm} d_{ikm} \right)$$

$$- \sum_{i \in w} \sum_p L_{ip} d_{ip} - \sum_{i \in w} PN_i d_i \geq 0 \tag{13}$$

Commodity balance restraint, each consuming region:

$$\sum_k \sum_{i \in j} \left(\sum_m X_{ikm} cy_{ikmc} + \sum_n Y_{ikn} cy_{iknc} + \sum_m Z_{ikm} cy_{ikmc} \right)$$

$$\pm \sum_{i \in j} \sum_p L_{ip} cy_{ipc} \pm \sum_{t \in j} T_{tc} \pm E_{jc} -$$

$$- \sum_{i \in j} PN_i \, cy_{ic} - EL_j \, cy_{jc} \geq 0 \qquad (14)$$

National commodity balance restraints, for cotton, sugar beets and spring wheat:

$$\sum_i \sum_k \left(\sum_m X_{ikm} cy_{kimg} + \sum_n Y_{ikn} cy_{ikng} + \sum_m Z_{ikm} cy_{ikmg} \right)$$

$$- \sum_i PN_i \, cy_{ig} - EX_c \geq 0 \qquad (15)$$

National export restraints:

$$\sum_j E_{jc} \geq EX_c \qquad (16)$$

National import restraints:

$$\sum_i E_{ic+e} \leq IM_{c+e} \qquad (17)$$

Nonnegativity restraints:

$$X_{ikm}, Y_{ikn}, Z_{ikm}, L_{ip}, DWH_i, IWH_i, DPP_i, IPP_i, FLG_i, FP_i, EL_i,$$

$$WB_w, WT_w, WI_w, WD_w, WX_w, WE_w, PN_i, T_{tc}, E_{jc}, E_{icre} \geq 0 \qquad (18)$$

The subscripts and variable s for the above equations are defined in the following section.

Subscripts and Variables of the Model

The subscripts and variables relating to the equations in the text are as follows:

Subscripts
c = 1, 2, ..., 15 for the endogenous commodities in the model
e = 1, 2, ..., 5 for the exogenous livestock alternatives considered
g = 1, 2, 3 for the commodities balanced at the national level

h = 1, 2, ..., 19 for the exogenous crop groups considered
i = 1, 2, ..., 223 for the producing areas of the model
j = 1, 2, ..., 30 for the consuming regions of the model
k = 1, 2, ..., 9 for the land classes in each producing area
m = 1, 2, ..., for the dryland crop management systems on a land calss in a producing area
n = 1, 2, ..., for the irrigated crop management systems on a land class in a producing area
p = 1, 2, ..., for the livestock activities defined in the producing area
t = 1, 2, ..., 458 for the transportation routes in the model
w = 1, 2, ..., 51 for the water supply regions in the model

Variables and Parameters

a = amount of land used by the associated activity from the land base as indicated by the subscripts

AIPP = number of acres of irrigated permanent pasture available in the subscripted producing area

ADWH = number of acres of dryland wild (noncropland) hay available in the subscripted producing area

AFLG = number of acres of forest land available for grazing in the subscripted producing area

ADPP = number of acres of dryland permanent pastures available for use in the subscripted producing area

AIWH = number of acreas of irrigated wild (noncropland) hay available in the subscripted producing area

ASL = per-acre allowable soil loss subscripted for land class, producing area, and activity

b = units of nitrogen-equivalent fertilizer produced by livestock, subscripted for producing area and activity

cy = interaction coefficient (yield or use) of the relevant commodity as regulated by the associated activity and specified by the subscripts

d = per-unit-of-activity water use coefficient as regulated by the associated activity and specified by the subscripts

DPP = level of use of dryland permanent pasture in the subscripted producing area

DWH = level of use of dryland wild (noncropland) hay in the subscripted producing area

E = level of net export for the associated commodity in the associated region as specified by the subscripts

EC = level of exogenous crop production by subscripted region

EL = level of exogenous livestock production consistent with the subscripted region
EX = level of national net export for the subscripted commodity as determined exogenous to the model
f = units of nitrogen-equivalent fertilizer required by the associated activity and specified by the subscripts
FLG = level of forest and land grazed in the subscripted producing area
FP = number of pounds of nitrogen-equivalent fertilizer purchased in the subscripted producing area
IM = level of national net imports for the subscripted commodities as determined exogenous to the model
IPP = level of use of the irrigated permanent pasture in the subscripted region
IWH = level of use of the irrigated wild (noncropland) hay in the subscripted region
L = level of the livestock activity with the type and region dependent on the subscripts
LD = number of acres of cryland cropland available for use as specified by the region and land class subscripts
LR = number of acres of dryland cropland available for use as specified by the region and land class subscripts
PN = level of population projected to be in the subscripted region
q = units of pasture in hay equivalents, consumed by the associated livestock activity and specified by the subscripts
r = units of aftermath or regular pasture, in hay equivalents, produced by the associated cropping or pasture activity and identified by the subscripts
SL = level of soil loss associated with any activity over the range $m + n$ in the region and land class designated by the subscripts
T = level of transportation of unit of the commodity either into or out of the consuming region designated by the subscripted
WB = level of water purchase for use in the water balance of the water supply region designated by the subscript
WD = level of desalting of ocean water in the water supply region designated by the subscript
WE = level of water to be exported from the water supply region subscripted
WI = level of movement of water in or out of the water supply region through the interbasin transfer network
WO = level of water requirement for onsite used such as mining, navigation, and estuary maintenance in the water supply region subscripted
WX = level of water use for the exogenous agricultural crops and livestock

in the water supply region subscripted
UC = cost per activity unit (unit cost)
X = level of employment of the dryland crop management system, rotation, in the region and on the land class as designated by the subscripts
Y = level of employment of the irrigated crop management system, rotation, in the region and on the land class as designated by the subscripts
Z = level of employment of the dryland crop management system, rotation, on the land class in the region as designated by the subscripts when the land has been designated as available for irrigated cropping patterns

Illustration of Results

Solution of the model provides indication of optimal land and water use in each of the 223 producing regions and each of the 2,007 land resource groups at prescribed levels of environmental quality restraints, consumer demand and distribution, export levels, and other policy, or market and technology parameters. Our illustration is in the case where the only environmental restraint is soil loss as a means of lessening no point pollution and in affecting water and land allocation. It also designates the level of production in each region and the optimal flows of commoditites to consuming regions and export markets.

For most purposes of illustration, we refer to solutions where (1) soil loss is not restricted, and (2) soil loss is restricted to 5 tons/acre/year for each soil resource group and exports are at a given level. Some comparisons are made, however, for 3-ton, 5-ton, 10-ton, and unlimited soil loss per acre. Although land use could be mapped or indicated by each of the 2,007 land resource groups, we illustrate on the basis of the 223 producing regions only. The model indicates not only land devoted to each crop use in each region and group, but also can indicate technologies for each such as dryland or irrigated, alternative rotations, conventional or reduced tillage methods, and other factors that affect land and water use and sedimentation.

Figure 3 indicates an optimal distribution of dryland and irrigated crop acreage among the 223 producing regions with no restraints in soil loss. Figure 4 illustrates the optimal distribution under a restraint of 5 tons/acre/year.

Soil loss

Although land use, tillage methods and soil loss are generated in the models by producing regions and land resource groups, we summarize results

Improving Water Quality and Nonpoint Pollution

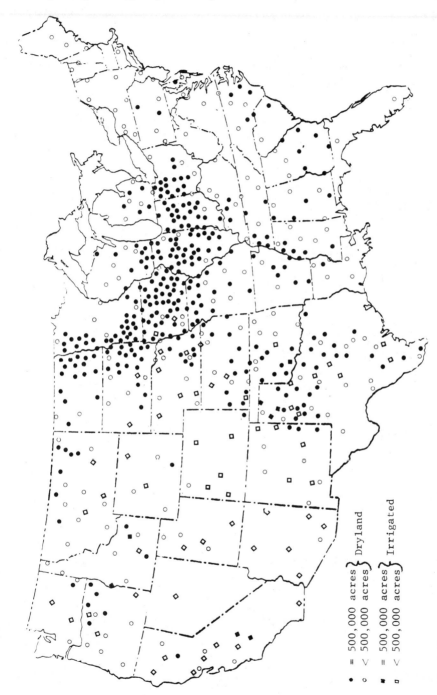

Fig. 3 Dryland and irrigated crops distributed among 223 producing regions, with no soil loss restraint.

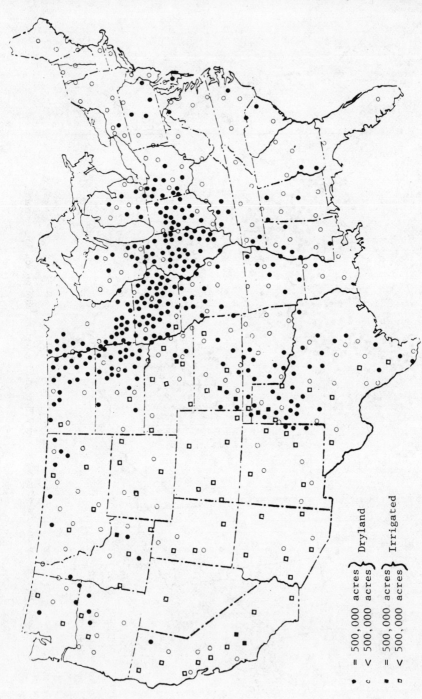

Fig. 4 Dryland and irrigated crops distributed among 223 producing regions, with 5-ton soil loss restraint.

only for seven major geographic regions of the United States because of space limits.

Restricting water runoff and soil loss per acre to 5 tons distributes land use and technologies interregionally to reduce national soil loss to 727 million tons. Without the restriction, interregional land use allocations and technologies to meet export demands generate a projected national soil loss 3.5 times greater, or 2.677 million tons. As Table 1 indicates, the reduction in average per-acre soil loss, as a source of sedimentation, would be extremely large on land classes V-VIII, which are most erosive. Although we do not do so here, our models allow indication of soil loss changes by each individual region.

Regional variation in reduced soil loss per acre is great. Largest reductions take place in the South Atlantic (18.2 tons) and South Central (11.5 tons) regions, where land and current land use methods give rise to sediment transport to streams (Table 2). The reduction in soil loss under a 5-ton-per-acre limit is attained especially by a switch from conventional tillage-straight row farming to contour, strip cropping, and terraces (Table 3). There also is a significant shift to

Table 1 National Soil Loss Total and Average per Acre by Land Resource Groups for Two Levels of Soil Loss Restriction, 2000

Item	Unrestricted soil loss	5-ton soil loss
Total tons (million tons)	2,677	727
Average tons per acre		
Class I and II land	6.2	2.7
IIIE & IVE land	17.8	3.1
Other III & IV land	15.6	2.8
V-VIII land	28.5	1.5
National Average	9.9	2.8

Table 2 Average Per-Acre Soil Loss by Major Region for Two Levels of Soil Loss Restriction Models, 2000

Region	Unrestricted soil loss	5-ton soil loss
National	9.9	2.8
North Atlantic	9.0	3.5
South Atlantic	21.5	3.3
North Central	9.2	2.8
South Central	15.1	3.6
Great Plains	3.2	1.5
Northwest	2.3	1.7
Southwest	3.3	2.6

reduced tillage farming practices to attain the environmentally attained soil loss of 5 tons/acre. Acres receiving reduced tillage practices increase from around 21 million in the unrestricted solution to near 58 million acres under the 5-ton solution. Conventional tillage practices decline from 248 million acres under the unrestricted soil loss to 201 million acres under the solution for 5-ton soil loss. Within the conventional tillage group, straight-row farming is nearly halved. Contouring is tripled, and strip cropping-terracing practices are increased 1,000% to meet soil loss restrictions (Table 3). Whereas reduced tillage nearly triples and very large increases occur in contouring, terracing, and strip cropping, straight-row methods of reduced tillage do not increase significantly.

To attain the soil loss reduction, there is a reduction of 16.5 million acres used for grain crops but a corresponding increase of only 5.5 million acres in hay on cultivated lands (Table 4). Hence, not all of the land shifted from crops is planted to hay and pasture. Part of the production required to meet national demands comes from an increase in noncropland roughage production (perma-

Table 3 Thousand Acres of Cultivated Land by Conservation-Tillage Practices for Two Levels of Soil Loss Restriction, 2000.

Conservation tillage	Unrestricted soil loss	5-ton soil loss
Conventional tillage	247,894	201,238
Straight row	233,475	129,120
Contoured	11,254	37,116
Strip cropped & terraced	3,165	35,002
Reduced tillage	21,219	57,644
Straight row	21,219	24,822
Contoured	0	18,902
Strip cropped & terraced	0	13,920

Table 4 National Production of Row Crops, Close-Grown Crops, and Rotation Roughage Crops for Two Levels of Soil Loss Restriction, 2000[a]

Land use	Unrestricted soil	5-ton soil loss
Acre cultivated (000)	269,113	258,882
Row crops (000)	148,226	136,035
Close grown crops (000)	75,535	73,478
Rotation roughage crops (000)	45,352	49,369
Nonrotation roughage crops (000)	303,060	310,697

[a]Demand levels are based on projected per-capita food consumption levels, 284 million people in 2000, and international trade of grains equal to the 1969-1971 averages.

nent hay and pasture). More of the reduced acreage required to meet the demand for agricultural products results because of the shift in production to the higher-cost, and higher-yielding, erosion-control practices. Also, a shift in acreage between land classes puts the grain crops on the higher-yielding and less-erosive lands.

Cost of production, in conjunction with the transportation network and the soil loss restrictions imposed, determine the national equilibrium prices for the commodities. Table 5 indicates the relative equilibrium prices of the commodities generated by the model under the two levels of soil loss and a single export level. Soil loss restrictions have the largest effect on prices for commodities that concentrate on land with high soil loss potential. Compared to absence of soil loss restrictions cotton and soybean prices increase by 12 to 15%, whereas wheat and hay crops increase by 1 to 3%. The increase in grain prices result in corresponding increases in cattle prices. In evaluating the effect of any environmental policy alternative, the effect on the desired parameter and the change in farm price of agricultural products are two easily observed change in our models.

Changes summarized at the national level do not, of course, reflect the effects in particular regions and on individual enterprises. These, however, are all available from our models. The shift in production from one region to another results in income repercussions on the rural community affected. The effect of such a shift depends on the degree of multiple-level resource use. As grain production shifts, livestock activities also may drift away from the region and underemployment of resources may occur in one region as prosperity is promoted in another rural area.

The data in Tables 1 through 5 indicate that American agriculture has great capacity and flexibility in adapting to certain environmental quality goals. By shifting land and water use among the many producing regions, water basins,

Table 5 Relative Farm-Level Prices for Some Agricultural Commodities with Two Levels of Soil Loss Restriction, 2000

Commodity	Unrestricted soil loss	5-ton soil loss
Corn	100	107
Wheat	100	103
Soybeans	100	115
Cotton	100	112
Hay	100	101
Cattle	100	104
Hogs	100	105
Milk	100	100

and land resource groups in terms of their comparative advantage in yields, commodity costs, location, and transportation, national and regional demands can be met without large increases in food prices and costs for consumers at the export level examined. The level of exports per se may have greater impact on consumer food costs than does a relatively wide adaptation of agriculture and land use to environmental quality goals. We shall, however, provide quantitive analysis of these possiblities, along with other environmental quality practices, in upcoming presentations.

Shifts in Technology and Trade-Offs in Variables

The reduction in soil loss from the agricultural sector is accomplished through changes in farming methods and cropping patterns and in reallocations of water and crops among regions. These changes have implications on the inputs required by the agricultural sector. The dollar value of total machinery inputs declines slightly as the allowable soil loss level is reduced (Fig. 5a). The decline results from the reduction in acres cultivated, as the per-acre expenditure for machinery is an oversimplification of the total impact. The machinery complement will change drastically to reflect the shift to a higher proportion of the lands being farmed under conservation tillage techniques (Table 3). The increase in per-acre machinery expenditures reflects the increased time required as farmers shift to the smaller fields associated with the contour, strip cropping, and terracing methods of farming.

The shift in farming method to a more intensive level of the conservation or reduced tillage methods is accompanied by an increased application of pesticides (Fig. 5c). The greatest increase is associated with the row crops which, on a per-acre basis, more than double their use of the pesticides as average soil loss is reduced from near 10 tons/acre to near 2 tons/acre (Fig. 5d). This increase results from the greater requirement for herbicides to control competing plant growth and insecticides to control the insects that would have been destroyed by the tillage operations under conventional farming alternatives.

The shift to the more productive regions and land classes increases the per-acre use of fertilizers (Fig. 5f). The increased use of pesticide and fertilizers compound environmental concerns with respect to their use. The model could have been formulated such that a penalty condition was incorporated to discharge an increased use of pesticides or fertilizers above the base model. Such a formulation would have changed the impact of the soil loss analysis as the new farming techniques used would be encouraged to reflect a lower use of these possible environmental contaminants.

The total impact on the farming sector is reflected in their production technologies and subsequent input requirements. The consequences of such action on commodity prices represents the impact on the consumer, generally

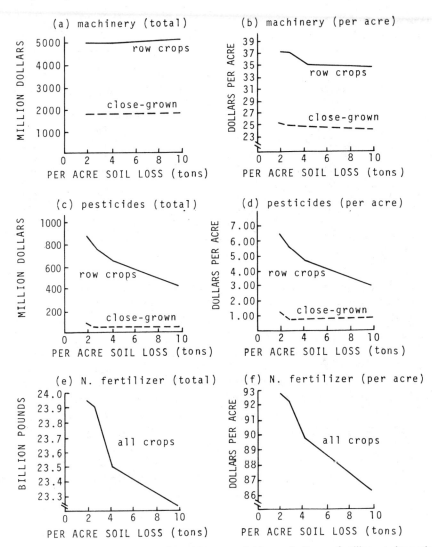

Fig. 5 Total and per-acre use of machinery, pesticides, and nitrogen fertilizer at alternative levels of per-acre soil loss in 2000 by crop type.

the group which is expressing the environmental concern. The impact on commodity prices is relatively minor at the 10-ton/acre level (Table 6). However, as the soil loss restriction level declines below 10 tons, changes in the prices of the commodities occur consistent with their regional production location and their susceptibility to erosion. Soybeans and cotton have the largest price increases, reaching 20 to 25% at the 3-ton restriction level, reflecting their

Table 6 Indication of Relative Farm-Level Prices for Some Agricultural Commodities for Alternative Soil Loss Restrictions and Export Levels in 2000

Commodity	Unrestricted ave. 69-71	10-ton ave. 69-71	Solution 5-ton ave. 69-71	3-ton ave. 69-71	5-ton 3* ave.[a]
			(unrestricted model = 100)		
Corn	100	100	107	106	126
Wheat	100	99	103	103	158
Soybeans	100	101	115	121	219
Cotton	100	100	112	125	107
Hay	100	99	101	106	132
Beef	100	100	104	105	119
Pork	100	100	105	104	122
Milk	100	100	100	102	111

[a]Export levels tripled in 2000.

general lack of residue and production location in the high-erosion regions. The livestock products increase in price relative to their consumption of the higher-valued feed commoditites and the value ot their feed costs relative to their total cost of production.

Increasing exports in conjunction with the 5-ton soil loss has a much greater impact on commodity prices than any of the soil loss alternatives considered (Table 6). Cotton prices do not change from the 5-ton alternative to the double-export alternative. This results as cotton exports are not increased and cotton, being the higher-valued crop, determines the resource rent in areas where it is grown and the higher demand for feed grains, soybeans, and wheat does not affect the opportunity cost of land (land rent) in cotton-producing areas. Soybeans have the largest increase in price because of their increased demand and susceptibility to erosion, causing soybeans to compete for the highly productive low-erosion lands. Once again, livestock prices increase relative to their feed costs.

Implication of the Analysis

The analysis, done by considering only a few alternatives in what is essentially a continuum of environmental dimensions in agriculture, indicates that the agricultural sector can reduce the level of gross field loss of soil substantially with only minor effects on the price of food, assuming levels of exports of agricultural commodities near the 1969-1971 average annual level. If higher export levels are considered in conjunction with the soil loss restraint, increases in farm-level prices for agricultural products are more pronounced.

Imposing a soil loss restriction on the cultivated lands in agriculture affects the regional production advantage directly in relation to the lands susceptibility to erosion. Farmers in high-erosion lands must incorporate higher-cost practices to continue production. This reduces the regions' and the farms' competitive positions. Farmers with lands not susceptible to water runoff and erosion receive a larger return to their fixed resource (land) as prices increase. This will increase capital requirement discrepancies among lands of varying erosion susceptibilities.

The indicated changes in machinery use will require farmers to replace their capital stock at a more rapid rate. Credit availability to those farmers required to make the change may be limited, and legislation setting the soil loss restriction could provide for loans, technical assistance, or, if deemed necessary, subsidies to aid in compliance. The major equity considerations include landowners suffering capital losses as their land priced on the historic production pattern now may have its price based on a less intensive use and the consumers affected most as their food costs increase.

The shift to reduced tillage compounds other environmental problems. That is, the use of increased amounts of pesticides to control weeds and insects introduces more of these compounds into the environment, thereby increasing the possiblity of environmental contamination. Similarly, the increased use of fertilizers may cause some environmental concern. Each alternative must be considered for its total impact, and the use of models such as these will enable the policy maker to understand better the directions of any shifts in production prices and farming methods occurring if a proposed policy is implemented.

The model is one of water allocation and distributes this resource among the 51 irrigation regions (Fig.2) in an optimal manner. As soil loss limits on all lands are made more severe, a greater transfer of water occurs among upstream water regions to downstream regions. In no case, however, with interregion transfers of water allowed, does the nation have a shortage of water for irrigated agriculture.

Extension of Model to Sediment and Chemical Delivery

The model reported here also has been extended to handle more directly sediment delivery to waterways. Plans include further extension of it to include chemical delivery to waterways and the subsequent improvement of water quality. In many parts of the United States, delivery of sediment from farms to streams is highly correlated with the simultaneous delivery of chemicals. To an important extent sediment serves as the vehicle for transporting excess nitrogen, phosphate, and even pesticides applied to farm crops. In other crops, especially that of irrigated agriculture, chemicals are carried with return flow to streams unaccompanied by sediment. Increased salinity results as surplus irrigation water picks up soil chemicals or those applied as fertilizers and pesticides and deposits them at other locations or as return flow to streams. We shall summarize the ongoing and planned extension of the model to include sediment and chemical delivery.

Sediment Delivery

Soil particles moving as sediment cause substantial damage in U.S. waterways. Erosion, which introduces soil into streams, is caused by natural forces. Hence, we developed a linkage among agricultural production, land use, and stream sediment loads. The relationships indicated in Equs. (2) and (10) refer only to gross soil loss; the amount of soil moved from one location on the land. The amount of sediment reaching rivers is something less than the gross soil loss as defined by Eq. (2). The sediment loads are built up from net amounts of sediment reaching the nation's river system.

Sediment delivery is based on the concept and reality that only a proportion of erosion or soil loss actually arrives in the major rivers. This is illustrated in Eq. (19), which includes the sum of sources controlled by the production or use activities of the land. Delivered sediment is available to be transported downstream. Thus, the following relationship between sediment sources and sediment delivered exists for each producing area:

$$D_i \left(\sum_j \sum_k S_{ijk} \cdot X_{ijk} + s_i^* + \sum_j S_{ij}^A \cdot X_{ij}^A \right) = X_i^D \tag{19}$$

where

X_{ijk} = acres of production activity k on land quality class j in producing area i

S_{ijk} = tons per acre of gross soil loss from activity k on land quality class j in producing area i

s_i^* = tons of gross soil loss in producing area i for land uses not endogenous to the model

X_{ij}^A = acres of idle cropland in land quality class j

S_{ij}^A = tons per acre of gross soil loss from idle cropland of class j

D_i = proportion of gross soil loss that reaches the stream in producing area i

X_i^D = tons of suspended sediment delivered from producing area i

Sediment Transport

The delivered quantity of sediment X_i^D, is assumed to behave according to the machanics of the stream. That is, sediment once in the stream moves with the water and "obeys" the properties of the stream. This characteristic is expressed by the following equation:[2]

$$T_i \left(\sum_{\ell \to i} X_\ell^D + \sum_{k \to i} X_k^T \right) = X_i^T \qquad (20)$$

where

X_ℓ^D = sediment delivered from the ℓth upstream producing area directly into producing area i

X_k^T = sediment transported through an upstream producing area k into the stream system of producing area i

T_i = proportion of sediment moved to producing area's i boundary that is transported through producing area j

X_i^T = sediment transported through producing area i

Basin Sediment Accounting

The total sediment load, $\overline{X}_i$, at the point of river basin outflow is

$$\overline{X}_{i'} = X_i^T + X_i^D \qquad (21)$$

or the sum of sediment delivered from the last (or ith) producing area in the river basin and the sediment transported through the ith producing area.

Equations (19), (20), and (21) form the basis for the linear programming model and the policy variations that can be analyzed. Put into the more familiar linear programming format, these equations are (22), (23), and (24):

$$\sum_j \sum_k S_{ijk} A_{ijk} + \sum_j S_{ij}^A X_{ij}^A - \frac{1}{D_i} X_i^D \leq -s_i^* \qquad (22)$$

$$\sum_{\ell \to i} X_\ell^D + \sum_{K \to i} X_K^T - \frac{1}{T_i} X_i^T \leq 0 \qquad (23)$$

$$X_i^T + X_i^D = \overline{X} \qquad (24)$$

---
[2] The symbolism $\ell \to i$ and $k \to i$ means "for all ℓ (or k) contributing directly to i."

The variables D_i and T_i are constants depending on the physiographic and hydrologic makeup of the production area. For this analysis the quantity on the right of Eq. (22) is a constant depending on the geomorphology of the area and the use of land/noncropland.

Chemical Delivery

Extension of the model to include chemical delivery will parallel the framework outlined above for sediment delivery. First, the coefficients must be estimated associating chemical delivery with sediment delivery. Next, estimates must be completed of the amount of chemical moving into solution in the water from both applied chemicals and salts residing in the soil. In areas of level irrigated land, chemical delivery will be from the last two sources alone. Hence, equations paralleling (19) and (20) also will exist for chemicals. (Relationships can be expressed for chemicals in aggregate or for individual categories. Our interests are largely in nitrogen, phosphate, and in irrigated areas, those compounds associated with salinity.) The total chemical load will be a function of the amount of chemicals applied, the type of soil and its chemical and physical properties, the amount of rainfall, the source and amount of water used for irrigation, the types and yields of crops, and the conservation-tillage practices used on the land. Interest is in concentration of these chemicals in the land in other producing areas or locations and in the amount deposited in rivers to be transported downstream. Developing data sets for this analysis is a large task. However, it is only slightly more complex than development of the data sets for either the gross soil loss or the net sediment loads carried by streams.

Acknowledgements

This work is from a project supported by the **RANN** Program of the National Science Foundation.

12

Cost Sharing and Pricing for Water Quality

Gideon Fishelson   George S. Tolley

Tel-Aviv University,   *University of Chicago,*
Tel-Aviv   *Chicago*

Introduction

The economics of cost sharing and product pricing have been widely studied with regard to private enterprises. However, there are hardly any similar studies of public enterprises — specifically, water quality projects. This chapter accordingly illustrates the economic principles of cost sharing and pricing of projects aimed at improving water quality. These principles are directed toward the economic efficiency of the project's scale, its operation, the waste discharged, and the usage of clean water.

What is water quality? Various economic activities result in by-products called waste. The concept of waste implies a product with a nonpositive market value. In a world of limited assimilative capacity, a continuous flow of waste beyond a certain rate results in waste accumulation which is a nuisance; people feel discomfort — the psychic costs — or, some of their activities become conditional upon expenditures for prior treatments and supplementary devices — the economic costs. The sum of the corresponding psychic and economic costs is the social cost of the environmental degradation; water quality is an index that represents these costs.

The discussion of water quality is frequently centered on water salinity;[1]

---

[1] Salinity per se is almost irreducible by conventional water treatment processes, except at very high costs. The only other feasible way to lower industrial salinity is by activity changes or dilution (including temporary storage or production rescheduling). The most damaging effects of salinity are to irrigated agriculture and to industries that use water for cooling, boiler feed, and processing.

indeed, water salinity is the main concern of water quality studies in the agricultural literature. Actually, any reduction of water quality can be defined as water pollution; for example, thermal pollution from electric power plants should also be considered to be water pollution. Thus, in this study, the terms salinity and water pollution are used interchangeably.[2]

The sources of pollution are either natural or man-made. Statistics for the Colorado River Basin indicate that they share equal responsibility (47 and 53%) for increased salinity. However, it is not necessarily the general case. Along rivers flowing in industrial areas (e.g., the Ohio and the Delaware), the man-made sources are usually responsible for over 80% of waste; floods and natural drainage make up the rest. The effects (damages) of man-made sources are called externalities. The economics of externalities and possible solutions to its inefficiency are presented in the second part of this study.

The objective of the policies aimed at improving water quality is to increase social welfare. The term "social" implies that the joint well-being of polluters, water users, and third parties is the target. The "social" concept ignores political borders (state and countries), since it deals with the population and economy of the entire water basin and frequently beyond it. This water basin might extend over a small spring two miles long, or, over several states, e.g., the Delaware River Basin. In principle, there is no difference between these two cases; it is only a matter of the scale and complexity of interactions, cross effects, joint purposes, and multiple projects involved.

Two topics that complement our study are the measurement of benefits from improved water quality and of the costs required for the improvement. The major advance in the first area has been made in the agricultural sector, where crop responses have been measured with respect both to salinity accumulation in the soil and to water quality.[3] However, very little has been done to evaluate municipal and industrial benefits. The main forms of these benefits are the savings in preusage treatment costs and operating costs, and the increase in the useful life of water-related equipment [see the Orange Country Survey (Wesner, 1974)]. Recreation benefits and wildlife effects are other significant benefits. The technology available and the corresponding costs of controlling slainity and waste vary widely. For point sources (which should be our main concern), this

---

[2] Salinity usually refers to the concentration of total dissolved solids. The definition of water waste is much broader, and includes nondissolved solids.

[3] The agricultural sector was traditionally concerned with water quality. In the United States, the problem is geographically concentrated in the West, where agriculture is irrigated. The history of irrigated agriculture in the United States is only about 100 years old. However, experiences of old cultures that vanished because of soil salinity problems is keeping U.S. agriculture on alert. The record until now indicates that matters do get worse. For a list of damage studies, see Yaron (1974). For air quality improvements see Cohen *et al.* (1974).

technology includes water treatment, desalting, diverting, evaporation ponds, plug wells, deep injection, and dilution. These means can be applied to natural sources of salinity (e.g., a salty spring such as the Blue Spring in Arizona), and to man-made sources (e.g., a refinery). A specific group of control technologies is that related to management and operation practices [e.g., irrigation scheduling, irrigation technology, groundwater management, basin management, waste water reclamation, and flood control (Maletic, 1974)]. Each technology (including the managerial ones) can be used at different scales, with different intensities and in combination with other technologies.

When the construction of a treatment plant is under consideration, the economic decisons are related to its capacity, location, technology, and methods of operation and finance. In this paper, economic efficiency is on decisions related to the sharing of the construction costs and to the pricing of its services.[4]

The financial arrangements within the society for project cost sharing (investment cost and operating cost), whether reached by mutual agreement or imposed by law, have obvious implications on the distribution of income. Matters related to income distribution are frequently referred to as issues of equity and fairness. On these two issues, as on optimal income distribution, the economist usually has very little to say (unless the social welfare function is well defined). This is the reason underlying the approach outlined by Harberger (1971): projects should be valued independently of whom the benefits or costs accrue. Cost sharing and output pricing extend out of the mere distributional aspects when they affect efficiency. In these cases, an optimal cost sharing policy or an optimal pricing policy is a prerequisite for overall efficiency.

The plan of this chapter is in line with the ideas presented above. The next section discusses the optimal project scale and the optimal cost sharing and pricing arrangements for projects aimed at improving naturally low water quality (sea water). The analysis is based on typical cases such as a predetermined project scale, a variable project scale, and the case where the quantity of water is limited. Then we discuss improving the quality of water which has been polluted by man-made waste. The discussion starts with a simple example of externalities. Solutions for externalities are outlined; the emphasis is on a waste treatment plant. The questions of cost sharing and pricing are discussed for various situations — such as increasing economies of scale in plant construction,

---

[4] Policies practiced to effect cost sharing in water projects of various types are summarized in the report of the National Water Commision (1974, Chap. 15). The recommendations of the Commission call for increasing the role of economic considerations in cost sharing. In particular, the Commission emphasizes uniformity of rules and pricing equity which is based on payments that are proportional to received benefits. The target is efficiency, and "appropriate cost sharing policies should provide incentives for the selection of efficient projects" (p. 496). [See also Marshall (1974)].

peak load pricing, and the financial constraint that revenues must equal costs.[5] The following section shows that land is the factor of production that bears the cost of improving water quality, and the final section summarizes the conclusions.

Improving Natural Sources of Water

The question of how to share the costs of a public water project is not a new one. However, major economic studies of water projects have not devoted much space to the question. They have concentrated on such aspects as project benefit, project cost, and project effieciency; or, on theoretical issues such as what items to include in the benefit-cost calculations and which interest rate is appropriate for project evaluation − for example, the social versus the private discount rate (Arrow, 1966; Mishan, 1967, Seagraves, 1970; and Feldstein 1974).

Cost sharing is a division of the financial responsibility of a project among groups of individuals with different purposes and interests. It centers on the equitable distribution of project costs and effects. As shown below, efficiency and cost sharing are, in most cases, independent. However, the absence of charges might induce a wasteful use of services by beneficiaries and increase the pressure for a project scale that is larger than is socially justifiable.

The traditional questions related to such projects concern their optimal size, location, capital financing, operating procedure, and financing of operating costs. The answers usually appear under a large set of *ceteris paribus* conditions − that is, each question is answered independently, assuming that all other questions are answered optimally. For example, Ciriacy-Wantrup (1954) addresses only the question of optimal pricing. He concludes that "it appears economically justified and politically equitable that beneficiaries from public resource development pay for the benfits received provided such benefits are practically assessable and provided that enough incentive is left for beneficiaries to participate in resource development" (p. 115).

Regan (1958) does not discuss the issue of market efficiency but emphasizes that in order to encourage the utilization of the developed resource, the charges and assessments should be less than a full value of the benefits received and in line with the ability to pay. Only Regan (1961) makes the specific point that "the marginal benefits from each purpose accruing to each participant (must) be sufficient to cover the corresponding marginal costs" (p. 239). Hence, when pricing is used, the charges imposed must be such that there will be neither waste nor under utilization implying that for each user equating marginal cost with marginal benefits is a necessary condition for efficiency.

---

[5] The cases discussed in the paper are static in the sense that all periods are identical. Two dynamic examples are available from the authors upon request.

Most studies on the economics of water projects (e.g., Krutilla and Eckstein, 1958; Eckstein, 1958; McKean, 1958; Hisrchliefer *et al.*, 1960, Maas et al., 1962 Marglin, 1967; Hamilton *et al.*, 1969 and many others) do not treat cost sharing and pricing explicitly. They concentrate on aspects of economic efficiency, although there has been some discussion on optimal finance (federal versus private) – for example, Krutilla and Eckstein (1958). A more general study by Feldstein (1974) on the financing of public projects emphasizes the role of pricing of private goods provided by the public sector as a source of revenue and as signals to consumers.

The Decision on Optimal Project Scale

Cost-sharing arrangements affect local decisions with regard to project type, technology, and scale. The problem is to make local groups select projects that are efficient and optimal from the national point of view. Figure 1 presents a simple example where the sharing affects project scale. MC^s is the social marginal cost, and MB^s and MB^p are the social and private marginal benefits.

The optimal project scale is p^*. However, the private groups may apply pressure toward a scale that is either larger or smaller than the one which is socially optimal. One way to alleviate the pressure is through a cost-sharing arrangement. The marginal cost charged to the private group MC^p is "tailored" such that

$$MC^p = MC^s \cdot \frac{MB^p}{MB^s} \tag{1}$$

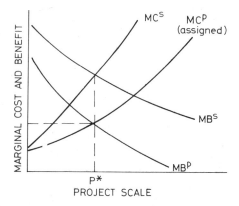

Fig. 1 Marginal cost assignment to alleviate pressure.

according to which the regions or other distinguishable beneficiaries represented in MB^P are to be charged. [This is, in principle, the idea in Marshall (1970).] Under these terms, the optimal project scale from the private point of view is identical to that from the social point of view.

When joint multipurpose projects are being considered and costs must be shared among the beneficiaries of each purpose (assuming they are distinct), the same working equation can be employed. Figure 2 is an example of three joint purposes. $\Sigma\,MB$ is the total marginal benefit — a vertical sum of MB^1, MB^2, and MB^3 (Tolley and Hastings, 1960). The optimal project scale is p^*, at which point $\Sigma\,MB = MC^s$ (Marshall, 1973).

In order to alleviate the pressure of local groups, we define

$$MC^i = \frac{MC^s \cdot MB^i}{\Sigma\,MB} \tag{2}$$

Thus, at p^*

$$MB^i = MC^i \qquad i = 1, 2, 3 \tag{3}$$

This practical solution implicitly employs the concept of fairness — namely, that costs are assessed in proportion to benefits.

An extension of the above problem is when a water project generates positive benefits to one region and negative benefits to another. A simplifying assumption is that local net benefits are measured by the value added to local businesses. The improvement of water quality in a region transfers recreationist from one region to another. The national account may show only a small change, whereas, on the regional level, a large benefit has been shifted.

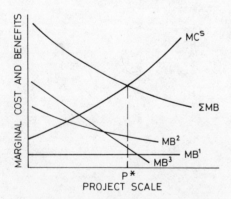

Fig. 2 Cost sharing in joint-purpose projects.

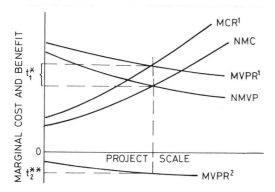

Fig. 3 Project scale, interregional benefits, and cost transfers.

NMC is the marginal (with respect to scale) direct cost to society, $MVPR^1$ is the marginal benefits to region 1, and $MVPR^2$ is the marginal benefit (which is negative) to region 2. Thus, the total social cost is represented by MCR^1, where $MCR^1 = NMC - MVPR^2$, while the net social benefits is defined by NMVP, where $NMVP = MVPR^1 + MVPR^2$. Following Fig. 3, the optimal project scale is p*; region 2 is to be compensated at the margin by t_2^* per unit of project scale, whereas region 1 is to be taxed by t_1^* per unit of project scale.

Rules for Cost Sharing and Pricing for Specific Project Types

The following analyses concentrate on specific types of projects for improving natural conditions. The focus is on the sharing of investment cost and the implications for water pricing.

Project Scale Is Predetermined

Water quality in various parts of the country is affected by salty springs. The technical solution is to divert the salty springs. It is helpful to consider the problem under the following assumptions. The investment costs of the project are the only relevant costs (there are no operating costs). More than one firm (farm) benefits from the diversion, and the firms are distinguishable. The problem then is to divide the project costs among the firms, assuming that the total project cost (C) is independent of the number of water users. The net benefits to the users are $B_1, B_2, \ldots, B_n$, where n is the number of potential users and the B's are present values. Let $\sum_{i}^{n} B_i > C$. Obviously, any amount charged to firm i should be below B_i (i = 1, ..., n): otherwise, user i would vote against the project (assume that unanimity is required). Following these conditions, any cost sharing is efficient if it does not affect the quantity of water used, the firms' outputs, and the quantity of other inputs. Such a cost-sharing arrangement enables the economy to obtain the maximum benefits from the project. The targets of equity, fairness, and maintenance of income distribution are to be achieved by other means.

Project Scale Is a Variable

The previous case is generalized by letting the quantity of salty water to be diverted be a decision variable. The cost of the project is a function of the quantity of water diverted. Water quality in the main river is also a function of the quantity of water diverted. The farms that utilize the water may have different production functions, but water quality and quantity are factors of production in all of them. Each farm may use different quantities of water, but each is forced to use the same quality. The problem that society faces is

$$\underset{DW}{\text{Max}} \sum_t \beta^t \left(\sum_i P_x^i (f^i (Q_t, W_t^i, V_t^i)) - P_w \cdot W_t^i - P_v \cdot V_t^i \right) - C(DW) \tag{4}$$

where $Q = h(DW)$. DW is the quantity of salty water diverted, $C(DW)$ is the associated cost, β is the discount factor, W^i is the quantity of water used by farm i, V^i is the quantity of other inputs used by farm i, and P_x^i is the price of output of farm i. P_w and P_v are the cost of pumping water and the price of purchased inputs, respectively, which are assumed to be equal for all farms. A single water quality, Q, is determined, and is equal for all users; (i.e., water quality is a public good). For each level of Q, different users have different marginal value products, and thus are willing to pay different prices. For simplicity, is is assumed that for *all* users the f^i function is separable in Q; i.e., the cross derivatives with Q are zero (the marginal product of water quality f_Q^i is independent of the quantity of other inputs, and vice versa). One obtains the obvious solution that the optimal Q has to fulfill the requirement

$$h_{DW} \sum_t \beta^t \sum P_x^i f_{tQt}^i = C_{DW} \tag{5}$$

where C_{DW} is the marginal cost of water diversion. Denoting MB_Q^i as the present value of the marginal value product of water quality in production, for firm i, then

$$\sum_i MB_Q^i = C_{DW} \tag{6}$$

Hence, the *vertical* sum of the marginal value products of diverted water should equal its marginal cost.

Cost Sharing and Pricing for Water Quality

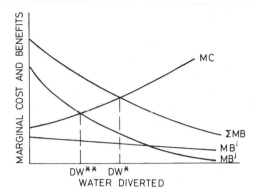

Fig. 4 Optimal water diversion. Farmer i is the nth farmer. The n − 1 farmers are denoted by j.

First, it is assumed the $C_{DW} > 0$ and $C_{DWDW} > 0$ (Fig. 4). DW* is the optimal quantity to be diverted. Total project cost is

$$C(DW) = \int_0^{DW^*} C_{DW}\, dDW \tag{7}$$

and total social benefit is

$$B(DW^*) = \int_0^{DW^*} \left(\sum_i MB^i \right) dDW \tag{8}$$

How should the costs be distributed among users? The economic answer is that equity and fairness are irrelevant as long as efficiency is unaffected. Hence, these targets must be reached by other means. However, if the farmers who are using the water are to finance the project, any coalition of n-1 participants will argue that if the nth farmer were not there, the optimal DW and, thus C(DW), and C_{DW} would be lower. When the nth farmer joins in, the total social cost of the project increases. Therefore, each user is to be charged the difference C(DW*) − C(DW**), given that it is less than B_i (DW*). C(DW**) is the total cost of an optimal quantity of diverted water when only n − 1 users are considered and the ith user is excluded; correspondingly, each user is to be considered the marginal user, and is to be charged accordingly.

The sum of payments is

$$nC(DW^*) - \sum_{i=1}^{n} C(DW_i^{**}) \tag{9}$$

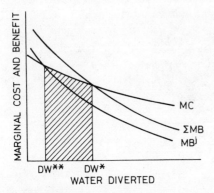

Fig. 5 Optimal water diversion with economies of scale. Farmer i is the nth farmer. The n − 1 farmers are denoted by j.

which, given an upward-sloping marginal cost curve, exceeds C(DW*). The redistribution of the difference is to be such that efficiency is unaffected (i.e., a lump-sum transfer).

Second, when C_{DW} is decreasing (Fig. 5) — that is, when there are economies of scale in water diversion — an additional (to that of $C_{DW} = \Sigma\, MB_i$) requirement for an optimal solution is that

$$\frac{d}{dDW} \Sigma\, MB_i > \frac{d}{dDW} C_{DW} \qquad (10)$$

The social optimum is reached at DW*, where *net* social benefits are positive. However, if the cost-sharing mechanism is as before, not enough is collected. The cost sharing thus has two parts. Each user pays his marginal contribution to social costs (the area DW**·ABDW*) and the residual is somehow divided among the users or, paid by the federal government or by any other institution. One constraint must be met: no user must pay more than the benefit he derives from the project. In order not to distort the efficiency of resource allocation and production, the payment is to be independent of output and quantities of inputs (including water).

Relaxing the separability assumption of the production function causes the quantity of water pumped and other inputs to be dependent on water quality, and vice versa. However, the conclusions on cost sharing do not change.

The conclusions in the two examples are based on a common characteristic: the quality of water changes, while the quantity of water remains unlimited, and the only limit to its usage is the pumping cost. The next case introduces a water quantity constraint.

Water Becomes a Limited Factor Because of the Project

Before the diversion project was in operation, the toal quantity of water pumped (per unit of time) was less than the quantity available. However, due to the improvement in water quality, the demand for water increased ($f_{QW}^i > 0$), and the new demand exceeds the quantity available. Thus, a mechanism is required to allocate the limited supply among the users. It is easy to show that the mechanism is a single price, equal for all firms. At this price, each firm will demand water until its marginal value product of water equals the price.

Water price is determined simultaneously with the decision on the optimal scale of the project. The water price is the shadow price on the water constraint. The social problem (discounting excluded) is

$$\text{Max} \sum_i \left[P_x^i f^i(Q, W^i, V^i) - P_w \cdot W^i - P_v \cdot V^i \right] - C(DW) \quad (11)$$

Subject to $Q = h(DW)$

$$\sum_i W^i \leq \bar{W} - DW$$

The decision variables are V^i, DW, and W^i. In the first-order conditions the same shadow price of water (λ) appears for each i:

$$P_x^i f_w^i(\cdot) - P_w - \lambda = 0 \qquad i = 1, 2, \ldots, n \quad (12)$$

The solution for λ is

$$\lambda = P_x^i f_x^i(\cdot) - P_w = P_x^j f_w^j(\cdot) - P_w \qquad i, j = 1, 2, \ldots, n \quad (13)$$

The price charged for water can be used for financing the water quality project. As before, the receipts may be more or less than the project costs. The distribution of the positive or negative differential is up to the policy maker and his viewpoint on income redistribution.

Quantity and Quality of Improved Water Are Variable

A specific example of changing natural conditions is a water desalination plant. The output is a "joint" product — a specific quantity of water of a given quality. Both the quantity and quality are decision variables. The quality is the same for all the water produced (one can consider a nonfixed quality plant producing different qualitites on different days of the week, supplying different customers each day). As in the previous example, the cost sharing for quantity and quality should be done jointly by charging a certain price for water. The price would be the marginal costs of producing the optimal quantity of water of the optimal quality. Optimality is defined by the joint equality of the demands for quality and quantity with the corresponding marginal costs.

This solution does not change if either or both the quantity and quality of water have downward-sloping supply curves. As already noted, the downward-sloping marginal cost generates a deficit in the budget (the upward-sloping cost generates a surplus); its disposal is efficient as long as it is unrelated to water consumed, output produced, or any other decision variable in the production process.

Conclusions

The general conclusion for the determination of project scale, cost sharing, and water pricing is that the main issue is efficiency. Thus, when there is no need for a mechanism to allocate output, no pricing of output is necessary. The cost sharing of the investment can take any form as long as it is independent of the quantity of water utilized. Pricing is necessary when output is to be allocated among users or when the variable cost is a function of output. For the latter case, the price should equal the marginal cost; for the former, the price should be such that when all of the output is allocated, the marginal benefit (product) of water in all uses is identical.

Externalities

Externalities occur when decisions and actions taken by one group of people have repercussions on others. The market system does not provide proper incentives to induce a decision maker to consider these repercussions. Hence, a free market fails to provide a maximum for social welfare.[6] One solution to this is a compensation scheme to be legislated that would force via economic means the consideration of externalitites. Since a bargaining solution is impractical when a large group of damaged people and many periods of time are involved, the government appears as the representative of the large group. The government utilizes its budget and its power to tax and to regulate in order to improve social welfare. In the context of water quality, government intervention can take various forms to induce the construction of water treatment plants.

---

[6]The number of studies dealing with the externality problem is huge. Some of the recent ones are Coase (1960), Buchanan and Stubblebine (1962), Davis and Whinston (1965,1966), and Wellisz (1964).

Cost Sharing and Pricing for Water Quality

The Basic Theory of Externalities with Water Quality as an Example

The equalization of price and social marginal cost in all sectors is a necessary and sufficient condition for a Pareto optimum (the state in which any action cannot improve the welfare of one individual without reducing that of another). This equality assures that the marginal rate of transformation between any two produced goods is the same for all firms; this transformation rate is the opportunity cost of production. However, when production externalities are present, the marginal cost to the firm (i.e., its opportunity cost) does not equal the social opportunity cost. Thus, decisions that are optimal from the firm's point of view are not best from the social point of view. Water pollution is a classic example of a production externality (e.g., the effect of upstream firms on downstream firms). Usually only external diseconomies are considered, although external economies can also be found (e.g., when the salinity of intake water is higher than that of effluent water due to water treatment before its usage).

The following is a mathematical representation of the economics of externalities. Firm A produces good x by employing input, b, and water, w. The level of output is determined by the relation

$$x = f(b^x, w^x) \tag{14}$$

where b^x and w^x denote the inputs used in producing x. The production function is assumed to have the traditional property that the marginal product of each input is positive and declining. The price of input b, P_b, is given to the firm, as is the price of x, P_x. Water is a "free" good, so its price to the firm equals the pumping cost, P_w. (f_b denotes the derivative of f with respect to b and hence the marginal product of b, and similarly for f_w.) Profit maximization by the firm leads to the use of b and w until the marginal value product of each equals its price:

$$P_x f_b = P_b \tag{15a}$$

$$P_x f_b = P_w \tag{15b}$$

A by-product of the production process is the flow of waste, z, which is dumped into a river. The quantity of waste is determined by the quantity of inputs used in production,

$$z^x = g(b^x, w^x) \tag{16}$$

where $g_b > 0$ and $g_w > 0$.

The waste, z^x, has no market. Dumping it into the river is the least expensive solution for firm A. Firm B, located downstream of A, produces a good, y, using input d and water, w. However, the quality of water used affects B's production cost. Thus, the effluent from Firm A, z^x, enters (after some natural dilution) into the production process of B. The output of firm B is thus determined by the relation

$$y = h(d^y, w^y, z) \quad \text{or} \quad y = h^*(d^y(z), w^y(z)) \tag{17}$$

where

$$z = f(z^x) \qquad f_z x > 0.$$

The marginal contributions to output of d and w are positive; that of z is negative. Just as for firm A, the profit-maximizing conditions for B are

$$P_y f_d = P_d \tag{18a}$$

$$P_y f_w = P_w \tag{18b}$$

Since z is defined as a nuisance, it is obvious that

$$f_d(d_0^y, w_0^y, 0) > f_d(d_0^y, w_0^y, z) \tag{19}$$

and

$$f_w(d_0^y, w_0^y, 0) > f_w(d_0^y, w_0^y, z) \tag{20}$$

for $z > 0$ and for any d_o^y and w_o^y. Hence, the presence of pollution in the water reduces the marginal products of the economic inputs and is therefore, definitively, a social cost. Furthermore, a maximization of the total output of society can show that the above independent solutions for firms A and B are not optimal from the social point of view. For simplicity, society is assumed to be composed of only the two firms. Hence,

$$\text{Max}\left[P_x \cdot x + P_y \cdot y - b^x \cdot P_b - d^y \cdot P_d - (w^x + w^y) \cdot P_w \right] \tag{21}$$

subject to

$$x = f(b^x, w^x) \tag{22a}$$

$$y = h(d^y, w^y; z) \tag{22b}$$

$$z = g(b^x, w^x) \tag{22c}$$

If the value of the joint output is greater than the sum of the outputs when each firm operates independently, it is obvious that the latter is inefficient. The maximization conditions are now

$$P_x f_b - P_b + P_y g_b h_z = 0 \tag{23a}$$

$$P_y h_d - P_d = 0 \tag{23b}$$

$$P_x f_w - P_w + P_y g_w h_z = 0 \tag{23c}$$

$$P_y h_w - P_w = 0 \tag{23d}$$

These conditions differ from the ones used by each firm independently. The additional term in (23a) compared to (15a) is negative; hence, f_b should have been larger, and b^x smaller. Similarly, the additional term in (23c) compared to (15a) is positive; thus, f_w should have been smaller, and w^x larger. Hence, because of decisions that did not account for the externality, the polluting input b was overused while the waste-reducing input w was underused in comparison to the socially optimal usage.

In a world without externalities, the goods price ratio, P_x/P_y, equals the social marginal rate of transformation and the ratio of consumers' marginal utilities. Indeed, this is the definition of a Pareto optimum

$$\frac{P_x}{P_y} = \frac{P_b/f_b}{P_d/h_d} = \frac{P_w/f_w}{P_w/h_w} = \frac{U_x}{U_y} \tag{24}$$

Now in the presence of production externalitites, the left-hand and right-hand terms do not change (they are determined in the goods market), but the terms in the middle have change. For Pareto equilibrium (social optimum) to hold, the condition is

$$\frac{P_x}{P_y} = \frac{(P_b - P_y g_b h_z)/f_b}{P_d/h_d} = \frac{(P_w - P_y g_w h_z)/f_w}{P_w/h_w} = \frac{U_x}{U_y} \tag{25}$$

The additional element in the middle terms denotes the value to society (through firm B) of the externality generated by using the nonoptimal quantities of b and w — that is, the externality leads to a distortion of resource allocation. (If h_z is zero, so that there is no external effect, the equations of independent decision making are kept.) The *social cost* of b and w used by firm A is not P_b and P_w, but rather $P_b - P_y g_b h_z$ and $P_w - P_y g_w h_z$. Therefore, the marginal cost of a unit of x is not P_b/f_b, being higher by a value of $P_y g_b h_z/f_b$, which is the difference between the social and the private marginal costs.

Policies for Optimization When Externalities Are Present

The above example of water quality externality and its economic inefficiency calls for policies that would correct the system toward a Pareto optimum. Indeed, the social inefficiency due to externalites has long interested economists (Pigou, 1932). The literature recently examined this topic from the viewpoint of environmental economics (Dales, 1968; Upton, 1968; Baumol and Oats, 1971; Baumol 1972; Mills, 1974; and Fishelson, 1974).

The suggested policies are examined below. The first five policies are internalization policies imposed upon the polluting firm, and the sixth is a public action policy.

1. Adjust the price of b to equal the social costs through a tax-subsidy scheme.
2. Tax the effluent z^x emitted by firm A.
3. Tax the output of firm A.
4. Impose water quality standards on firm A.
5. Sell pollution rights — that is, fix the price per unit of pollution and let firm A buy any quantity it wants, or (when there are many firms) fix the quantity of rights and let the price per unit be established in the market for rights (obviously, firm B can also participate in this market).
6. Construct a treatment plant that removes z from the effluent of firm A.
7. Induce the merger of firms A and B (not dealt with here).

From the cost-sharing and pricing points of view, the internalization policies are not relevant. The cost-sharing and pricing decisions become relevant when policy 6 is followed although the internalization policies do affect income distribution. When the demand facing firm A for x is infinitely elastic (as is assumed) and the supply of b is also infinitely elastic, then the consequences of any tax-subsidy policy and effluent or production constraints are borne entirely by firm A. The effect of internalizing externalities by any policy will be revealed in the site value of the location; for example, the site value of firm A (upstream) will decline, while that of B (downstream) increases. The decline and increase in land values equal the present value of the difference in profits due to the policy.

The fifth policy is very attractive on theoretical grounds. The direct intervention of the policy maker in the production decisions is minimal, and the equilibrium is left to market forces. In principle it is identical to both an effluent tax and a standard, but it has the advantage of flexibility: rights change hands when there is an economic justification for it. This characteristic is of significant importance in a dynamic world.

One possible response of the polluting firm to any of the internalizing policies is to construct a treatment plant or, if it operates on a small scale, to introduce a waste control device into its production system. This alternative does not involve cost-sharing or pricing decisions. Leaving it to the firm, the assumption of profit maximization is sufficient to assure economic efficiency (under the policy constraint) if no other externalities are generated (e.g., transformation of water waste into air pollution). Thus, cost-sharing and pricing decisions are relevant only when the public constructs a treatment plant.

Cost-Sharing and Pricing of Public Projects

Recent studies of water quality projects discuss cost sharing within the main framework of economic efficiency (e.g., Kneese, 1964; Cleary, 1965; Kneese and Smith, 1966; Kneese and Bower, 1968). Kneese (1964) argues that "the extremely large potential of industrial process adjustment and waste reclamation practices in reducing waste discharges and the important effect of location decisions on system costs emphasizes the importance of cost distribution as an element of system design" (p. 129).

Kneese provides a detailed example of cost sharing from the Ruhr water system (pp. 169-187) (see also Kneese and Bower, 1968, pp. 244-253). The initial costs of the drainage system were divided equally between the mines and the cities, based on the fact that about half of the costs were not attributable to mining activity; this was a marginal cost assessment. Later costs were partly borne by the mine responsible for them. Projects not needed in order to achieve the organization's (*Genossenschaft*) objectives have their costs assessed to the member asking for them. The purpose of the effluent charges is to distribute the costs of water quality in proportion to the quality and quantity of the effluent; thus, the marginal cost principle is applied. In practice the charges are based on effluent measurements, not on damage. Thus, unless the damage-effluent concentration relationship is linear, the system misses the theoretical optimum (Fishelson, 1974) even though total social costs (including administrative costs) may still be minimized. Examples of sewer surcharge formulas for the Winnipeg Sanitary District and the Allegheny County Authority are found in Kneese and Bower (1968, pp. 166-170).

The National Water Commission (1974) does not detail the cost-sharing arrangements for waste water, but does cite some interestimg figures: "HUD may grant up to 50 percent of waste water collection project costs, or, under need criteria, up to 90 percent for communities of less than 10,000 people" (p. 489). "The Environmental Protection Agency (EPA) provides grants for up to 75 percent of such (interceptor sewers and sewage treatment) project cost" (p. 490). Are these figures an outcome of economic efficiency, or are they equity targets? Without having a sound set of principles we do not know. Some such principles are established below by describing various scenarios for the construction of treatment plants.

Polluters Liable for Pollution Damages and Forced to Treat Waste Water

Within a river basin there are n polluting firms. The basin water is used by downstream firms and households directly or indirectly (fishing, recreation). The upstream firms are found liable for the damages. They are forced to treat the waste so that the quality standard is met. The question is how the cost of a treatment plant be shared by the polluters. The cost of the treatment plant is a function of plant size, measured in millions of gallons per day (MGD). Plant size is a function of the demand for its services. The objective is to minimize the cost subject to a water quality standard.

The quantity and quality of the waste water from each firm are decision variables, and the demand function for treating the water for each polluting firm is denoted by $W^i = D(P)$, where P is the treatment price charged (per MGD of treated water). Elliott (1973) and Ethridge (1970) showed that $D_p < 0$: the lower the charge, the more waste water is sent to be treated. ΣW^i is the aggregate demand for treatment services, and MC(W) the long-run marginal cost of treatment services. The optimal scale of the treatment plant and the price per unit of water treated are both determined by the intersection of the supply and demand curves. The price is the same for all polluting firms; each firm is considered to be the marginal firm generating the observed marginal treatment cost.

The above example is the setting for two other cases. In the first, the quantity of water to be treated is assumed to be fixed. In the second, the size of the regional plant is assumed to be fixed (at least in the short run), and no effluent discharge is allowed. In the first case, each firm is allowed to build its own treatment plant or, to form a coalititon with other firms. Obviously, when there are diseconomies of scale to start with, it is optimal for each firm to treat its own waste. Thus, the interesting case concerns economies of scale over some range. When there are many polluters (n), the potential number and forms of coalitions is very large. For the case of n = 4, each firm can treat its own waste (A, B, C, D); two firms can join efforts (AB and CD, or AB and C and D, or CD and A and B, etc.); three firms can join together (ABC and D, or ACD and B, etc.). The firms would form a coalititon that minimized their (private) treatment costs. From the social point of view, if economies of scale are continuous the optimal coalition is of all firms. Thus, the cost sharing of that plant has to be made attractive to all participants, compared to those when forming alternative, partial coalititons. The solution for the cost sharing is not necessarily unique. Also, the cost sharing has nothing to do with the socially optimal plant size, since the total quantity to be treated is assumed to be fixed.

Cost Sharing and Pricing for Water Quality

A technical device for getting one feasible sollution is the linear programming model. The basic assumption is that the cost for each possible coalititon is known. The largest coalition (ABCD) is identical with the regional authority. The example below is again for a region with four polluters. The linear programming problem is

$$\begin{aligned}
\text{Min } Z = {} & X_1 + X_2 + X_3 + X_4 \\
\text{subject to } & X_1 + X_2 + X_3 + X_4 \geq \bar{X} \\
& X_1 \leq X_1^0 \\
& X_2 \leq X_2^0 \\
& X_3 \leq X_3^0 \\
& X_4 \leq X_4^0 \\
& X_1 + X_2 \leq X_{12}^0 \\
& X_1 + X_2 + X_3 \leq X_{13}^0 \\
& X_1 + X_4 \leq X_{14}^0 \\
& X_2 + X_3 \leq X_{23}^0 \\
& X_2 + X_4 \leq X_{24}^0 \\
& X_3 + X_4 \leq X_{34}^0 \\
& X_1 + X_2 + X_3 \leq X_{123}^0 \\
& X_1 + X_2 + X_4 \leq X_{124}^0 \\
& X_1 + X_3 + X_4 \leq X_{134}^0 \\
& X_2 + X_3 + X_4 \leq X_{234}^0
\end{aligned} \qquad (26)$$

where $\bar{X}$ is the project costs when all four polluters participate, X_{ijk}^0 is the cost for a partial coalition, and X_j is the payment by firm j.

This approach is similar to the one suggested by Giglo and Wrightington (1972) under "cost sharing based on bargaining (including the Regional Authority)" (pp. 1140-1143), except that they minimize the payoffs by the regional authority. The regional authority can be added to our model explicitly as a fifth partner, in the sense that if a regional plant is built a government subsidy of a certain amount is avialable (e.g., equal to the value of positive externalities accruing to the nation). X_5 is then added to the objective function and to the first constraint. An addition constraint, $X_5 \leq S^0$, is added where S^0 is the maximum federal subsidy to be paid for a regional project. The subsidy assures the feasibility of the cost-sharing program, but leaves a wide interval over which to bargain.[7]

The above cost-sharing model can be further extended by introducing a policy-maker bias with regard to cost distribution among the participants: a dollar paid by participant i is valued differently from a dollar paid by j. The only technical difference is that the coefficients in the objective function ($X_1 + X_2 + X_3 + X_4 + X_5$), instead of being identically unity, will take values $\gamma_1, \gamma_2, \gamma_3, \gamma_4$ (in example n =4); where the γ_i for those relatively favored is higher than for the nonfavored (for them let $\gamma_j = 1$). If a dollar paid by polluter 2 is valued twice as much as that paid by 1, 3, 4, and the government, the problem is

$$\text{Min } Z^1 = X_1 + X_2 + X_3 + X_4 + X_5$$
$$\text{subject to } \quad X_1 + X_2 + X_3 + X_4 + X_5 \geq \bar{X} \tag{27}$$

etc. as in (26)

The objective function now measures the subjective social cost.

It is our contention that this way of supporting polluter 2 is inefficient, and that the government should use other ways of transferring income. For example, firms 1, 3, . . . , n might form a coalition in which each pays less than in the project that is economically efficient. The model, however, can be used for distributing costs when firm sizes (in terms of water waste) are widely dispersed. Then the weights will represent relative wastes (the larger the waste, the smaller the corresponding γ in the objective function). Also in this case, the model is just a device for solving for a feasible cost sharing. Its economic principle is that each firm minimizes its cost, but an overall efficiency criterion is not present in the sharing procedure.

The second extension is the case when the treatment plant capacity is fixed. Each firm pays the marginal treatment cost, and the payment per unit is the same for all polluters. The question that arises once full capacity is reached and no additional quantity can be treated concerns the price newcomers should pay. They should pay the same price charged to previous users (assuming no long-run fixed price contracts). As demand increases, and the price per unit treated exceeds a certain level, one of the following will take place in the medium or long run: (1) Some firms will build their own treatment facilities; (2) the regional authorities will expand the existing plant or build a new one; (3) firms will leave the basin because of high treatment costs; or (4) the input mix and technology

---

[7]When a government subsidy is available, the case of diseconomies of scale becomes relevant again. When joining the social project, the polluters may end up paying less than when each builds his own treatment plant. The government can also define a different subsidy level for different sizes of coalitions if the formation of water treatment conditions is considered to be desirable for its own sake.

Cost Sharing and Pricing for Water Quality

will change. Non of these actions affects efficiency so long as the same price is charged to all users and this price is equal to the marginal cost of operating the treatment plant at full capacity.

Joint Treatment of Pollutants

Thus, far, it has been implicitly assumed either that there is one kind of waste or that the waste mix is constant. Correspondingly, the cost of treating waste or abating the effluent has been defined as a unique function of the quantity of waste. Actually, however, industries discharge more than one pollutant, and the pollutants mix varies over industries. Treatment costs vary for different pollutants, and in may cases are not independent of the mix; the cost of removing BOD, for example, is positively related to the concentration of nitrates in the water. Therefore, when a downstream water user treats the water before using it and removes two or more types of pollutants, each of which is the effluent of a different upstream polluter, a problem of cost allocation arises. Sometimes the treatment costs are nonseparable (by treating one residual, the other is treated at no extra cost); the treatment cost is then a function of the quantity of water treated and the overall level of cleanliness. In this case, any cost sharing or pricing between the polluters is arbitrary (with the upper limit being the polluters' own treatment cost). When the jointness of costs is not perfect, treatment cost is a function of the quantity of water treated (W) and the specific reduction of each pollutant (R^1, R^2):

$$C = C(W, \bar{R}^1, \bar{R}^2) \tag{28}$$

$$\text{Max} \left[P_x f(R^1) + P_y h(R^2) + P_z g(W, R^1 - \bar{R}^1, R^2 - \bar{R}^2) - C(W, \bar{R}^1, \bar{R}^2) \right]$$

$$f_1 > 0, \ f_{11} < 0, \ h_1 > 0, \ h_{11} < 0, \ g_1 > 0, \ g_{11} < 0, \ g_2 < 0,$$

An optimal cost sharing between pollutants is feasible through pricing. First, it is assumed that $C_1 > 0$, $C_2 > 0$, $C_3 > 0$, $C_{11} < 0$, $C_{22} > 0$, $C_{33} > 0$, and $C_{23} > 0$. The transformation between residual effluent (E^i) and its concentration in the treated water (R^i) is monotonic. $g(\cdot)$ denotes the production function of the downstream plant producing the good Z. The social problem is

$$g_3 < 0 \tag{29}$$

where $f(\cdot)$, $h(\cdot)$, and $g(\cdot)$ are the production functions of the firms emitting the waste and the firm that uses the polluted water.

The first-order conditions are

$$P_x f_1 + P_z g_2 = 0 \tag{30a}$$

$$P_y h_1 + P_z g_2 = 0 \tag{30b}$$

$$P_z g_2 - C_2 = 0 \tag{30c}$$

$$P_z g_3 - C_3 = 0 \tag{30d}$$

$$P_z g_1 - C_1 = 0 \tag{30e}$$

The procedure is to solve for W, R^1, $\bar{R}^1$, R^2, and $\bar{R}^2$, and to introduce the values $\overset{*}{W}$, $\overset{*}{R}^1$, $\overset{*}{R}^2$ into C_2 and C_3 to obtain the optimal charge to be levied on each polluter. Charges of $C_2(\overset{*}{W}, \overset{*}{R}, \overset{*}{R}^2)$ and $C_3(\overset{*}{W}, \overset{*}{R}^1, \overset{*}{R}^2)$ leave a surplus for the treatment plant. The polluters can be reimbursed by a lump-sum refund if the questions of fairness are raised.

However, if $C_{22} < 0$ and $C_{33} < 0$, the payments to the treatment plant according to optimal pricing will not cover the treatment costs. The difference is to be met by a lump-sum payment in order to preserve optimal resource allocation. The cases of jointness with a declining or increasing marginal cost of treatment are important when long-run behavior and dynamics are considered. The situation also gives rise to possibilities of "bribing" to gain membership in the watershed industrial community when the C_{22}, C_{33}, and C_{23} are negative and to stay out when C_{22}, C_{33}, and C_{23} are positive.

Peak-Load Pricing for Water Treatment

The quantities of effluent to be treated may vary over the year. One reason may be a seasonal effect on the production process of the polluting plant; another is the different assimilative capacity of "nature." The search is for the optimal plant, given that a treatment plant is needed in order for water quality to stay above a fixed level (specified exogeneously). Treatment pricing and the distribution of costs among the users of the treatment plant are discussed below.

The peak-load pricing problem is typical of regulated public utilities. Traditionally, only the telephone, electric, and gas industries have been considered. These utilities and treatment plants also have decreasing average costs over a wide range of output.

Two cases are discussed below. The first assumes that the only alternative to waste treatment is dilution. The river flow determines the maximum quantity that can be diluted without the water quality falling below a predetermined standard. Waste cannot be stored, and the production technology requires a year-round operation, so the demand for treatment services is inelastic in each time period.

The social problem is to design a treatment plant with the lowest possible cost. Dilution costs are assumed to be zero. The daily quantities to be treated are denoted by $T_i, \ldots, T_n$ (MGD), each over $\alpha_1, \ldots, \alpha_n$ fraction of the year. The treatment costs are the sum of capital cost (r·k) and operating cost (OP_t). The treatment function is $T(k, OP_t)$, where $T_k > 0$, $T_{kk} < 0$, $T_{OP_t} > 0$, and $T_{OP_t OP_t} < 0$. The $T(k, OP)$ function can be of increasing returns to scale, but in the context in which the problem is presented, this does not make any difference. The formulation of the problem (k is investment cost and r capital recovery factor) is

$$\text{Min } Z = rk + 364(\alpha_1 OP_1 + \alpha_2 OP_2 + \ldots + \alpha_n OP_n)$$

subject to

$$T_1 \leqslant T(k, OP_1)$$
$$T_2 \leqslant T(k, OP_2)$$
$$\vdots$$
$$T_t \leqslant T(k, OP_t) \tag{31}$$
$$\vdots$$
$$T_n \leqslant T(k, OP_n)$$

The optimal plant size k^* and the minimum operation costs OP_*, $t = 1, \ldots, n$, are determined from the cost-minimization first-order conditions. Any cost sharing is efficient due to the assumption of inelastic demand for treatment ($T_t = T_t^0$ for all t). For accounting purposes, the marginal cost for each increment can be calculated as follows. First the optimal plant size and operating cost for the smallest amount (T_1) to be treated year-round is calculated.

$$Z_1^* = rk_1^* + 364(OP_1^*) \tag{32}$$

Then, for a plant that has to treat T_1 MGD for an α fraction of the time and T_2 MGD for the rest, the lowest costs are

$$Z_2^* = rk_2^* + 364[\alpha OP_1^* + (1-\alpha) OP_2^*] \tag{33}$$

The ratio $(Z_2^* - Z_1^*)/(T_2 - T_1)$ is the marginal cost per marginal unit treated in the second period. This procedure is to be continued until T_n is treated per day $(T_n > T_{n-1})$, in α_n days per year. The total cost is

$$Z_n^* = rk_n^* + 364(\alpha_1 OP_1^* + \alpha_2 OP_2^* + \ldots + \alpha_n OP_n^*) \tag{34}$$

The marginal cost for the last increase in output is

$$MC_n = \frac{Z_n^* - Z_{n-1}^*}{T_n - T_{n-1}} \tag{35}$$

The price schedule by level of demand is

Level 1 $\quad \dfrac{Z_1^*}{T_1}$

Level 2 $\quad \dfrac{Z_1^*}{T_1} + \dfrac{Z_2^* - Z_1^*}{T_2 - T_1}$ (36)

$\vdots$

Level n $\quad \dfrac{Z_1^*}{T_1} + \ldots + \dfrac{Z_{n-1}^* - Z_{n-2}^*}{T_{n-1} - T_{n-2}} + \dfrac{Z_n^* - Z_{n-1}^*}{T_n - T_{n-1}}$

This price schedule is efficient because the treatment costs are minimized for each demand pattern while the polluters' behavior is not affected by the prices.

The pricing problem becomes relevant once the assumption of inelastic demand for the treatment plant services is relaxed. This is the second case. A plausible assumption is that the demand for treatment is a downward-sloping function of treatment charges. The height and slope of the demand function are determined by river flow and other alternatives for waste treatment that are available (e.g., settling ponds).

Figure 6 illustrates a regional annual demand pattern. Each demand equation is effective over an α_t fraction of the year. D_1 is the daily demand at the high-flow period, D_t at an intermediate-flow period, and D_n at the low-flow period. The regional authority must determine the optimal plant size and the pricing policy for the treated water. The treatment plant is similar to a power plant in the sense that it cannot inventory output. Consumers can inventory

Cost Sharing and Pricing for Water Quality

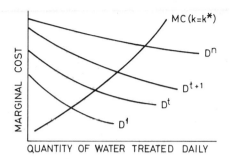

Fig. 6 Monthly demand for water treatment.

waste through ponds and storage tanks, but only to a limited extent and with relatively high costs. (This possibility is already embodied in the demand for treatment services.)

The pricing rule for optimal resource allocation is identical to that for the conventional peak-load problem: It is the traditional marginal cost pricing. One of the characteristics of this solution is that the revenue is not required to equal the cost (or some fraction of the cost), implicity assuming (as we have all along) that the first priority is economic efficiency and that the government has other means for income redistribution. However, when income redistribution is not feasible and the self-financing of the project is an effective constraint, the marginal cost pricing is replaced by pricing that is inversely proportional to the elasticity of demand (Mohring, 1970; Baumol and Bradford, 1970; and Vickrey, 1973).

The optimal marginal price is the solution to the problem of maximizing social rent, defined as the difference between the willingness to pay (the area under the demand curves) and treatment cost. $D_t(T)$ denotes the maximum price the polluter is willing to pay per unit treated in period t. Hence the problem is

$$\max_{k, OP_1, \ldots, OP_n} z = 364 \left[\alpha_1 \int_0^{T_1} D_1(T) \, dT + \alpha_2 \int_0^{T_2} D_2(T) \, dt \right.$$

$$+ \ldots + \alpha_n \int_0^{T_n} D_n(T) \, dT \qquad (37)$$

$$\left. - \left(rk + 364 \sum_{t=1}^{n} OP_t \cdot \alpha_t \right) \right]$$

Subject to

$$T_1 \leq T(k, OP_1)$$
$$T_2 \leq T(k, OP_2)$$
$$\vdots$$
$$T_n \leq T(k, OP_n)$$

The solution is not detailed here. It consists of values of k^*, $OP_1^*, \ldots, OP_n^*$, from which $T_1^*, \ldots, T_n^*$, are determined. The latter are introduced into the corresponding demand for treatment functions, $D_1, \ldots, D_n$, to evaluate $P_1, \ldots, P_n$, the optimal prices per unit of treatment in each time period.

As noted, the solution $T_1^*, \ldots, T_n^*$ and $P_1^*, \ldots, P_n^*$ does not assure that

$$\gamma \left(rk^* + 364 \sum_{t=1}^{n} \alpha_t OP_t^* \right) = 364 \sum_{t=1}^{n} T_t^* P_t^* \tag{38}$$

where γ is a policy variable and $\gamma = 1$ is a specific case at which annual revenue equals annual cost. Equation (38) is referred to below as the cost-revenue constraint. In the following section, this constraint is introduced.

Pricing of Water Treatment Under a Cost-Revenue Constraint

In this section, we don otn repeat the Baumol-Bradford inverse elasticity results (1970, pp. 268-272), but take another approach to the same problem. The basic problem formulation was presented in the previous section. [Equation (38) is added to the n constraints of problem (37) after removing the asterik from OP_t^*, T_t^*, and P_t^*.] It is obvious now that the price schedule is an integral part of the problem, and prices must be solved simultaneously with k and OP_t (t = 1, ..., n).

The financial constraint is

$$\gamma \left(rk + 364 \sum_{t=1}^{n} \alpha_t OP \right) - 364 \sum_{t=1}^{n} \alpha_t P_t D_t(P_t) = 0 \tag{39}$$

Here, the prices are explicitly introduced $[D_t(P_t)]$. The decision variables k, $OP_1, \ldots, OP_n$ are solved from the first-order conditions of (37) and (39). Accordingly, $T_1^*, \ldots, T_n^*$ are calculated and introduced into the demand equations to find $P_1^*, \ldots, P_n^*$. The value of γ appears explicitly in the solution for $T_1, \ldots, T_n$ and $P_1, \ldots, P_n$. γ might reflect the a priori determined share of public investment in the treatment project $\gamma < 1$) or a tax imposed upon the treatment plant due to odors emitted ($\gamma > 1$).

Cost Sharing and Pricing for Water Quality

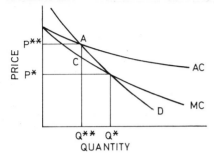

Fig. 7 Optimal pricing with and without a cost-revenue constraint.

The power of the cost-revenue constraint is illustrated in an example of one commodity in one period, where the marginal cost is negatively sloped (Fig. 7). The demand for the product is

$$P = a + bQ \qquad a > 0, \; b < 0 \tag{40}$$

The marginal cost of production is

$$MC = c + dQ \quad \text{such that} \quad c > 0, \; a < c, \; |b| > |d|, \; \text{and} \tag{41}$$
$$|bc| > |ad|$$

The average cost of production is correspondingly

$$AC = c + dQ/2 \tag{42}$$

Under the marginal cost pricing rule, equating the price to marginal cost gives

$$Q^* = \frac{a - c}{d - b}$$
$$P^* = \frac{ad - bc}{d - b} \tag{43}$$

When imposing the cost-revenue constraint, the optimal quantity produced is determined by the intersection of average cost and demand. Figure 7 shows the differences in solutions. The optimal quantity and price are now

$$Q^{**} = \frac{a - c}{d/2 - b}$$
$$P^{**} = \frac{ad/2 - bc}{d/2 - b} \tag{44}$$

The relative difference between the marginal cost and the price $(P^{**} - MC^{**})/P^{**}$ is equal to

$$\frac{-d(a - c)}{ad - 2bc} \tag{45}$$

The elasticity of demand at Q^{**}, P^{**} is

$$E = \frac{ad - 2bc}{2b(a - c)} \quad \text{(note that } E > 1\text{)} \tag{46}$$

Hence, following the result obtained by Baumol and Bradford (1970, Eq. 4), we find that for $\gamma = 1$ (cost equals revenue),

$$\frac{1 + \lambda}{\lambda} = \frac{d}{2b} \tag{47}$$

where λ is the shadow price of the cost equals revenue constraint. Thus,

$$\lambda = \frac{d}{2b - d} \tag{48}$$

The constraint is less costly (γ is smaller) the larger (in absolute terms) is b and the smaller (in absolute terms, and $|d| < |b|$) is d.

The social welfare loss due to the constraint is the triangle ABC (Fig. 7). In order to save this loss while maintaining the cost-revenue constraint, a two-part payment system can be used. The fixed part is for the right to purchase. The variable part is proportional to the marginal price. When consumers are identical, the fixed charge per customer equals the difference (positive for a downward-sloping MC) between total cost and total revenue divided by the number of customers. When consumers are not identical (small and large firms in our case), the fixed charged can be calculated identically — but then it becomes a regressive tax. Hence, a pricing rule that balances efficiency and equity is needed.

The Incidences of Cost Sharing and Pricing Policies

We argued above that any internalization policy would, under the specific conditions, be paid by the firm on which it is imposed. The same obvious rule applies to cost sharing and pricing policy. However, in this section we show that the incidence of costs of improving water quality will finally fall on the immobile factors of production. Whereas the fixed capital will get lower returns in short

Cost Sharing and Pricing for Water Quality

run, in the long run, the entire incidence is on land, and landowners will benefit from improved water quality. Hence, in the long run, internalization policies and cost-sharing and pricing policies result in income redistribution among landowners. The secondary effects of these policies is a different input mix in production and in the long run a different geographic distribution of industries. This does not necessarily imply a conflict between economic efficiency and income distribution.

Some Remarks on Income Distribution

Income redistribution may be justified on various ethical grounds. On economic grounds, a project is regarded as increasing social welfare only when those whose welfare has been improved by the project have gained more than enough to compensate the losses suffered by others. Water projects have been used as instruments for attaining social goals in the United States and elsewhere (Israel is a perfect example). They were intended to provide intangible assets such as employment, settlement opportunities, and conditions for development of underdeveloped and depressed regions. In the context of this paper, public funds must be used to finance the project because a collective good which is nonmarketable is produced. Any attempt that disregards economic efficiency in order to shift the burden of this investment to those who generated the social problem or benefit from the project will lower the project's total welfare, regardless of its income distribution effects.

Krutilla and Eckstein's (1958) treatment of the income distribution aspect is an attempt to identify the differences between privately and publicly financed hydroelectric system. For the Willamette Basin, they estimate the portion of opportunity cost not financed by the government, and the amount of shifted tax liabilities in connection with federal development. Then they identify the incidence of the income transfers by income class and by regions (Chap. VII). They find that the distribution of incidences of increased federal tax liabilities differ from the distribution of increased benefits. Hence, the cost-sharing and pricing do generate an intra-regional and inter-regional income redistribution.

In the following section, a similar conclusion is reached. When demand for the product is elastic the incidence of the cost of the treatment plant is on the landlords, while the benficiary is the society as a whole. When the demand for the region's output is downward-sloping for each single producer, it is still horizontal. The final equilibrium, regardless of whether a cost-sharing or water quality price is imposed, will be on the demand curve at a higher price and lower quantity. Thus, the returns to the immobile factor will decline by less than for the infinitely elastic demand. The consumers of the product will "pay" part of the costs of improving water quality.

The Incidence of Costs

The underlying propositon is that any production cost increase – whether explicit in the form of input prices or taxes, or implicit in the form of constraints or cost-sharing arrangements – is eventually paid by the immobile factor, land.

The production function is assumed to be homogeneous of the first degree in all inputs, including land. Because of the limited quantity of land at any specific location, the marginal cost of production is upward sloping. The demand for the product the region is facing is infinitely elastic. Given that the industry is in long-run equilibrium, the cost of the immobile factor can be built into the long-run average cost LRAC (Fig. 8). The area bounded by the demand curve (D) and the long-run magrinal cost curve (LRMC) denoted by G is the annual return to the immobile factor (its annual rent). This value is built into the LRAC. Clearly, the price of the immobile factor is G/r, where r is the interest rate.

Additional costs are now imposed upon the producing firms. The first is a cost-sharing one which is independent of the level of output produced; the other is a cost which stems from the pricing of the services of the water treatment plant, so it is positively related to output. The cost sharing can be regarded as a fixed annual cost when perfect capital markets are assumed. An additional fixed cost will not affect LRMC or D. Hence, the long-run equilibrium will be set at A. LRAC now contains the return to the immobile factor and the fixed cost. Thus, by definition, the sum of the two equals the return to the immobile factor before the fixed cost was imposed. An interesting case is when the additional fixed costs are larger than G. The implied annual rent is negative; unless the demand (D) shifts upward, production in the area will cease.

Water-improving costs related to output level cause an upward shift of LMRC. It is obvious that G, the annual rent, declines. The new equilibrium will

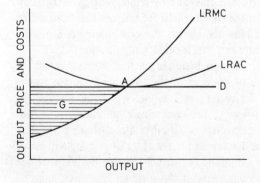

Fig. 8 Firm equilibrium and the returns to the immobile factor.

be on D, but at a point to the left of A. The new LRAC (LRAC′) embodies a smaller quantity. The intensity of usage of the immobile factor declines as before.

The only case where consumers of the product pay the whole improvement cost is when demand is inelastic. Landlords enjoy the same returns as before the treatment plant was constructed. Output level does not change, nor does factor intensity.

Conclusion

The distribution of the incidences of the cost sharing and service pricing depends on the relative elasticities of the supply curve of the product and the demand for the product. The frequently expressed concern that labor is the factor whose returns will diminish has no foundation unless labor is the immobile factor. We regard land as being the immobile factor. Thus, landlords are the ones among which the costs are distributed.

Conclusions

Cost-sharing arrangements are formulated in order to pay for the investment costs of a water project or a waste treatment plant. The pricing of the project and plant services should be aimed at financing the operating costs. The two costs must be kept separate since any attempt to include the cost sharing in the price will distort the optimal resource allocation (i.e., the optimal usage of plant services). The leading relationship for pricing of the services is that the price equals the marginal cost of producing and providing the services. This price will eventually make the marginal benefits to the users of the services equal the marginal cost of production. These marginal benefits may indirectly stem from the constraints the society imposes on the effluent of water waste or the tax imposed on its emitters.

The question of who ultimately pays for cleaning up the rivers and waterways is less crucial than might be supposed. In the long run, landlords of the land on which pollution is generated will pay the cost, while landlords currently receiving low rents because their land is adjacent to polluted water will reap the benefits. Hence, the role of public action in building treatment plants and forcing polluters to share in the investment and operating costs is merely an income redistribution among landlords. Thus, in general, society should not be concerned about the income distribution consequences of water quality policies. It should, however, be on the alert for economic efficiency in cost sharing and pricing decisions.

References

Arrow, K. J. (1966), Discounting and public investment Criteria, In A. V. Kneese, and S. C. Smith (eds.), *Water Research*, The Johns Hopkins Press, Baltimore.

Baumol, W. J. (1972), On taxation and control of externalities, Amer. Econ. Rev., *62*: 307-322.

Baumol, W. J., and Bradford, K. F. (1970), Optimal departure from marginal cost pricing, Amer. Econ. Rev., *60*(2): 265-283.

Baumol, W. J., and Oates, W. E. (1971), The use of stands and Prices for protection of environment, Swedish J. Econ, *73*(1): 42-54.

Buchanan, J. M. and Stubblebine, W. C. (1962), Externality, Economica *29*(4): 371-384.

Ciriacy-Wantrup, S. V. (1954), Cost allocation in relation to western water policies, J. Farm Econ., *36*(1): 108-129.

Cleary, E. J. (1965), *The Orsanco Story*, The Johns Hopkins Press, Baltimore.

Coase, R. H. (1960), The problem of social costs, J. Law Econ., *3*:1-44.

Cohen, A. S., Fishelson, G., and Gardner, J. L. (1974), *Residential Fuel Policy and the Environment*, Ballinger Publishing Co., Cambridge, Mass.

Dales, J. H. (1968), *Pollution, Property and Prices,* University of Toronto Press, Toronto.

Davis, O. A., and Whinston, A. B. (1965), Some notes on equating private and social cost, Southern Econ. J., *32*(1) 113-126.

Davis, O. A., and Whinston, A. B. (1966), On externalities information, and the government assisted invisible hand, Economica, *33*(3): 308-318.

Eckstein, O. (1958), *Water Resource Development, The Economics of Project Evaluation,* Harvard University Press, Cambridge, Mass.

Elliot, R. D. (1973), Economics study of the effect of municipal sewer surcharges on industrial wastes and water usages, Water Resources Res., *9(5):* 1121-1132.

Ethridge, D. E. (1970), An economic study of the effect of municipal sewer surcharges on industrial waste, Ph.D. Dissertation, Dept. of Economics, North Carolina State University, Raleigh, N. C.

Feldstein, M. S. (1972a), Distributional equity and the optimal structure of public prices, Amer. Econ. Rev., *62*(1): 32-36.

Feldstein, M. S. (1972b), Equity and efficiency in public sector pricing the optimal two part tariff, Quart. J. Econ., *86*(2):176-187.

Fishelson, G. (1974), Taxing emissions, a theoretical note, Air Pollution Control Association *24*(1): 44-47.

Giglio, R. J., and Wrightington, B. (1972), Methods for apportionment of costs among participants in regional systems, Water Resources Res., *8*(5): 1133-1144.

Hamilton, H. R., Goldstone, S. E., Milliman J. W., Peigh, A. L. III, Roberts, E. R., and Zellner, A. (1969), *System Simulation for Regional Analysis, An Application to River Basin Planning*, M.I.T. Press, Cambridge, Mass.

Harberger, A. C. (1971), Three basic postulates for applies welfare economics, An interpretive essay, J. Econ. Lit., 9(3): 785-797.

Hass, J. E. (1970), Optimal taxing for the abatement of water pollution, Water Resources Res., 6(2): 353-365.

Hirschliefer, J., Devhaven, and Milliman, J. W. (1960), *Water Supply Economics Technology and Policy*, University of Chicago Press, Chicago, Ill.

Kneese, A. V. (1964), *The Economics of Regional Water Quality Management*, The Johns Hopkins Press, Baltimore.

Kneese, A. V., and Bower, B. T. (1968), *Managing Water Quality, Economics, Technology Institutions*, The Johns Hopkins Press, Baltimore.

Kneese, A. V., and Smith, S. C. (1966), *Water Research*, The Johns Hopkins Press, Baltimore.

Krutilla, J., and Eckstein, O. (1958), *Multiple Purpose River Development, Studies in Applied Economic Analysis*, The Johns Hopkins Press, Baltimore.

Loughlin, J. C. (1970), Cost sharing for federal water resource programs with emphasis on flood protection, Water Resources Res., 6(2): 366-382.

Loughlin, J. C. (1971), A flood insurance model for sharing the costs of flood protection, Water Resources Res., 7(1): 236-244.

Maass, A. Hefschmidt, M. M., Dorfman, R., Thomas, H. A., Marglin, S. A., and Faer, G. M. (1962), *Design of Water Resource Systems*, Harvard University Press, Cambridge, Mass.

Maletic, J. T. (1974), Current approaches and alternatives to salinity management in the Colorado River Basin. In J. E. Flack and C. W. Howe (eds.), *Salinity in Water Resources*, Merriman Publishing Company, Boulder Colo. pp. 11-29.

Marshall, H. E. (1970), Economic efficiency implications of federal-local cost sharing in water resources development, Water Resources Res., 6(3): 673-682.

Marshall, H. E. (1973), Cost sharing and multi-objectives in water resource development, Water Resources Res., 9(1): 1-10.

Marshall, H. E. (1974), Cost sharing and efficiency in salinity control. In J. E. Flack, and C. W. Howe (eds.), *Salinity in Water Resources*, Merriman Publishing Company, Boulder, Colo. pp. 139-152.

McKean, R. N. (1958), *Efficiency in Government through System Analysis with Emphasis on Water Resource Development*, John Wiley, New York.

Mills, E. S. (1974), User fees and the quality of the environment. In W. L. Smith and J. M. Culbertson (eds.), *Public Finance and Stabilization Policy*, North Holland Publishing Co., Amsterdam, pp. 53-70.

Mishan, E. J. (1967), Criteria for public investment, Some simplifying suggestions. J. Political Econ., 75(2): 136-146.

Mohring, H. (1970), The peak load problem with increasing returns and pricing constraints. Amer. Econ. Rev., 60(4): 693-705.

National Water Commission (1974), *Water Policies for the Future*, Water Information Center Inc., Port Washington, N.Y.

Pigou, A. C. (1932), *Economics of Welfare*, 4th ed., Macmillan, London.

Regan, M. W. (1958), Sharing financial responsibility of river basin development, J. Farm Econ., 40(5):1690-1702.

Regan, M. M. (1961), Economically desirable institutional arrangements and cost sharing requirements, In G. S. Tolley and F. E. Riggs (eds.), *Watershed Planning*, Iowa State University Press, Ames, Iowas, pp. 230-245.

Seagraves, J. A. (1970), More on the social rate of discount, Quart. J. Econ., 84(3): 430-450.

Tolley, G. S., and Hastings, V. S. (1960), Optimal water allocation, the north Platte River, Quart. J. Econ., 74(2):279-295.

Upton, C. (1968), Optimal taxing of water pollution, Water Resources Res., 4(5): 865-876.

Vickrey, W. (1973), Direct testimony before the Postal Rate Commission, U.S. Postal Service, Washington, D.C.

Wellisz, S. (1964), On external diseconomies and the government assisted invisible hand, Economica 31(4): 345-362.

Wesner, G. M. (1974), The importance of salinity in urban water management, In J. E. Flack and C. W. Howe (eds.), *Salinity in Water Resources*, Merriman Publishing Company, Boulder, Colo., pp. 108-119.

Yaron, D. (1974), Economic analysis of optimal use of saline water in irrigation and the evaluation of water quality. In J. E. Flack and C. W. Howe, (eds.), *Salinity in Water Resources*, Merriman Publishing Company, Boulder, Colo. pp. 60-85.

13

Legal and Institutional Approaches to Salinity Management

Ralph W. Johnson

School of Law
University of Washington, Seattle, Washington

The salinity problem in U.S. rivers and elsewhere worsens year by year. Not considered serious prior to World War II, especially in the United States, this problem now demands attention worldwide because the cumulative impact of irrigation and other water uses is raising salinity to damaging levels.

There are no simple legal or institutional solutions to salinity problems. They generally occur in basins that are hydrologically and legally highly developed, where farming and industrial investments are well established, where vested interests are aggressively protected, and where cultural attitudes are firmly fixed. The onset of damaging levels of salinity is ordinarily gradual, the cumulative result of thousands of small individual farming and reservoir operations. Solutions are even more difficult here than for the normal industrial and municipal pollution problems, where sources can be located, identified, tested, and monitored. Agricultural salinity comes from diffused, nonpoint sources, which makes it technologically difficult as well as economically impractical to test, monitor, and establish controls for individual sources.

The legal and institutional approaches to pollution control that work elsewhere are considerably less effective in controlling these nonpoint salinity sources. The special relationship of the land, water, population, crop distribution, and legal systems that exists in each basin means that solutions must often be tailored uniquely to meet each problem. Finding principles of general applicability is especially difficult. However, we believe that some such general principles do exist, and that they can best be examined and illustrated through a case study of the agricultural salinity problems in a particular basin.

An examination of the literature concerning water management problems in different parts of the world reveals a paucity of writing on the legal and institutional aspects of salinity management. One of the basins where this problem has been critical has been the Indus, where 6.5 million acres of actually or potentially culturable land were seriously affected by waterlogging and/or high salinity, and several million additional acres were somewhat less adversely affected.[1] A number of authors have described the Indus problem, its impact on agricultural production, and proposed engineering solutions.[2] (The master plan for solving the problem in Pakistan, for example, called for the installation of 31,500 tube wells, 7,500 miles of major drainage channels, and 25,000 miles of supplemental drainage channels.[3]) However, little writing has occurred on the legal-institutional actions taken to resolve the Indus problem.

A careful analysis of the legal-institutional approach to the salinity problems in the Colorado River Basin might be of unique value because information has been published and the remedial actions are visible. At the end of this

---

[1] White House Department of the Interior Panel on Waterlogging and Salinity in West Pakistan, Report on Land and Water Development in the Indus Plain, p.62 (1964).

[2] Salinity is one of the major problems accompanying the irrigation of lands in West Pakistan. One solution is to establish a "salt balance," i.e., remove the salt left in the soil after leaching by proper drainage and additional leaching so that the salt is not concentrated in the root zone. Karpov, Indus Valley—West Pakistan's life line, Proc. Amer. Soc. Civ. Eng. *90*: 228-229 (1964). To solve the problem and reclaim salinated soil, it is essential that the Indus Valley provide permanent drainage facilities and undertake leaching as a means to reduce the salt content of the soil. *Ibid.*, p. 241.

According to a 1964 White House study, the saline and waterlogged areas were increasing at the rate of 50,000 to 100,000 acres/year (or 0.2 to 0.4% year). This rate was expected to slow after a while. White House Department of the Interior Panel on Waterlogging and Salinity in West Pakistan, Report on Land and Water Development in the Indus Plain 63 (1964). The study notes that waterlogging and salinity are only two of several problems and cannot be solved in isolation.

See also Hasan, WAPDA's master plan for the eradication of salinity and waterlogging, Indus, pp. 4, 506 (February 1965); Fowler, The Indo-Pakistan water dispute, Yearbook of World Affairs 101 (1955); Khosla, Development of the Indus River system, an engineering approach, India Quart. *14*: 233 (1958). For a discussion of Soviet attempts to deal with the salinity problem, see Mubashir, Recent Soviet advances in waterlogging and salinity control, Indus, p. 23 (October 1962).

[3] Baxter, The Indus Basin, in The Law of International Drainage Basins, pp. 443, 474 (1967).

The India-Pakistan Indus Waters Treaty of 1960 has made more water available for all parties and has wiped the state clean of all existing rights. Foreign Policy of India 1947-64, p. 63 (1966). The International Bank for Reconstruction and Development played a major role in mediating disputes between the two countries. The success of the Bank's efforts was based partly on its funding power. In the future, others may find it necessary to entrust disputes to a third party for conciliation (as the Bank in fact did), although it may be more difficult if no financial ties are involved. See Baxter, The Indus Basin, pp. 467-477.

chapter we shall note the applicability of some of the concepts used in the Colorado Basin to other river basins and especially to basins in the developing part of the world.

The Colorado is one of the larger rivers of the world that has been both substantially appropriated and subjected to extensive legal and institutional actions to control salinity. Aside from the salinity controls, the Colorado long has been subjected to an intensive legal control primarily concerned with water quantity allocation. Some understanding of this legal background is essential in order to understand the complexities involved in designing salinity control remedies. A full discussion of the so-called law of the river for the Colorado fills many volumes.[4] We shall give only the barest summary here.

The seven states[5] in the basin all apply the prior appropriation doctrine, which holds that water users who are "first in time are first in right." California also applies the riparian system. Nearly all the streamflows within the basin have been appropriated under these water law systems. Groundwater withdrawals within the basin are not controlled so extensively, different states having adopted one or a combination of some four legal systems applicable to such waters, including:

1. the rule of correlative rights
2. the rule of reasonableness
3. the rule of absolute ownership
4. the rule of prior appropriation

Statutory groundwater codes, although adopted more widely in recent years, tend to be far less comprehensive than surface water laws.

The Colorado River Compact of 1922 (which became effective in 1929) apportioned beneficial uses of 7.5 million acre ft to the Upper Basin and 7.5 million acre ft to the Lower Basin. The Boulder Canyon Project Act of 1928, combined with certain California state legislation, limited California's consumptive use to no more than 4.4 million acre ft/year of the total 7.5 million acre ft allocated to the Lower Basin states plus not more than half the surplus.[6] The

---

[4] Clark, Waters and Water Rights (1967); Trelease, Arizona v. California, Allocation of Water Resource to People, States, and Nations, 1963 Supreme Court Rev. 158; Hutchins, Water Rights Law in the Nineteen Western States (1974); Farnham, Water and Water Rights(1904); Meyers, The Colorado River, Stanford L. Rev. *19*: 1 (1966).

A brief statement of the law of the river for the Colorado can be found in National Academy of Sciences, Water and Choice in the Colorado Basin (1968).

[5] Arizona, California, Utah, Nevada, Wyoming, Colorado, and New Mexico.

[6] A more complete statement of this allocation is as follows:

> According to Articles III(a) and III(b) of the compact, each basin was allocated 7,500,000 acre-feet and the lower basin was given permission to increase

Boulder Canyon Project Act also authorized the Secretary of the Interior to enter into contracts, which he has subsequently done, to deliver 2.8 million acre ft consumptive use to Arizona and 300,000 acre ft to Nevada. The Supreme Court's 1964 decree, implementing its 1963 decision in *Arizona v. California*,[7] interpreted the Boulder Canyon Project Act to mean that it apportioned only the waters of the mainstream, not including diversions from tributaries such as the Gila River in Arizona. Later the Colorado River Basin Project Act of 1968 gave California's 4.4 million acre ft/year a priority over the Central Arizona Project. Thus, the Colorado River Compact, the Boulder Canyon Project Act, the Supreme Court decree in *Arizona v. California*, and the 1968 Colorado River Basin Project Act all combine to apportion the water supplies among the Lower Basin States.

The 1963 decision in *Arizona v. California* also held that in the event of a shortage the Secretary of the Interior had discretionary power to allocate that shortage among the Lower Basin States, and even within the states after satisfaction of certain rights antedating 1929. In 1948 the Upper Basin States agreed to an apportionment of their share of Colorado River water in a compact that gave 51.75% of total beneficial use to Colorado, 11.25% to New Mexico, 23% to Utah, 14% to Wyoming, and 50,000 acre ft/year to Arizona, which has a small part of the Upper Basin within its borders.

In 1944 the United States and Mexico entered a treaty that guaranteed Mexico 1.5 million acre ft to be increased in years of surplus to 1.7 million acre ft and reduced in years of extraordinary drought in proportion to the reduction of consumptive uses in the United States. This treaty has been supplemented by three agreements embodied in the minutes of the International Boundary and Water Commission, created by the 1944 Treaty. These minutes were designed

---

its uses by an additional 1,000,000 acre-feet. But neither of these allocations was guaranteed. They merely indicated how the water would be divided if it were available. There was another provision of the compact, however, which, for all practical purposes, does seem to represent a guarantee. Article III(d) stipulates that the upper basin "will not cause the flow of the river at Lee Ferry to be depleted below an aggregate of 75,000,000 acre-feet for any period of 10 consecutive years." This would essentially mean an annual delivery of 7,500,000 acre-feet. Presently, the upper basin contributes about 13,700,000 acre-feet annually to the mainstream. Consequently, to fulfill its compact obligation, it must limit its consumptive uses to 6,200,000 acre-feet, some 1,300,000 acre-feet less than anticipated. Indeed, it may have to curtail its uses even more in order to satisfy its share of the Mexican burden if return flow should eventually prove unacceptable under the treaty, and if it should be determined that the 75,000,000 acre-feet does not include its contribution to Mexico. [Norris Hundley, *Dividing the Waters*, Univ. of Calif. Press, pp. 181-182 (1966)].

[7] *Arizona v. California*, 373 US 546 (1963).

Approaches to Salinity Management

to provide a solution to the problem of increasing salinity of the water flowing in the Colorado across the border from the United States into Mexico.

Why does the salinity problem arise now? Several reasons combined to bring this problem to the fore in the 1960s. The level of salinity of the river had been increasing gradually ever since the first irrigation projects were built. However, prior to 1961 there were no major problems because the salinity of the water delivered to Mexico was generally within 100 ppm of the water on the U.S. side at Imperial Dam. In 1961 the United States commenced operation of a series of drainage wells as part of the Wellton-Mohawk project,[8] and these wells discharged highly saline water into the Colorado below the last U.S. diversion point but above the Mexican diversion, raising the salinity of the Colorado River water flowing to Mexico from an average of about 800 ppm in 1960 to 1,500 ppm in 1962. At about the same time the total flow of water reaching Mexico was sharply reduced by the United States as water was held back to fill the reservoir behind Glen Canyon Dam.[9] This exacerbated the salinity problem. Mexico objected strenuously to these actions.

After the winter of 1961-1962 the United States took provisional measures to minimize the impact of the high-salinity drainage from the Wellton-Mohawk project. At the same time the United States entered negotiations with Mexico to arrive at a longer-term solution and reconstituted the Committee of Fourteen representing the basin states to advise the Secretary of State in connection with the salinity problem.[10]

In 1965 the two governments reached a five-year agreement set out in Minute 218 of the International Boundary and Water Commission. According to this Minute the United States agreed to three actions: (1) to construct an extension of the Wellton-Mohawk Drain so that these highly saline waters could, at times chosen by Mexico, be carried directly into the Gulf of Mexico rather than into the Colorado above the Mexican diversion points; (2) to engage in selective pumping of wells in the Wellton-Mohawk Project to alleviate salinity at times most critical to Mexico; and (3) to release about 50,000 acre ft/year from behind Imperial Dam to replace the water flushed down the Wellton-Mohawk drain extension into the Gulf of California.

These actions, which cost the United States about $11 million, reduced

---

[8] Gila Reclamation Project Relocation Act, July 30, 1947, ch. 382, 61 Stat. 628 (1947), 43 U.S.C. §§ 613-613e (1970).

[9] This loss of dilution water can be emphasized by two figures: for the 10-year period from 1951 to 1960 the average delivery to Mexico at the Northerly International Boundary was 4.2 million acre ft/year, whereas for the succeeding 10-year period from 1961 to 1970, the flow averaged only 1.5 million acre ft/yr.

[10] Holburt, International problems of the Colorado River, Natural Resources J.*15*: 11, 14 (1975).

the average annual salinity of waters delivered to Mexico from 1,500 ppm in 1962 to 1,245 ppm in late 1971. Mexico also took certain unilateral action during this period which further reduced the average salinity to about 1,160 ppm.

The above actions were not considered a permanent solution to the problem, and negotiations continued throughout 1971 and 1972 toward such a permanent solution. Meanwhile Minute 218 was extended for a year, until November 1971. A new, temporary agreement, embodied in Minute 241, was agreed to in July 1972; it was to run only until December 31, 1971. This was estimated to result in a reduction in salinity of about 100 ppm, from 1,242 in 1971 to 1,140 in 1973. In addition, Mexico requested the United States to bypass the balance of the drainage, about 100,000 acre ft annually, from the Wellton-Mohawk Project so that the water would be carried directly to the Gulf of California. This water counted against the Mexican entitlement of 1.5 million acre ft. However, it further reduced the salinity of Colorado River waters delivered to Mexico to slightly below 1,000 ppm.

During the state of negotiations when the United States and Mexico were trying to work out a permanent agreement, consideration was given to presenting the issue to the International Court of Justice, or to a special arbitral tribunal convened for this purpose. However, both sides preferred a bilateral, negotiated settlement, and that is what occurred.

The main issue dividing the two countries was "the difference in quality between the water available to U.S. users below Imperial Dam (about 850 ppm for most) and the water delivered to Mexico at the northerly international boundary, in compliance with the Water Treaty (then averaging about 1,140 ppm under the operation of Minute 241)[11]" These and certain subsidiary issues were finally set aside in favor of an agreed overall objective, that is, to deliver to Mexico water of a quality suitable for agricultural purposes. More specifically, the agreement reached was to deliver water to Mexico at Morelos Dam of a quality that is 115 ± 30 ppm greater than the average annual salinity of the water arriving at Imperial Dam.

While these negotiatons were going on, the United States was becoming increasingly aware of its own salinity problem and that, without reference to Mexican claims, action would have to be taken to protect water users on the Lower Colorado in the United States from excessive salinity. This concern had initially surfaced several years before, in the 1950s, resulting in the authorization of studies of the salinity problem in the 1956 Colorado River Storage Project Act and subsequent legislation on major projects in the Upper Basin.

---

[11] Brown and Eaton, The Colorado River salinity problem with Mexico, American J. International Law 69: 255, 259 (1975).

Approaches to Salinity Management 311

The active program of the Colorado Basin States began about the time[12] of the publication of the Colorado River Board of California Report, "Need for Controlling Salinity of the Colorado River," in August 1970. In 1971 the EPA released an eight-year study, "The Mineral Quality Problem in the Colorado River Basin." Projections were made of salinity levels at four key locations on the Lower Colorado River for unlimited and limited development conditions.

The EPA projections are summarized below[13]:

	Unlimited development (mg/liter)			Limited development (mg/liter)	
	1960 (base)	1980	2010	1970	1980 & 2010
Hoover Dam	697	876	990	760	800
Parker Dam	684	866	985	760	800
Palo Verde Dam	713	940	1,082	800	850
Imperial Dam	759	1,056	1,223	865	920

The EPA defined limited development as completion of all currently authorized projects, with no new developments. The projections of other agencies differ numerically, but they all show the same trend[14]:

	At Imperial Dam (mg/liter)			
	1980	1990	2000	2030
California Colorado River Board	960	1,080	1,210	—
U.S. Bureau of Reclamation	930	1,115	1,169	—
Water Resources Council	1,260	—	1,290	1,350

---

[12] John T. Maletic, Chief, Water Quality Office, Engineering and Research Center, Bureau of Reclamation says:
> It is difficult to pinpoint the initiation of the "active program." The historic trace of studies done by other agencies goes back many years. The Environmental Protection Agency and its predecessor agencies have worked on the problem for 8 years—covering most of the decade of the 60s. Also, various aspects of the salinity conditions were studied by the U.S. Geological Survey and the Bureau of Reclamation. I believe that meetings between the basin state representatives and Interior Department officials in 1971 just prior to the enforcement conference were instrumental in establishing the active salinity control program within the Bureau entitled "The Colorado River Water Quality Improvement Program." [Letter from J.T. Maletic to Ralph W. Johnson, November 25, 1975.]

[13] U.S. Environmental Protection Agency, The Mineral Quality Problem in the Colorado River Basin, p. 21 (1971). [Hereinafter cited as EPA Report.]

[14] U.S. Department of the Interior, Bureau of Reclamation, Colorado River Water Quality

In a subsequent 1974 study,[15] a special Department of the Interior study team analyzed the impact on water use that is projected to result from energy development in the Upper Colorado Basin till the year 2000. The report says:

> Salt loading due to energy development will be insignificant in comparison to the concentrating effects of the consumptive use projected for these [energy] developments with the possible exception of salt leaching from spent disposal piles. [The relative impact of each development is dependent upon a number of variables; comprehensive studies have not yet been made.] However, as an example, it has been estimated that a shale production of 1 million barrels per day would result in an increased salinity concentration of 9 mg/liter at Imperial Dam if surface water were the sole source and there were no return flows....

This energy study, dated July 1974, will result in changes in the 1972 projections contained in the earlier Bureau study on anticipated salinity levels in the Colorado, although the exact amounts of the changes have not yet been published. Obviously, many variables are involved in any such projections, and one especially important variable is the existence of the energy crisis, which began after 1972 and which has resulted in a national policy designed to make the United States increasingly independent of Mideast oil sources. The continuation of the energy shortage, and the policy of national energy independence could well accelerate the development of energy sources in the Upper Colorado Basin and exacerbate the salinity problem.

In view of the projected increases in salinity, even in 1971, the EPA recognized the need for action.[16] The Report recognized that implementation of a basinwide salinity-control program based on salt-load reductions would face several legal and institutional constraints. Some of these would be imposed by existing state water laws,[17] although there is no evidence that these particular constraints posed any major roadblocks for the salinity control program.

Improvement Program, p. 9 (1972). The figures for Bureau of Reclamation estimates were provided by J. T. Maletic, November 25, 1975.

[15] U.S. Department of the Interior, Report on Water for Energy in the Upper Colorado River Basin, p. 56 (1974).

[16] Earlier, in 1968, a cooperative salinity control program for the Colorado River Basin was proposed by the Director, Pacific Southwest Region, Federal Water Pollution Control Administration, and the Chief Engineer, Bureau of Reclamation. The proposed program, however, was caught in the change of administration and, as a result, was not put into effect until the above-mentioned meetings were held between the states and Interior Department officials. Letter from J.T. Maletic to Ralph W. Johnson, November 25, 1975.

[17] EPA Report, p. 53.

Approaches to Salinity Management

Improvement of water transportation and irrigation efficiency with consequent reductions in water use and return flows raises other possible conflicts with state water rights laws.

The EPA report noted the lack of any overall planning and implementation entity for the basin:

> An important institutional factor for consideration is the lack of a single entity with basinwide jurisdiction to direct and implement a salinity control program. In addition, water quality and water quantity considerations are generally under the jurisdiction of different agencies at both the state and federal levels. This split in jurisdiction poses coordination problems to all interests affected by a salinity control program.[18]

Other authors have also regretted the lack of existence of a comprehensive management entity for the basin. "Most forms of salinity require the creation of a regional agency to administer them."[19] "A regional agency may provide water quality management for the basin, or it may provide river basin management in all respects including water quality."[20]

The EPA study identified two preexisting objectives that it concluded should not be altered by the salinity control program: (1) full development of the water supply allocated to each state by applicable water laws and compacts, and (2) expansion of the regional economy.[21] Beyond this the study recommended the prompt setting of numerical salinity limits in the water quality standards for the Colorado, the creation of a single institutional entity with basinwide jurisdiction responsible for both planning and implementing a control program, legislative authorization and funding of salinity control projects, a systems analysis using "a refined water quality simulation model and updated economic evaluation"[22] to evaluate the quality and economic aspects of the problem and alternative solutions, and additional research on salinity control technology.

One basic, underlying attitude held uniformly by the Colorado Basin states was that salinity control for the benefit of Mexico should be considered a national (not a regional) responsibility and financial obligation[23]; indeed, this philosophy has been implemented in congressional action, discussed below.

---

[18] EPA Report, p. 55.
[19] Gindler and Holburt, Water salinity problems: Approaches to legal and engineering solution, Natural Resources J. 9: 329, 400 (1969).
[20] Gindler and Holburt, ibid., p. 343.
[21] EPA Report, p. 57.
[22] EPA Report, p. 57.
[23] For an analysis of the Regional Coalitions that powered this attitude, see Mann, Politics

The EPA study provided the basis for a major Conference, begun in Las Vegas in February 1972, and concluded in Denver in April 1972. The topic of the conference was "In the Matter of Pollution of the Interstate Waters of the Colorado River and Its Tributaries," and the meetings provided important background information for consideration by the negotiators of Minute 242, with Mexico, setting forth the final settlement on the international salinity problem.

During the period of negotiations toward Minute 242, the United States was under considerable pressure to provide a permanent supply of water to Mexico with acceptable salinity values. At the same time it was under pressure from a strong coalition of the seven basin states to adopt a solution that (1) imposed no costs on them for salinity control remedies, and (2) included federally financed solutions to their own intra-U.S. salinity problems. Although the final agreement and 1974 implementing legislation is very favorable to the basin states, they nonetheless objected to the fact that they were not more directly included in the decision-making process with Mexico[24] and that some of their arguments were "ignored."[25]

In 1974 Congress enacted P.L. 93-320, the Colorado River Basin Salinity Control Act, authorizing the Secretary of the Interior to build and modify various projects in the basin for the protection of both U.S. and Mexican water quality. Title I of the Act establishes a program for improving the quality of the water reaching Mexico. The costs of this program are totally nonreimbursable and are to be borne by the United States as a national obligation. Title II of the Act provides salinity control for the benefit of the Colorado River Basin States. The costs for this part of the program are split, 75% to be paid by the federal government, and to be nonreimbursable in view of the federal responsibility for an interstate stream and international county with Mexico and because of federal ownership of federal lands which contributed heavily to the salinity problem; the other 25% to be paid from the sale of hydroelectric power (generated by the federal power system in the basin) to users within and outside the basin.

The final international agreement and the congressional legislation implementing that agreement can be summarized as follows[26]: construction of a 120 million gallon per day desalting complex for treatment of the heavily saline

---

in the United States and the salinity problem of the Colorado River, Natural Resources J. *15*: 113, 127 (1975).

[24] Dregne, Salinity aspects of the Colorado River, Natural Resources J. *15*: 43, 51 (1975).

[25] Goslin, Outline History of Colorado River Development, p. 50. (Paper presented at the annual meeting of the American Association for the Advancement of Science, February 28, 1974).

[26] Mann, Politics in the United States, pp. 123-125.

draw-water from the Wellton-Mohawk Project with all costs nonreimbursable; acceleration of a program to improve irrigation efficiency on the Wellton-Mohawk Project, with the district bearing part of the cost; acquisition by the Secretary of the Interior of an initial 10,000 acres of the 75,000 acres in the Project for the purpose of reducing return flows, to be paid for by a reduction in repayment obligation for the Project and an offset for any increased operating costs; acquisition of additional lands above Painted Rock Dam for temporary flood storage; construction of a new canal or lining of the Coachella Canal for a length of 49 miles for the purpose of conserving water presently lost to seepage with the cost of this construction to be repaid by the Coachella Valley Country Water District in 40 years except that the payment period would not begin for several years; construction and operation by the Secretary of the Interior of a well field near the Mexican border, the water to be delivered to Mexico under the Treaty obligation with the cost of the well field nonreimbursable; authorization to construct four salinity control projects at (1) Paradox Valley, Colorado, (2) The Grand Valley Basin, Colorado, (3) Crystal Geyser, Utah, and (4) Las Vegas Wash, Nevada, and various other salinity control projects; authorization of investigation of 12 other sources of salinity; and creation of a Colorado River Basin Salinity Control Advisory Council composed of representatives of each basin state to advise on the salinity control program.

It is appropriate to note here that one additional solution considered, and rejected, at least for the present, was the augmentation of the Colorado by importing water from either northern California rivers, or the Columbia River in the Pacific Northwest. These alternatives would bring into the Colorado from 1.7 to 5.9 million acre ft annually with resulting reductions in annual salinity concentrations at Hoover Dam ranging from 100 to 300 mg/liter. However, the flow augmentation proposals were estimated to be more costly per unit of salinity reduction than other solutions. These solutions would also take longer to implement and would raise many issues not related to the salinity problem, not the least of which is the continuing hostility of both areas of origin, California and the Pacific Northwest states, to any such proposals. And the 10-year statutory moratorium on the study of any major interbasin transfer from the Columbia River to the Southwest is still in effect.

One of the concerns of EPA has been to assure that numerical quality standards on allowable concentrations of salinity are adopted in the Basin (although the overall objective of maintaining salinity below 1972 levels was adopted at the 1972 enforcement conference). Thus on December 18, 1974, acting under authority of the Federal Water Pollution Control Act Amendments of 1972, the EPA issued regulations[27] requiring the states in the basin to adopt

---

[27] 40 CFR 120; 39 Fed. Reg. 43721 (1974).

water quality standards for salinity, consisting of minimum salinity criteria and a plan of implementation for salinity control. The standards were to be submitted to EPA for approval on or before October 18, 1975, and are to treat the salinity problem as a basinwide problem. The goal continues to be to maintain Lower Colorado River "salinity at or below 1972 levels while the basin states continue to develop their compact apportioned waters."[28]

The seven basin states are utilizing the Colorado River Basin Salinity Control Forum as a voluntary interstate planning entity for drawing up standards and plans of implementation. The Forum has indicated[29] that it will adopt numeric criteria at key locations on the Colorado as follows:

Below Hoover Dam	723 mg/liter
Below Parker Dam	747 mg/liter
Imperial Dam	879 mg/liter

The plan of implementation being considered by the basin states is essentially to accept what the federal government proposes to do in the Colorado River Salinity Control Act of 1974[30]; that is, the states plan to take only modest action on their own. Thus the Forum's plan will (1) urge prompt construction and operation of the initial four units authorized by the federal act (federal action), (2) urge construction of the 12 other units listed in that Act after receipt of favorable planning reports (federal action), (3) recommend the placing of stringent effluent limitations on new industry (the states have agreed on a "no salt return concept" for new power and industrial installations) (state action,)[31] and (4) recommend the reformulation of previously authorized, but unconstructed, water projects to reduce salt loading affect (state action).

The plan recommends further state action in urging the use of saline water for industrial purposes whenever practical, programs by water users to cope with the river's high salinity, improvements in irrigation systems and management to reduce salt pickup, studies of means to minimize salinity in municipal discharges, and studies of future possible salinity control programs.[32]

---

[28] 39 Fed. Reg. 43723 (1974).
[29] Proposed Water Quality Standards for Salinity Including Numeric Criteria and Plan of Implementation for Salinity Control: Colorado River System. Prepared by Colorado River Basin Salinity Control Forum, June 1975, p. i.
[30] Letter from Joseph C. Lord, Interstate Streams Engineer, State of Wyoming, to Ralph W. Johnson, April 16, 1975; letter from Felix L. Sparks, Director, Colorado (State) Water Conservation Board, to Ralph W. Johnson, April 15, 1975. It is estimated that if all the programs recommended by the Forum are implemented, about 80% of the salinity reductions would be due the federal program and 20% to nonfederal actions.
[31] Letter from Arthur E. Williamson, Administrator, Water Quality Division, Department of Environmental Quality, State of Wyoming, to Ralph W. Johnson, April 29, 1975.
[32] Letter from Arthur E. Williamson, ibid.

Approaches to Salinity Management

The plan does not recommend mandatory controls on irrigation or land use.[33] According to one state official,[34] any "zoning, irrigation efficiency or other such controls...would be utterly impossible of enforcement and otherwise barren of production results." Neither the plan, nor EPA, recommends placing emphasis on the permit system under the Federal Water Pollution Control Act Amendments of 1972 for controlling agricultural salinity sources,[35] although it does urge this system of control for industrial salinity. Instead it "proposes increased educational activity in an effort to promote irrigation methods which would decrease salt loading".[36] Holburt describes it this way:

> The major emphasis for salinity control from irrigation sources is on improving irrigation management and on improving control of water flow in canals, laterals, and drainage systems. The ways that these will be accomplished on existing irrigation projects are through the Bureau of Reclamation's Irrigation Management Services (IMS) and Water Systems Improvement (WSI) Programs.
>
> The IMS Program's objective is to increase on-farm irrigation water efficiency. Benefits projected to be derived from the program include increased crop yields, water savings, reduced leaching of salts, and reduced drainage requirements. It is anticipated that these benefits to the farmers will exceed the costs to the farmers. The IMS Program will be accomplished mainly through an education program and will be on a voluntary basis. Programs are currently under way in the above mentioned irrigation projects.
>
> Research is being carried out on the use of sprinkler and drip irrigation systems, with very low leaching fractions to precipitate harmless salts in the soil profile. It is not clear at this time how expensive on-farm improvements would be funded. Some financial assistance is anticipated through Federal programs such as Rural Economic Assistance Programs.
>
> The WSI Program involves changes or additions to structures in water conveyance systems in order to reduce seepage into the

---

[33] Letter from Vernon E. Valentine, Assistant Chief Engineer, Colorado River Board of California, to Ralph W. Johnson, May 7, 1975; letter from Joseph C. Lord, Note 30.

[34] Letter from Felix L. Sparks, Note 30.

[35] "At present, a practicable and effective technology for controlling salinity in irrigation return flows cannot be defined. Consequently, the initial NPDES permit program will be one which is designed to provide data upon which informed judgements may be made as to the type of control measures which should be taken."

Proposed Water Quality Standards for Salinity, Colorado River System, Prepared by Colorado River Basin Control Forum, June 1975.

[36] Letter from Joseph C. Lord, Note 30.

ground and subsequent drainage and salinity pickup problems. The structural changes studied under this program include lining of canals and laterals, installation of field drainage systems. The measures should result in a reduction in water losses from the irrigation system and into deep percolation, thereby reducing water contact with high saline soils, shales, and ground water aquifers.

The results of studies recently conducted by Colorado State University for the Environmental Protection Agency in Grand Valley, Colorado, indicate that a 50% reduction in return flow will result in 30% to 70% reduction in salt load in that portion of Grand Valley.[37]

The plan, which covers the period 1974 through 1990, does not include augmentation of the Colorado River from any other river basin, although it does include some additional use of California State Project water in the South Coastal Plain of Los Angeles in lieu of a portion of the Los Angeles Metropolitan Water District's allocation of Colorado River water, which would be used for power plant cooling in the Colorado desert.[38]

National Aims and Salinity Control

In considering these various solutions to the salinity problems of the Colorado, we must keep in mind the national aims or goals that have traditionally provided guidance in planning water projects, especially in the Southwest. As described by the National Academy of Sciences, [39] these aims can generally be classified in five groups.

1. National economic efficiency, that is, bringing a net increase in national wealth
2. Income redistribution, for example, limiting the availability of water from federal projects to small (320 acres or less) family farms; subsidizing the economic growth of the Southwest to assure it continues at about the same rate as in the past; rescuing areas threatened with floods such as Imperial Valley, or diminishing water supplies such as Phoenix
3. Political equity, for example, expressing international goodwill through the Mexican Treaty of 1944; meeting Indian water rights claims

---

[37] Letter from Myron B. Holburt, Chief Engineer, Colorado River Board of California, to Ralph W. Johnson, May 30, 1975.
[38] Letter from Vernon E. Valentine, Note 33.
[39] *Water and Choice in the Colorado Basin,* Nat. Acad. Sci., Washington, D. C., 1968.

Approaches to Salinity Management

4. Controlling the natural environment, that is, the view that places a premium on technical proficiency in regulating volume and quality and reflects the notion that an uncontrolled resource is a wasted resource and that if man has the capacity to control and completely utilize the waters of a river he should do so
5. Environmental protection, preservation, and esthetics, for example, the view that a free-flowing river through natural canyons may be of greater value than the power and irrigation benefits resulting from reservoirs partially filling those canyons

In the context of these goals and aims we can see that the ones that have dominated the Colorado salinity control program are as follows:

It is clear that national economic efficiency has not been the dominant, or even a highly important goal in the design of the Colorado Basin salinity management program. This goal has been subordinated to the twin goals of political equity and income redistribution as described below.

The income redistribution aspect of the recent salinity control measures is apparent in the "rescue" approach. The Colorado River water users have got themselves into an awkward position. Their use of the waters of the river has produced a gradual but persistent increase in salinity (on top of a high natural salinity level) projected to get worse in the future. They could, of course, be required to take care of the problem themselves, but instead have persuaded Congress to subsidize a substantial part of the rescue operation through the use of federal funds from general revenues (in the same way that industrial and municipal polluters have persuaded Congress to allocate massive subsidies to help clean up the nation's waters under the 1972 Federal Water Pollution Control Act Amendments). The Colorado Basin states have also convinced Congress that they are entitled to federal support to continue the past rate of "expansion of the regional economy."[40]

The aim of political equity is served through the agreement with Mexico to provide agriculturally usable quality water at the border, and the decision to spend federal dollars to implement this through construction of the Wellton-Mohawk drain extension and the Wellton-Mohawk desalination plant. Similarly, the basin states have been persuasive in arguing that the allocation of water among the basin states under compacts, Supreme Court decisions, and federal laws should be recognized and affirmed by the federal government and that the national as well as regional objective should be "full development of the water supply allocated to each state by the applicable water laws and compacts."[41]

---

[40] EPA Report, p. 57.
[41] EPA Report. The past, and continuing, objective for the development of the basin's water resources should be "full development of the water supply allocated to each state by

The basin states have also consistently supported international comity arguments to the effect that the obligation to provide Mexico with usable quality water is a "national" obligation and should not be the burden solely of the basin states.

Analysis

The problem of excessive salinity ordinarily does not arise until a river is already intensively appropriated for irrigation, power, industrial, and municipal uses. By that time facilities are in place, rights are established, and investments made. The problem ordinarily arises from the cumulative impact of many, many users. This makes solutions exceptionally difficult. The optimal solution will generally be found by considering the problem on a basinwide basis. In fact, the solution should often reach beyond the basin, for example, where water is exported from or imported into the basin.

In most cases the optimal solution is one that would result in the least total cost of salinity reduction. However, the decision on an optimal solution is especially difficult when there exists a multitude of public and private entities within the basin, each with economic, legal, political, and social investments in existing projects, facilities, and waters, and where any particular plan of implementation will fall unevenly.

At the very least the optimal solution requires extensive voluntary cooperation among all water users. Realistically, however, such voluntary cooperation can hardly be expected to produce and implement an optimal plan.

In theory, what is needed is some basinwide, regional, or larger entity with sufficient jurisdiction and power to (1) gather the needed data, (2) design the optimal plan, and (3) implement (or insist upon implementation) of that plan. Such an entity should also have the capacity to arrive at and implement a program of cost allocation for the actions taken. The management entity should make a complete systems analysis of the problem. It should be able to consider, and implement if appropriate, the widest possible array of alternatives, and it should provide opportunity for full discussion and revelation concerning the allocation of costs.

---

applicable water laws and compacts." See also comments of A. E. Williamson, Director of Sanitary Engineering Services, Department of Health and Social Service, Wyoming, who noted the "ground rules" for planning salinity control actions are: (1) in no way would water quality standards or such ever be used to circumvent the allocation of waters as laid out in the Compact; and (2) "in no way would we infringe on a state's right to use their allocated share of water"; Letter of Arthur E. Williamson to R. L. O'Connell, Environmental Protection Agency, June 1, 1971 (A. E. Williamson is currently otherwise employed, see note 31.)

It should, however, be remembered that this theoretically neat decision structure may, in the real world, prove impractical, and come with too high a political cost to justify adoption. This appears to be true in the Colorado Basin. Nor will it ordinarily be wise simply to add another layer of bureaucracy to an existing governmental structure for the purpose of solving the salinity problem, for although salinity management is important, it still is a lesser variable among a host that must be considered in determining the total management structure for a given river. Nonetheless, the theoretical model offers sufficient advantages to justify sighting on it as a goal as salinity management problems are approached.

One can expect that the larger contributors to the salinity problem will especially object to bearing the burden of cleanup in proportion to their contribution to the problem on the theory that when they made their water use investments the rules of the game did not require salinity control and they should not be singled out to bear those costs now that the rules are being changed.

One might argue, in theory at least, that no one has a legal right to pollute, and that is the premise on which most national pollution control programs are based, both in this country and abroad.[42] But the salinity problem is far more complex than this statement implies, for it results from a wide variety of causes, including out-of-basin diversions, structure, location, and regulation of reservoirs, phraetophyte control, evaporation control, and so on,[43] and is not produced simply by agricultural uses of a river. Also, it is widely recognized, and accepted in pollution abatement programs everywhere, that financial assistance in the form of subsidies should be available to help historic polluters through the period of transition to less polluting methods. This argument has been persuasive in the Colorado River Basin, and can be expected to arise elsewhere.

The recommendations for a comprehensive basinwide or regional management entity to plan and implement a salinity control program for the Colorado Basin have primarily produced, to date, a voluntary, cooperative interstate discussion group, the Colorado River Basin Salinity Control Forum. It must be noted, however, that in spite of the voluntary nature of this discussion group, it has been a dynamic force in approaching the Colorado Basin salinity problem, having established the plan of implementation, conducted comprehensive com-

---

[42] Johnson and Brown, *Cleaning Up Europe's Waters: Economics, Management, Policies,* Praeger Publishers, New York, 1976.

[43] See, for example, Pionke and Nicks, The effect of selected hydrologic variables on stream salinity, Bull. Int. Assoc. Sci. Hydrol. *XV*: 4 (12/1970); Pionke and Workman, effect of two impoundments on the salinity and quantity of stored waters, Water Resources Bull. *10* (0): 66 (Feb. 1974); Pionke, Nicks, and School, estimating salinity of streams in the southwestern United States, Water Resources Res. *8* (6): 1597 (Dec. 1972); Pionke, Effect of climate, impoundments, and land use on stream salinity, J. Soil Water Conservation, *25* (2): 62 (March-April 1970).

puter and other studies, and carried out complex interstate and federal-state negotiations on salinity issues. P.L. 93-320 has also been enacted, establishing the basis for a basinwide program based on the Bureau of Reclamation's 1972 report entitled "Colorado River Water Quality Improvement Program." This Act provides for cooperation between the Secretaries of the Interior and Agriculture and the EPA administrator. P.L. 93-320 also created the Colorado River Basin Salinity Control Advisory Council to act as liaison between and among the federal agencies and the states, and to make recommendations for appropriate studies. One can only conjecture whether time, and the gradual but predictable worsening of the salinity problem in the basin, will necessitate the creation of an entity with more substantial legal powers of management.

In the meantime the management decisions lie in the hands of the federal government. Fortunately, its legal powers are broad enough to design and implement a wide range of solutions if it has the political will to do so. The federal powers are found in the President's powers to make treaties[44] and empowering Congress to regulate interstate and foreign commerce,[45] administer public lands,[46] protect the Native American,[47] and provide for the general welfare.[48] Congress can override state laws if it wishes to do so and it has gradually moved in, and has become the immediate supervisor, if not actual implementer, of other pollution control programs throughout the United States under the Federal Water Pollution Control Act Amendments of 1972. It is not difficult to predict that the federal government will gradually continue to move into and assume ever-greater jurisdiction over the Colorado salinity management problem, either through an existing federal agency such as the EPA or the Department of the Interior, or through the creation of a more strongly empowered basin or regional entity.

It is possible that some changes in the appropriation system in the West might marginally help to alleviate the salinity problem. That system has seldom considered water use for pollution control a beneficial use, has often denied legal rights to instream uses of water, and has sometimes discouraged efficiency in water transportation and irrigation. However, it remains doubtful whether changes in the appropriate systems—which would have to occur in each state—would have more than a minor marginal impact on salinity levels in the river. Besides, any such changes could certainly not occur without substantial political cost in view of the legal, economic, and cultural attachment that Western farmers have for the appropriation system.

---

[44] U.S. Constitution, Art. II §2.
[45] U.S. Constitution, Art. I, §8.
[46] U.S. Constitution, Art. I, §3.
[47] U.S. Constitution, Art. I, §8.
[48] U.S. Constitution, Art. I, §8.

Approaches to Salinity Management

In theory, one might think of zoning or water use regulations as legal tools to be considered in a salinity management program. Thus it is conceivable that an area might be zoned, or water users regulated, or that only certain crops could be grown, only a limited quantity of water used for irrigation, or irrigation carried out only at certain times. Such an approach may in the future become necessary as the salinity problem worsens and as marginal remedies become more attractive. To date, however, these remedies have not been seriously tested, and there is no firm evidence as to how effective they would be if tried. They are, in any event, quickly rejected by Colorado Basin water users, who have opted for other solutions.

To date all irrigation efficiency and water use management programs have been voluntary and are based on education and persuasion rather than on legal regulation. And, indeed, most water managers and other experts in the Colorado Basin believe the voluntary program will, in the long run, be more effective. It is noteworthy, however, that a decision has been made to reduce by 10,000 acres the amount of the authorized irrigable acreage from the Wellton-Mohawk Project and to make further reductions in the irrigable acreage as appropriate, with the consent of the Wellton-Mohawk Irrigation District. Also several other planned, but unconstructed projects are being reevaluated and may possibly be redesigned (possibly with less land, or different land going under irrigation) in light of the goal of salinity control.

Salinity management poses complex challenges for legal and institutional structures. In general, private law systems will only partially be able to meet these challenges. Whereas the optimal solution, one that produces the optimal reduction in salinity for the least cost, may theoretically result from the creation or existence of a comprehensive basin or regional management entity with power to consider, plan, and implement the widest possible range of alternative solutions, real-world constraints, as on the Colorado and elsewhere, will often dictate otherwise. In view of the great complexity of the problems concerning physical causes of salinity, uncertain legal rights, potential for protracted litigation, and potential fallibility of new layers of bureaucracy, the best results will often be attained by working through and redirecting the priorities of existing institutions.

Applicability of the Colorado Basin Approach to Other River Basins

There is danger in too-easy generalization about the applicability of U.S. water management practices to other countries. Water problems are everywhere unique, and solutions must be designed in light of the prevailing geographic, hydrological, and agricultural situation, the political and legal history, religious beliefs, and economic conditions. Nonetheless, some of the lessons of the Colorado would seem to have a bearing on salinity problems elsewhere.

Salinity problems cannot be solved in isolation. The optimal solution will be found only after an analysis of the entire hydrological, economic, political, agricultural system, that is, after a full systems analysis. A team studying salinity and waterlogging in the Indus Valley concluded in 1962 that these problems "must be attacked within the context of a broad approach toward a large and rapid increase in agricultural productivity," and should be organized to permit a coordinated attack on all aspects of the agricultural problems."[49]

Solutions to salinity problems may require a wide range of actions, including construction of drainage ditches, drilling of wells, construction of desalination plants, taking land out of irrigation, reducing or increasing the amount of water used for irrigation, changing methods or timing of irrigation, changing crops, construction or alteration of the design of reservoirs, and changing the regulation of existing reservoirs. And the list goes on, as illustrated elsewhere in this volume. It must be remembered too that almost every action taken to resolve a salinity problem will (1) cost money, and (2) adversely impact some existing activity.

Where does this leave us? The lesson to be learned for the United States or elsewhere, is that a systems analysis is vital for achieving the optimal solution to salinity problems. In the highly developed United States, with its strong economic system and educated populace, a substantially voluntary approach has achieved some success in the Colorado Basin. Whether this approach will continue to be effective for that Basin in the future, as the intensity of competition for water increases, is not at all clear. What does seem clear is that in most places in the world such a voluntary program will not work, at least not as effectively. A more likely approach is one where the central government, or one of its agencies, will analyze the problem, consider and select among the multiple alternative solutions, and implement those solutions by direct and authoritative governmental action. Varying legal and institutional formulas will be used depending on the political situation in the particular country. The ultimate goal, however, should be the same everywhere, that is, to establish the legal-institutional machinery that has the authority to consider, and implement, a comprehensive systems approach to the solution of the salinity problem.

---

[49] Baxter, The Indus Basin, p. 474.

1.4

Desalination, A Review of Technology and Cost Estimates

Nathan Arad and Pinhas Glueckstern

Mekorot Water Co., Ltd., Tel-Aviv

Introduction

It is now commonly accepted that fresh water extracted from either seawater or brackish water[1] sources could be a reliable resource for water development. In some areas and situations the desalting of seawater or brackish water may provide an added resource for the development, but in others it may be the only means for subsistence. Present desalting technology, though more complex than conventional reservoir-aqueduct supply systems, provides reliable solutions for all uses, if users are able to pay for the water produced. Conversion of salt water into fresh involves rather intricate chemical processes, and complex machinery, vessels, control equipment, and large amounts of energy. It also requires highly skilled labor to operate and maintain the plants. Thus, the introduction of desalting presupposes a rather extreme transformation in water resources systems management and costing.

For many centuries man has tried to invent processes and develop the technology of extracting salts from sea and brackish waters. Many processes and process variations have been developed to bench-scale experimentation; fewer has reached the pilot-plant stage. Present technology is based on modifications of distillation desalting for seawater and on two membrane processes, electro-dialysis and reverse osmosis, for separating salts from brackish water. Plants based on these processes have been built and operated for some time, and are

---

[1] Brackish (salty) water is commonly defined as having a total dissolved salt (TDS) content of 1,000 to 10,000 ppm.

commercially available. The largest single distillation units currently sold rate at about 40,000 m³/day. Membrane facilities for removing salts from brackish water usually consist of modules installed in parallel. The basic electrodialysis module is of the order of 2,000 m³ product/day, that of reverse osmosis, 50 m³/day.

This chapter reviews current technology and the limitations of the most-favored processes. The systems and their applications are described, some typical investment and unit water costs are presented,[2] and some typical operational data are summarized. The reader may wish to consult sources of detailed information listed in the references.

Distillation Processes

Introduction

Distillation is by far the most common means for desalting seawater. Whether the energy source is solar, thermal, electrical, or mechanical, distillation processes allow seawater to change into salt-free vapor that is subsequently condensed to form fresh water. Process modifications involve the method of applying energy, the method by which seawater is made to change phase, and the methods by which the vapor is condensed and energy thus released or conserved. A given combination of these methods forms a given process, and usually predetermines its limitations.

Current technology for converting sea- into fresh water is developed to the extent that a reliable engineering solution is commercially available for all applications, be they of relatively small capacity for remote hotel sites, or large, for urban domestic use or even as blend water for irrigation.

The basic principles of commercially available processes are discussed first, and then the engineering systems based on these processes, and their applications, are described.

System Description and Applications

The multistage flash (MSF) evaporation process (Fig. 1) uses steam as heat energy, and electrical power or steam to drive the process pumps. Vapor is formed spontaneously as heated brine is made to pass from one flash chamber, in which equilibrium conditions exist, into an adjacent, lower-pressure chamber.

---

[2] All cost data relate to market prices prevailing in early 1975.

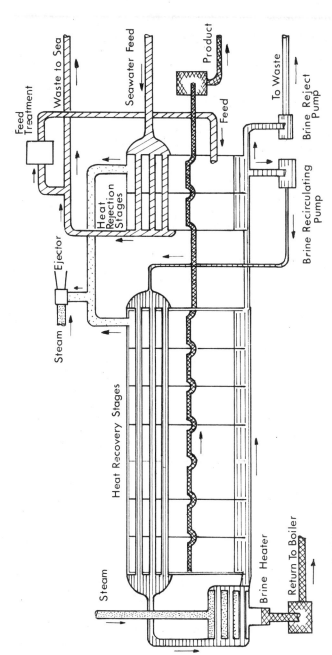

Fig. 1 Multistage flash distillation process.

The release of vapor creates equilibrium conditions in the lower-pressure chamber and the brine, partially cooled, continues to cascade through a number of subsequent chambers where the flashing process is repeated. The vapor released in each chamber flows to cooler condenser-tube banks upon which it condenses; its heat of vaporization being transferred to the seawater or brine flowing in the tubes while the distillate drips into troughs beneath them. The distillate then cascades from chamber to chamber, to be withdrawn as product water from the last chamber. From the last and coolest chamber, part of the brine is circulated back into the condenser tubes to recover the heat given up by the condensing vapor; the remainder is rejected, back into the sea. To replace the rejected brine, chemically treated seawater is fed afresh to the recirculated brine stream.

The multistage flash evaporator system is best suited to capacities of 1,000 m^3/day and up. Single units with capacities of 40,000 m^3/day have already been built. MSF evaporation is currently the most accepted seawater desalting system. Table 1 presents some typical features of the MSF system.

The multieffect distillation (MED) process (Fig. 2) uses steam as heat energy, and electrical power or steam to drive the process pumps. Prime steam from the heat source is introduced to the first effect (either inside or outside the

Table 1 Typical Features of Distillation Systems

	Multistage flash evaporation	Multieffect distillation	Vapor compression
Steam comsumption (kg/kg product)	0.08-0.13	0.08-0.13	–
Electrical energy consumption (kWhr/m^3 product)	3-5	1-3	14-20
Equivalent fossil fuel consumption for steam and electrical energy (kg/m^3 product):			
Single-purpose[a]	6-10	6-10	3.5-5.0
Dual-purpose[b]	4-6	4-6	
Ratio of rejected brine to seawater feed	1.7-2.0	1.7-3.0	1.7-3.0
Top brine temperature (°C)	90-120	70-120	30-100
Product water salinity (ppm TDS)	5-50	5-50	5-50

[a]For the production of water only.
[b]For the production of water and electrical power.

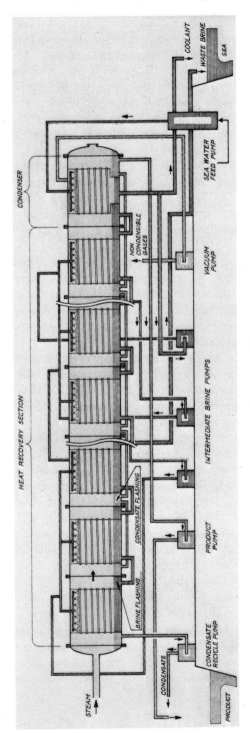

Fig. 2 Schematic cross section of a horizontal-tube multieffect distillation plant. (Courtesy of Israel Desalination Engineering, Ltd.)

condenser tubes, depending on specific design) to vaporize part of the heated seawater. In each subsequent lower-pressure and temperature effect, vapor from the higher-pressure effect is admitted, where it condenses on condenser tubes, transferring heat to vaporize brine at the chamber's temperature; brine from the higher-pressure effect is admitted, making up for that converted into vapor; brine is fed to the lower-pressure effect; brine is fed to a pump to be recirculated through the condenser. Distillate water cascades from one effect to the next, to be withdrawn from the last as product water. Vapor formed in the last effect is condensed in a final, seawater-cooled condenser.

Multieffect distillation plants are offered in capacities of 1,000 m^3/day and up. The largest units so far built (Schevchrenko, U.S.S.R.) are in the range of 4.5 million gallons per day (MGD) (17,000 m^3/day). Construction of a 3 MGD (11,400 m^3/day) module, with some experimental features, has been completed at Orange County, California. The modular nature of the plant will permit later expansion to a final capacity of 12.5 to 15 MGD (47,000 to 56,700 m^3/day).

Multieffect distillers are offered in different designs. The vertical tube evaporator (VTE) design allows for brine to flow inside the tubes while vapor condenses on their outside. This design is thus most adaptable to the use of enhanced heat-exchange tubes for promoting higher heat transfer rates. A novel improvement in VTE design is the incorporation of an MSF evaporator as a feedwater heater; the Orange County module noted above has one.

In the horizontal-tube, falling-film evaporator (HTE) system, brine is sprayed on the outside of the condenser tubes while vapor condenses inside them. The HTE system has some inherent features that make it superior to the MSF. Although there is less experience with constructed plants, the technology, especially in 4,000 m^3/day plants, may now be claimed proven. HTE plants will normally operate at a top brine temperature of about 70°C. This feature favors long-range plant performance, saves on seawater feed, and permits choice of construction materials to match process temperatures, such as using aluminium instead of copper alloys for condenser surfaces. Low-temperature operation will also significantly affect energy costs charged to plants by back-pressure steam from a power station.

The MED system design can be accommodated to variations in water demand or steam availability. It can thus operate efficiently at capacities between 40% and over 100% of nominal, by simply adjusting the steam supply rate. This flexibilities is not possible in MSF system design.

Table 1 presents some typical features of the MED system.

Figure 3 shows a 4,000-m^3/day HTE installation.

The vapor compression (VC) process (Fig. 4) uses electrical energy to drive a vapor compressor and process pumps. Heated, chemically treated seawater is inroduced into the VC vessel (either inside or outside the condenser tubes, depending on specific design). The part of the seawater vaporized is fed to the

Fig. 3 A 4,000-m³/day MED (HTE) installation. (Courtesy of Israel Desalination Engineering, Ltd.)

vapor compressor, the rest being either rejected back into the sea or, after partial brine rejection and seawater makeup, recirculated back to the condenser. The compressed vapor is now introduced into the condenser, where it condenses to form fresh product water while vaporizing across the condenser tubes containing newly fed seawater or recirculated brine.

Vapor compression distillation plants are suited for capacities from a few to 500 m³/day. Technology of large compressors limits the economic size to which VC units can be built. A battery of large VC units may sometimes be found more reliable and economical than MSF or MED desalting plants of the same capacity. The VC process has some inherent features that make its specific energy comsumption lowest of all distillation processes (this feature is later appraised in detail). It is very simple to operate and users report reliable, relatively uniform performance over time.

VC plants are ideal for remote locations, because they require only a modest, feedwater supply and do not require highly skilled manpower and technical services. The VC units are designed to operate either at ambient or higher temperatures (up to 100°C). Ambient-temperature operation allows the use of cheaper materials and savings on seawater feed treatment. Higher-temperature operation, although requiring special care in material selection and water treatment, has the advantage of compressing vapor of lower specific volumes.

Table 1 presents some typical features of the VC system.

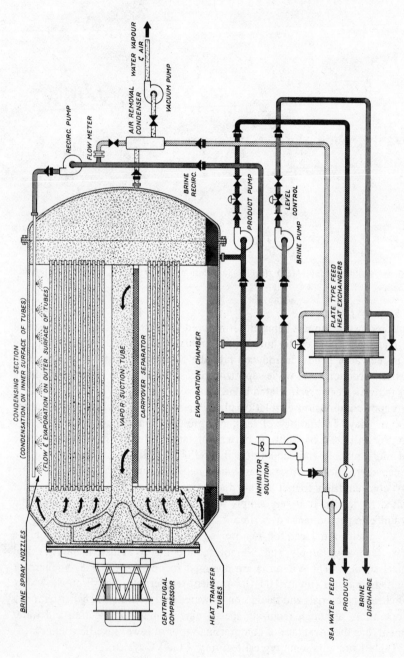

Fig. 4 Flow diagram for the vapor compression process.

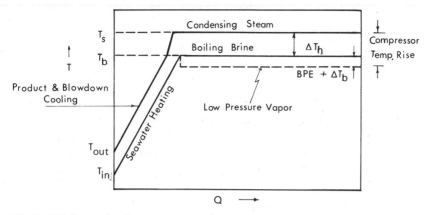

Fig. 5 T-Q diagram for the vapor compression process.

Review of the Technology and Its Limitations

Process-Temperature Driving Force

Separation of fresh water from a salt solution by distillation involves the transfer of a quantity of heat—to the brine and from the vapor. The quantity of heat transferred for each unit of water condensed is the same for all distillation processes provided that the temperature (and pressure) at which vapor is released and the product delivered are identical. Since distillation processes are essentially heat-transfer operations, they are most easily described by temperature-heat transfer (T-Q) diagrams similar to those describing simple heat exchangers. Here, the curve is of temperature plotted against quantity of heat transferred for each of two heat-exchanging streams. The vertical distance between the two curves represents the temperature driving force for heat transfer.

Figure 5 presents a typical T-Q diagram for the VC system. Fresh seawater is heated from inlet temperature T_{in} to the boiling brine temperature T_b by heat exchanged with the product water (condensate) and rejected brine. The dashed line shows the saturation temperature of the vapor released after allowing for boiling-point elevation[3] (BPE) and the saturation temperature decrease associated with pressure drops (ΔT_b) inside the evaporator vessel. The temperature difference between the boiling brine and the condensing steam is

---

[3] Boiling-point elevation is the temperature difference between the boiling point of a salt solution and that of pure water.

the heat transfer driving force ΔT_h. The heat-transfer temperature difference in the feed heating section ($T_{out} - T_{in}$) will normally have about the same value as ΔT_h. Energy to run the process is in the form of compressor work.

Process Thermodynamics

The conversion of saline water into fresh requires energy input. It can be shown that the theoretical minimum (ideal reversible) energy to convert a water of given salinity into fresh water is dependent only on the temperature at which the process takes place and on the extraction rate.[4] Thus, the minimum energy to convert normal seawater at a temperature of 25°C amounts to 0.7, 1.1, and 1.8 kWhr/m³ for extraction rates of 0, 50, and 80%, respectively.

Any practical process, however, involves, sources of irreversibility that increase the energy required. Thus, if ΔT_{rev} is the driving force corresponding to a reversible process and the driving force charaterizing the irreversibility of the system is given by ΔT_{irr}, the actual total driving force required will amount to

$$\Delta T = \Delta T_{rev} + \Delta T_{irr}$$

The energy consumption of a practical process, W, is therefore the sum of the reversible desalination energy, W_{rev}, plus the energy required to overcome the causes of irreversibility, W_{irr}.

For a reversible process

$$W_{rev} = \frac{\Delta T_{rev}}{T} L$$

where T is the process absolute temperature and L is the heat of vaporization for water at T.

Also, for a real process,

$$W = W_{rev} + W_{irr} = W_{rev}\left(1 + \frac{\Delta T_{irr}}{\Delta T_{rev}}\right)$$

The ratio W_{rev}/W is known as the thermodynamic yield, Y, and

$$Y = \frac{W_{irr}}{W} = \frac{\Delta T_{rev}}{\Delta T_{rev} + \Delta T_{irr}}$$

---

[4] Extraction rate is defined as the ratio of product output to feed rate.

For distillation processes ΔT_{rev} is usually in the range of 0.5 to 1.0°C, whereas for real systems, ΔT_{irr} is in the range of 4 to 10°C. The resulting thermodynamic yield for the current-technology distillation processes is therefore in the range of 0.05 to 0.20.

A real vapor compression process may operate at a thermodynamic yield of 0.2. When operating at an extraction rate of 50%, the system consumes about five times more energy than the theoretical minimum, that is, 5.5 kWhr/m³ of product water, provided that the vapor is compressed in an ideal adiabatic compressor. To calculate nonideal compressor energy, the ideal adiabatic energy must be divided by the compressor and motor efficiency. The actual energy consumption of a real system is therefore of the order of 14 kWhr/m³ of product water. If supplied from a modern, highly efficient power grid, this will entail the burning of about 3.5 kg of fuel at the power station.

The above analysis holds for the case where process work is provided as high-grade mechanical or electrical energy. When the energy input source is a low-grade process steam, only part of it is available to do work, that is, to be converted to electrical or mechanical forms of energy. At best, the available energy fraction of total steam heat energy is equal to Carnot efficiency, $(T_{source} - T_{sink})/T_{source}$, where T is the absolute temperature of an ideal heat engine operating between steam source and sink (usually seawater) temperatures. For a multistage or multieffect distillation process, thermodynamic yield is of the order of 0.15. If the process operates at an extraction rate of 0.5 (minimum ideal of 1.1 kWhr/m³), the actual high-grade energy consumption will be about 7 kWhr/m³. To effect the process with 7 kWhr, if the heat source is steam at 100°C (539.0 kcal/kg steam) and the sink temperature is 30°C, the Carnot efficiency is 0.188 (70/373 = 0.188), so that we must, at best, supply 37.2 heat kWhr (7/0.188 = 37.2), or 32,021 kcal for each cubic meter of product water. This requires the feeding of 59.4 kg (32,021/539 = 59.4) of source steam for each cubic meter of distilled water produced. Actually, for most plants, 80 to 120 kg of source steam are required for each cubic meter of distilled water produced. To produce 80 to 120 kg of steam of 100°C, 5.0 to 7.5 kg of fuel are required.

From the above analysis one tends to conclude that for real (nonideal) distillation systems energy consumption is high. This stems from the fact that real systems must have a thermodynamic yield smaller than 1. Many designs have shown that the thermodynamic yield for optimally designed systems must be in the range of 0.1 to 0.2; those designed for a value greater than 0.2 would entail heavy capital investment. If other losses are added, such as compressor efficiency losses in the vapor compressor system, or the losses of conversion of low-grade heat energy to work energy in the MSF or MED systems, the result is that real systems' energy consumption is about 12 to 25 times higher than the ideal (reversible) minimum.

Process Operating Temperatures and Heat-Transfer Rates

One method of improving process thermodynamic yield is to enhance heat-transfer rates. Since heat-transfer rates increase with temperature, enhancement is obtained by increasing operating temperatures, at which released vapor volumes are smaller and irreversible losses decreased.

Heat-transfer rates can be improved by enhanced heat-transfer surfaces. Fluted tubes installed in a VTE system, though not yet commercial, more than doubled heat-transfer coefficients (Alexander and Hoffman, 1971).

Dual-Purpose Installations

Heat energy supplied to MSF or MED systems may derive from a steam turbine exhaust. In dual-purpose installations producing power and desalting water, prime steam expands through the turbine to be condensed in the desalting system's brine heater. The condensing temperature is consistent with the system's top brine temperature. The total available steam energy is split between the turbine, to produce power, and the desalting system, to produce water. The share of the desalter in this split is substantially smaller (15 to 50%) than the prime-steam heat energy needed to produce steam to run a single-purpose plant (desalting only). The full-condensing turbine cycle used in plants for power production only is the basis for determining the total available energy split between the two products. The loss in power production caused by diversion of back-pressure steam still able to generate power to the desalter is the available energy consumed by the desalter.

The fraction of available energy consumed by low-temperature desalting processes is obviously smaller than that for higher-temperature processes. A desalter operating at a process temperature of 100 to 120°C will consume about 50% of the available energy in the prime steam; a process temperature of 60 to 65°C would consume only 15 to 20% of it. A low-temperature HTE desalter consumes heat costing roughly one-third of that consumed by high-temperature MSF and VTE processes. As fuel consumed to produce prime steam in dual-purpose plants is almost equal to that required to produce low-pressure process steam for single-purpose desalting plants, dual-purpose plants have the obvious advantage of substantially lower heat costs.

The dual-purpose concept has further advantages in capital cost savings from sharing in site development, buildings, and common structures and components, such as seawater intake and outfall facilities, and operation and maintenance savings—from joint plant management, operating staff, and maintenance facilities.

Dual-purpose plants may have some disadvantages, usually relating to different demand patterns for the two products, and to location of the power and water production plants. The problem of flexibility in production of power and water in dual-purpose plants has been recently investigated and various schemes and system modifications proposed (Glueckstern et al., 1972).

Feedwater Pretreatment

It has often been said that the art of desalting is the art of feedwater pretreatment. Without proper pretreatment, solid layers of mineral scale deposit on heat-transfer surfaces, causing operating difficulties and thermodynamic yield loss. Scale in distillation plants usually contains calcium sulfate, magnesium oxide, and calcium carbonate; it forms when the solubility limit of these compounds is exceeded.

The solubility diagram of calcium sulfate is shown in Fig. 6. The limits it dictates are decisively important to plant designers. The solubility limits of the hemihydrate and of gypsum must be avoided; that of the anhydrite may enter the region of supersaturation if crystallization seeds are absent.

Seawater is not saturated with magnesium hydroxide, but when evaporated at temperatures above 70°C, the point at which the solubility limit is exceeded is soon reached, and magnesium hydroxide precipitates. The solubility limit is a function of pH, so that deposition is controlled by regulating the seawater or brine pH.

A number of methods for preventing calcium carbonate scale have been proposed and tried. Modern practices suggest two methods: (1) acid addition to reduce makeup alkalinity; and (2) addition of compounds to retard scale deposition.

With acid treatment, calcium bicarbonate is converted to calcium sulfate according to the reaction:

$$Ca(HCO_3)_2 + H_2SO_4 = CaSO_4 + 2CO_2 + 2H_2O$$

Sulfuric or hydrochloric acids, if used in excess, may cause severe corrosion. Very carefully designed acid injection and neutralizers such as sodium hydroxide, forming a controlled pH system, are favored today. The pH in various parts of the feed and brine streams is controlled to within scale and corrosion limits. Such systems, if properly designed and maintained, give excellent service.

Polyphosphates and polyelectrolytes suppress precipitation thresholds. They are reliable and are known to provide good service if brine temperatures are kept below 90°C. The mechanism of precipitation suppression varies with the treatment. It generally depends on chelation, in which the alkali-metal ions causing scale are either tied up in un-ionizable "complexed" water-soluble salts or inhibited so that precipitation occurs during flashdown instead of during heating.

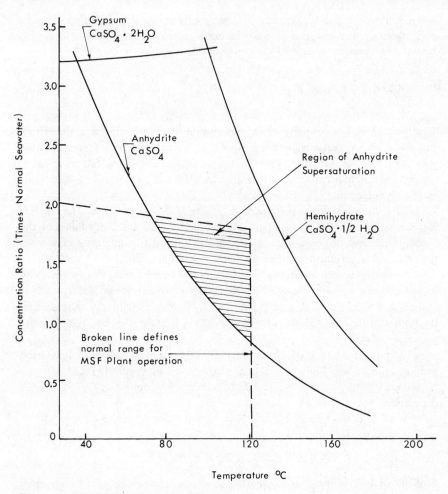

Fig. 6 Solubility diagram of calcium sulfate.

Scale retardation does not altogether eliminate its buildup. The performance of plants using scale retarders degrades over time, lowering the average thermodynamic yield, but designed performance is regained by periodic chemical plant cleaning. Scale retarders are considered safer for preventing corrosion than acid-injection systems in which there are always possibilities of malfunction or human error that may lead to acid overdosing and hazards of severe corrosion that, if uncorrected, could endanger plant integrity.

Selection of Materials

Materials for constructing distillation plants are designed to be compatible with service and exposure conditions, detailed as follows:

1. Condenser Tubes and Tube Bundles. These are usually the most expensive single items in distillation plants. The materials most commonly used are 90-10 copper-nickel alloy for high-temperature plants, aluminium-brass for temperatures up to 100°C, and special aluminum alloys for processes at temperatures below 70°C. A 0.75- to 1-in. diameter is usually specified for MSF plants, and a 1.5- to 3-in. diameter for MED and VC plants. Currently, the most specified tube-wall thickness is 0.049 in.; some designers of large plants prefer a 0.035-in. tube-wall thickness.

2. Evaporator Shells. All operating plants currently have evaporators with cylindrical or rectangular steel shells. As steel is corroded by oxygen and low-pH brines, the shells are usually coated with an inert material such as an epoxy paint. Plating with corrosion-resistant alloy is sometimes specified, especially for high-temperature stages.

3. Evaporator Internals. The evaporator internals include tubing supports; condensate product-water trays; brine-flow baffles and interstage orifices; noncondensable vent lines; and special demisters for separating entrained brine droplets from the vapor. Monel or stainless-steel wires are generally used for the demister. Steel with various plastic coatings or noble materials such as stainless steel or copper-nickel alloys are used for the other internals. Plastic materials are also used, especially for low-temperature processes.

Unit-Size Limitations

Theoretically, there is no limit to unit size. Manufacturers and users have, however, as yet avoided quotation of units larger than about 12 MGD (about 45,000 m^3/day). Plants with capacities much in excess of 10 MGD (such as the 180,000-m^3/day Hong Kong MSF plant) consist of multiple units in parallel, but evaporator chambers, piping, process pumps, and tube bundles become excessively large and heavy. Consequently, for current and near-term technology, economy of scale is not very significant above plant sizes of about 10 MGD (40,000 m^3/day). Unit-size plants will eventually be adjusted to market demand. It may be expected that orders for plants in the 100-MGD (about 400,000 m^3/day) range, if and when placed, will entail the design and construction of units in the 20- to 50-MGD range.

Membrane Processes

Introduction

Membrane processes comprise methods in which salts are separated from salt solutions by means of membranes, without phase changes. In electrodialysis (ED), salt ions are extracted by means of an electrical current and concentrated in special cells or compartments, separated by cation- and anion-permeable membranes. In reverse osmosis (RO), hydraulic pressure in excess of the osmotic pressure is applied onto a chamber containing salt solution, causing the passage of pure solvent, via a semipermeable membrane, into a second chamber.

Both ED and RO have developed to the extent that water-treatment systems are commercially available for any moderate need. The range is from very small units for home drinking water to several-MGD installations, both municipal and industrial. The Colorado River Project, presently being considered, has a projected 100-MGD capacity. Whereas ED is commonly used

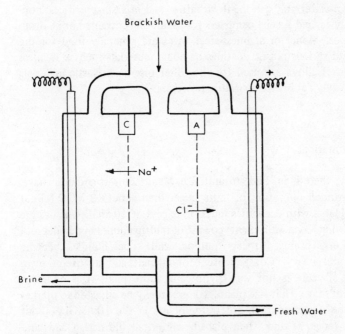

C — Cation Permeable Membrane
A — Anion Permeable Membrane

Fig. 7 Principle of electrodialysis.

for brackish water of up to 5,000 ppm TDS only, RO is being rapidly extended to separation of salts from seawater.

The aim here is to present the basic principles and the applications and limitations of commercially available systems.

System Description and Application

Electrodialysis

The principle of ED operation is shown in Fig. 7. Salts, when dissolved in water, are in the form of negatively and positively charged ions. When direct electrical current is applied to a brackish feedwater, positively charged ions such as sodium are forced through the cation-permeable membrane toward the cathode, whereas negatively charged ions, such as chloride, are forced through the anion-permeable membrane toward the anode. The water in the compartment between the membranes is depleted of salt, whereas the mineral content of water in adjacent compartments increases.

The membrane stack or ED process unit consists of several sets of anion- and cation-permeable membranes. Passage through one stack removes 30 to 50% of the entering salts. Additional stacks, in series, increase the degree of demineralization. Each added stack is a stage. Volume throughput is increased by stacks being added in parallel.

Figure 8 is a schematic flow diagram of the ED system. Feedwater, pumped through a pretreatment section and then through successive membrane stacks, is processed into product water. A portion of the concentrated brine is recycled to improve system performance. Concentrated waste brine is removed from the system, to be replaced by feedwater. In electrodialysis, the energy required to produce product water of a given solids content is proportional to the salinity of the brackish water being processed. Thus, ED produces cheaper product water from water of relatively low salinity than from feedwater of higher salinity.

Hundreds of ED plants, demineralizing brackish water for municipal, industrial, military, and other applications, are spread all over the world. Plants with output capacities above 1,000 m^3/day have been in operation for over a decade. Raw feed salinity generally does not exceed 5,000 ppm TDS, salinity of the product water being dictated by specific requirements and determined by the number of stages.

Reverse Osmosis

The phenomenon of osmosis, in which fresh water tends to diffuse through a semipermeable membrane into a salt solution, is reversed in the RO

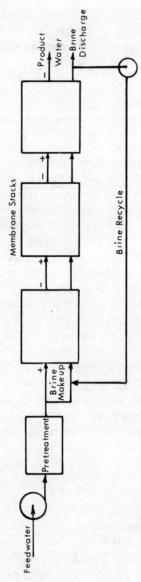

Fig 8 Schematic flow diagram of electrodialysis system.

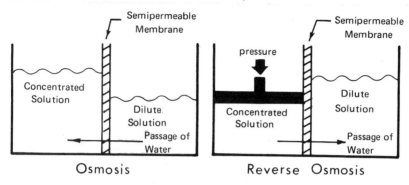

Fig. 9 Principle of reverse osmosis.

process. This is most attractive because of its simplicity and theoretically low energy requirement. Development of a strong, reliable, well-packed, pressure-resistant, semipermeable membrane was the precondition that allowed development of RO desalting. In the RO process, a membrane acts as a molecular filter, separating over 90% of the dissolved minerals from the water produced when pressure is applied to a saline feedwater. This pressure reverses the osmotic solvent flow through the semipermeable membrane, which also acts as a barrier to certain molecules, preventing their passage into the dilute solution. Brackish feedwater is thus demineralized by the passage of its water molecules through a semipermeable membrane, whereas its salt molecules are held back, or rejected (Fig. 9).

The RO deslating plant comprises membrane modules, high-pressure pumps, and auxiliary equipment. The basic flow diagram for RO systems is shown in Fig. 10. After a pretreatment to reduce suspended solids, possibly excess manganese and iron concentrations, and prevent biological growth, feedwater is raised to system-design pressure by the high-pressure process pump and pumped to the inlet of the RO membrane modules. A portion of the feedwater diffuses through the membrane as pure product water, which flows through porous backing materials or tubes into a collection trough. Collected product is lead to the inlet of the product pump, which delivers product water to the plant boundaries. As pure water diffuses through the membrane, the feedwater stream simultaneously becomes more salt-concentrated, to be finally discharged as rejected brine at a pressure somewhat lower than that of the feed. Most RO plants operate at 50 to 80% extraction rates, whereas 50 to 20% of the incoming feed is rejected, at two- to fivefold the initial concentration, respectively.

RO systems for water treatment are now commercially available for treating brackish water of up to 10,000 ppm TDS. Today, the basic RO module has moderate output capacities, of the order of 10 to 50 m³/day. A commonly

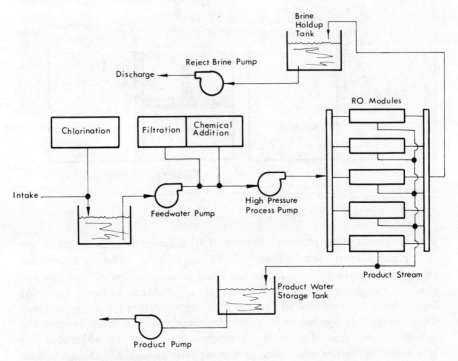

Fig. 10 RO plant process diagram.

accepted system design has many moduels in parallel, so grouped that several modules are serviced by one high-pressure pump. Thus, the factor limiting the size of a single RO desalting unit is the capacity of the high-pressure pump. The grouping of several RO units in parallel, however, results in plants with virtually unlimited desalting capacities.

Review of the Technology and Its Limitations

Electrodialysis Membranes and Other Basic Components

The basic ED device is a large number of alternating cation- and anion-selective membranes assembled between a pair of electrodes to form an electrodialytic stack. Between the membranes are narrow channels into which saline water is introduced. Flow of the process streams—both dialysate and concentrated brine— is contained and directed by spacers that alternate with the membranes. The thickness of the spacers ensures the distance between adjacent cation and anion membranes—to the order of 1 mm. This represents a compromise between direct-current heat energy losses and the pumping energy required, for the latter increases or decreases, respectively, with decreasing or increasing distance between the membranes. Each alternate membrane pair and spacer constitute a stack compartment or a cell pair. The assembly of membranes, spacers, and electrodes is compressed together by a pair of end plates. A typical stack may contain several hundred cell pairs with membrane surface areas ranging from 2,000 to 12,000 cm^2.

ED Membranes. The several types of membrane now available on the market are the result of intensive efforts to produce suitable material for ED. Their thickness ranges from less than 0.1 to 0.6 mm. Some are reinforced by fabrics supplying the mechanical strength required to withstand the pressure differences in adjacent electrodialysis compartments. The properties desired for selective ED membranes include: high electrical conductivity; high ion selectivity and permeability; homogeneous quality; mechanical strength and chemical stability; resistance to deflection under pressure; and dimensional stability in different solutions.

ED spacers are plastic sheets of a thickness equal to the distance desired between adjacent membranes. Spacers are also designed to serve as conduits and liquid distributors. Some consider the spacer the most critical component in the ED system, determining as it does both hydraulic capacity and limiting current density of the basic ED cell.

Electrodialysis Process Parameters

ED units use direct current (DC) electrical power to transport ions through the membranes. The applied voltage per cell pair must be at least equal to the countervoltage consistent with minimum energy requirements. Because of ion depletion in the liquid layers adjacent to the membranes, the electrical resistance of each cell pair increases as the opposing voltage is increased. This effect becomes more pronounced at higher current densities and is, in fact, the factor limiting the ED process. The current at which this becomes significant, reflected in a sharp increase in the stack's electrical resistance, is called the polarization current. During polarization, hydroxyl ion transfer through the anion-selective membranes will cause a high pH in the concentrated brine compartments and sbusequently cause scale formation and membrane damage. High currents and correspondingly high removal of dissolved solids thus cannot be effected in a single ED stack. In theory, it should be possible to operate at low voltages (0.1 to 0.2 V per cell pair), but at very slow rates. This would, however, entail extremely large and capital-intensive units. Overall economic evaluation shows that it is beneficial to operate at higher voltages (1 to 2 V per cell pair), since electric power losses caused by conversion of energy into heat are more than offset by savings in capital investment.

In accordance with Faraday's law, approximately 50 A of current is required to remove 100 ppm of dissolved solids from 1 m^3 of water per hour. The application of 1 to 2 V per cell pair will, accordingly, entail consumption of 50 to 100 Whr to remove 100 ppm of dissolved solids from 1 m^3 of water. For comparison of two stacks operating at the same salinity and identically removing salts, it is useful to measure, in each, the ratio of DC energy consumed per unit of fresh water produced, to stack production per total membrane surface area. Of two stacks identically removing salts, the one consuming less DC energy per unit of fresh water produced is superior, when both produce it at the same rate per total membrane surface area.

Pumping the streams through the stack, feed pumping and brine disposal, and other minor services add 0.5 to 0.8 kWhr/m^3 to the DC energy requirement.

Dissolved solids removal by an ED stack is affected by the temperature and the ratio of the monovalent ions to the total dissolved solids in the feedwater. The higher the feedwater temperature and the higher the ratio of monovalent ions to TDS, the better the DC energy rating of the stack. The DC energy rating improves, moreover, with increasing extraction rate (per given amount of salts removed) and stack flow.

Reverse Osmosis Membranes and Module Configurations

Many natural materials have semipermeable characteristics, but only two types of membranes are currently used in RO equipment. These are (1) asymmetric cellulose acetate (CA) membranes, and (2) membranes of asymmetric hollow fibers of aromatic polyamide (AP) and its derivatives. Development of these membranes and the modular equipment for packing them are the result of intensive research and development in the last two decades. Commercial CA and AP membrances presently operate only on brackish water, but membranes for single-pass seawater desalination are in development; some are indeed already on the market, but without guarantee on their lifetime.

CA membranes are usually packed in a spiral module or fabricated as tubes. The AP membrane, developed by Du Pont's Permasep Products, consists of hollow, hairlike fibers, packed in a shell, that accept brackish feed on one side and discharge product permeate on the other.

Fluxes presently achieved in practice with CA membranes are limited to 10 to 20 gallons per day per square foot (gfd) of membrane surface (or 0.4 to 0.8 m^3/day/m^2). The flux is governed by several factors: (1) the intrinsic permeability of CA to water; (2) the thinness of the film that can be manufactured without imperfections; (3) the excessive energy and capital costs associated with operating at higher pressures, if higher fluxes are desired; and (4) boundary-layer effects. The fluxes achieved with hollow-fiber AP membranes are limited to less than 2 gfd (0.08 m^3/day/m^2) by some of these factors, apart from the danger of fiber collapse if higher pressures were to be used.

Both CA and AP membranes have limited life spans. CA membrane is sensitive to: (1) degradation by hydrolysis if not operated in a pH range of 5 to 6; (2) membrane fouling or gross module plugging; and (3) biological attack. The lifetime of hollow-fiber AP membranes is limited mainly by (1) chlorine attack where feedwater chlorination is necessary; and (2) gross module plugging.

Of all the configurations for packing or modulizing RO membranes that have been explored and developed in recent years, three remain commercially available: tubular, spiral-wound, and hollow-fiber. Tubular design offers open-flow characteristics, giving it certain advantages over other RO designs. However, because of the limited packability (measured as square feet of membrane surface area per cubic foot of module pressure vessel) inherent in its design, the tubular concept has not gained wide acceptance.

The packability of the spiral-wound design is about 200 ft^2 of membrane surface area per cubic foot of the housing pressure vessel. If one notes typical CA flux to be about 15 gfd of membrane surface area, specific production rate for the spiral-wound modular design is about 3,000 gpd/ft^3 (400 m^3/day/m^3) of the module pressure vessel. The basic hydraulic configuration of a spiral-wound module element is depicted in Fig. 11. As noted, feed is introduced normal to the spiral-wound element, whereas product flow is led spirally inward to a

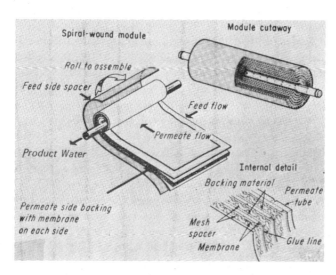

Fig. 11 ROGA spiral wound RO membrane module. (Courtesy of Ajax International Corp.)

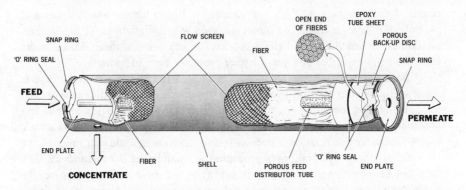

Fig. 12 Cutaway of Du Pont's Permasep permeator. (Courtesy of E. I. Du Pont de Nemours & Co.)

central product-water collector tube. Several spiral elements arranged in series are housed within a plastic-lined pipe with feedwater entrance and brine exit at its ends. A product water exit that is an extension of the central collector tube is situated at the center of one end of the pipe. The 4-in. module diameter has been scaled up recently to a diameter of 8 in., with an initial product water capacity of approximately 2,500 gpd (9.45 m^3/day). The first commercial spiral-wound module, using CA membranes, was introduced by the ROGA Division of Universal Oil Products Corp.

The specific production rate of the hollow-fiber design is approximately 10,000 gpd/ft^3 (1,300 m^3/day/m^3) of the module pressure vessel. The basic hydraulic flow patterns and components of the hollow-fiber system are shown in Fig. 12. Pressurized feedwater is introduced on the outside of the hollow-fiber bundles. Permeate—fresh water—diffuses through the fibrous membrane wall into the fibers' hollows. The fibers are sandwiched at one end inside an epoxy tube sheet and serve to collect the permeate. Concentrated brine flows from another opening in the module to ensure constant feedwater flow. The first commercial hollow-fiber module was introduced by Du Pont in 1967. It uses polyamide fibers, of 85 μm outer diameter, housed in 4-in.-diameter modules. These were recently upscaled to an 8-in. diameter with an initial product water capacity of 14,000 gpd (53 m^3/day) when operated with a feed of 1,500 ppm NaCl, at a pressure of 400 psi (28 atm), temperature of 25°C, and a 75% extraction rate. The polyamide fiber membranes have certain properties superior to those of cellulose acetate membranes in that they are more resistant to degradation over a wider pH range (i.e., 4.0 to 11.0) and can withstand slightly higher operating temperatures. A drawback of the polyamide fibers is their sensitivity to free residual chlorine.

Reverse Osmosis Process Parameters

Pressure and Energy. The barest minimum pressure requirement for the RO process is identical with the osmotic pressure—the pressure difference that exists between the dilute (product water) and the concentrated solution (brackish-water feed). The operating pressure of real RO systems are many times the osmotic pressure. The osmotic pressure at 25°C of brackish water with a 2,000 ppm salt content is about 23 psi (1.6 atm), but the operating pressures of RO units range from 250 to 800 psi (17 to 54 atm).

The effect of pressure on salt rejection is to cause it to rise steeply until the operating pressure exceeds the osmotic pressure by approximately 400 psi, when salt rejection levels off. Thus nominal salt rejection cannot be expected at low pressures. The effect of pressure on flux is approximately linear.

Temperature. Salt rejection of most RO membranes is essentially constant over the operating temperature range, but solvent flux increases with temperature. There is, in effect a 2 to 3% increase in production rate for each 1°C above the nominal operating temperature. Most manufacturers of RO equipment recommend a maximum feed temperature of 86°F (30°C). The rate of membrane hydrolysis and membrane creep or compaction increases at higher temperatures, shortening the useful life of the membrane.

Compaction. Compaction is the term used to describe the decrease of flux with time. The phenomenon, reduction in membrane thickness effected by pressure and time, is widely believed to be a membrane densification, increasing resistance to flow through the open pores and thus reducing the flux. Since compaction is a function of pressure, the only possibility of alleviating it is to operate at the lowest pressure acceptable. Current commercial membranes, operating at a pressure of 400 psi (27.2 atm), show a 20 to 30% flux degradation in 3 years. Since rated performance is usually guaranteed for 3 years, manufacturers must supply membranes with an initial flux 20 to 30% higher than rated by the guarantee. Some manufacturers, however, recommend operating the RO system at a constant production rate by varying operating pressures— from lower than rated at the start and increasing as the membrane degrades.

Another cause of flux decline is the progressive hydrolysis of cellulose acetate during membrane life. Cellulose acetate membranes are limited to an operating pH range of 3 to 7, outside of which rapid membrane degradation occurs; the optimum pH lies between 5 and 6.

Concentration Polarization. Concentration polarization is a boundary-layer phenomenon, reducing flux and salt rejection, caused by solute being allowed to accummulate on or near the membrane surface. This is brought about by inadequate flow rates or turbulence. The problem may be solved by maximizing the transport of solute away from the membrane surface, and into the feed solution, by any means commensurate with economical operating costs.

Feedwater Pretreatment

The extent of pretreatment required for RO and ED systems varies between module designs and is to a certain extent dependent on the hydraulic design configuration involved. For example, tubular RO units have a decided advantage, requiring minimum pretreatment because of their well-defined open-flow characteristics and ready accessibility for in-place, mechanical cleaning. Hollow-fiber RO systems intrinsically require the highest degree of feedwater clarity and place severe restrictions on feedwater turbidity. Spiral-wound RO units do not require as high a degree of feedwater clarity, and a turbidity level of 1.0 Jackson turbidity unit (JTU), or less, is acceptable. A turbidity level of 5 JTU or less is considered acceptable for ED units. The ability of ED systems to tolerate higher turbidity levels is partially due to the ease with which ED stacks can be disassembled for hand cleaning of the membranes. Another major factor affecting the degree of pretreatment is the brine concentration that the system can accept without forming calcium sulfate scale. In several brackish-water desalting systems, 10 to 20 ppm of sodium hexametaphosphate are commonly injected into the feed to supersaturate calcium sulfate. With this addition, desalting plants may be operated successfully in the range of 150 to 200% saturation in the brine exist stream. Beyond this level, the only effective way to control calcium sulfate scale is to soften the feedwater, through lime addition, lime soda softening, or ion exchange.

Iron and manganese, either alone or together, commonly contaminate brackish water. Membrane systems can generally tolerate a maximum combined iron and manganese level of 0.3 ppm, although some suppliers prefer that it be no higher than 0.2 ppm.

Acid is generally added to RO and ED systems, serving several functions. The pH depression ensures against the formation of calcium carbonate and magnesium hydroxide scale and tends to supress organic and biological fouling. Moreover, CA membranes not only hydrolyze at high pH levels, but membrane service life is maximized if a pH of about 5 is maintained.

Reverse Osmosis for Removal of Salts from Seawater

The problems encountered in using RO membranes for removing salts from seawater relate to the much higher osmotic pressures of seawater and to the greater risks of membrane fouling by biological deposits, particulate matter, and corrosion products. The osmotic pressure of seawater range from 370 to 470 psi (26 to 33 atm), corresponding to total salt content of approximately 35,000 ppm TDS for open oceans and up to 44,000 ppm TDS for more saline seas such as the Red Sea. These are theoretical figures for zero extraction rate, without allowing for the higher membrane-surface salt concentrations due to concentration polarization. Consequently, the real osmotic pressure for practical systems will be in the order of 500 to 600 psi (35 to 42 atm). Thus, for operating fluxes or production, operating pressures of not less than 800 to 1,000 psi (54.4 to 68.0 atm) must be applied.

The salt-rejection rate required of RO seawater membranes depends on whether a one-pass, or a two-pass system having two stages in series, is contemplated. The rejection rate of a one-pass system must be at least 99%; two-pass systems can tolerate membranes with rejection rates of about 90%. A one-pass system is, of course, more attractive from the operational and economic points of view, but although membranes with rejection rates above 99% were experimentally fabricated in the late 1960s, the first commercial membrane with a rated rejection of 98.5% was marketed only in September 1973 by Du Pont—for limited application. The 4-in.-diameter version uses aromatic polyamide hollow-fiber (92-μm outer diameter and 41-μm inner diameter) membranes to produce 1,500 gpd (5.67 m^3/day) at 800 psi (56 atm) operating pressure and 10 to 50% extraction rates. When fed with 30,000 ppm NaCl solution at 25°C and 30% extraction rate, the module will rejection 98.5% of the salt. The 8-in.-diameter version, not yet available commercially, will have characteristics similar to those of the 4-in. version, but with a production rate of 5,000 gpd (18.9 m^3/day).

A further recent development is that of the Fluid Systems Division of Universal Oil Products Corp., which has developed CA membranes with rejection rates of up to 99.7%, able to produce potable water from seawater in a one-pass system. The company prefers, however, to offer their commercial plants, of up to 5 MGD capacity, as two-pass systems. At operating pressures of 800 and 400 psi (54.4 and 27.2 atm) in the first and second stages, respectively, and 30 to 45% overall extraction rates, a demonstration two-stage system produced 15,000 gpd (56.7 m^3/day) of less than 500 ppm TDS product water.

Performance and Cost Data

Introduction

In 1974, the U.S. Department of the Interior Office of Saline Water (OSW) reported that, as of 1972, there were 812 land-based desalting plants of 25,000 gpd (100 m^3/day) capacity or larger in operation or under construction throughout the world. The total desalting capability of these plants is 348 MGD 1.3 $\times$ 10^6 m^3/day). Distillation processes constitute 93% of this capacity and membrane processes more than 6%. Since then, a substantial increase in plant capacity has been reported. These include a 48 MGD (180,000 m^3/day) MSF plant in Hong Kong; a 3-MGD (11,340 m^3/day) ED plant in Oklahoma; and expansion of the RO plant in Kashima, Japan, from 0.8 MGD to 1.4 MGD (5,300 m³/day).

This section surveys the performance of a number of sample plants and presents pertinent investment and operating costs. The sensitivity of desalted water costs to changing energy prices and investment costs, in ranges prevailing today and projected for the future, is also analyzed. All costs reported here are related to market prices in late 1975.

Plant Performance

Water is acutely short in many parts of the world. This has stimulated international interest in water desalting research, technology, and development, and many countries have embarked on programs to develop the desalting art. The program of the OSW is the largest and most elaborate. It sponsors research and development, constructs and operates plants to demonstrate the technology and economics of most known processes, and publishes detailed annual and other reports on all aspects of the technology and its economics. Several of these reports deal with the performance of existing plants (Newton et al., 1972; Rapier et al., 1972).

The main causes of reduced performance in desalting systems, both distillation and membrane, are failure of construction materials as a result of severe corrosion, and scale formation as a result of design or malfunctioning of feedwater pretreatment systems. In recent years, substantial experience has been gained in material selection for various conditions of service and exposure, while the art of feedwater pretreatment for corrosion and scale control has also developed. The present-day designer thus has better information on adequacy and expected life of construction materials for optimal plant design. The large-scale desalting systems of the future may thus be expected to provide optimal performance at minimal water costs, when compared with the small and medium-sized systems that pioneered the industry.

Table 2 summarizes the performance of 17 desalting systems at 11 different sites. The data for this table were compiled from data of Hornburg et al. (1973) and Glueckstern et al. (1975). Desalting units located at one site, with common feed, product delivery, and energy-supply systems, were grouped together. In these cases capacity and on-stream factors are presented as weighted averages (see notes to Table 2).

Investment and Water Costs

The recent rapid rise in oil prices has had significant impact on the prices of most materials and services, including materials for constructing seawater and brackish-water desalting plants. Since investment and energy costs account for a major portion of desalted water costs, the rise in oil prices was followed by a steep rise in these costs. This section presents an estimate of desalted water costs and analyzes the sensitivity of these costs to changing investment costs and energy prices, in ranges known today and projected for the future.

Three seawater desalting processes are considered here: multistage flash, multieffect, and vapor compression distillation. Apart from vapor compression, applicable to small plants only, the processes may be used in small (1 MGD range) and large (100 MGD range) plants. The two processes for desalting brackish water considered here are electrodialysis and reverse osmosis, both in the 1 MGD range. The RO process with technology projected for separation of salts from seawater (small plants only) is also discussed.

Investment Costs[5] for Small Plants (1-MGD Range)

MSF and MED. Overall investment depends on site conditions and on a raw material prices (mainly for copper and aluminium alloy tubes). Because of price instability this analysis considers the following range of unit investment costs: low, at \$2.5/gpd (\$2.3/m^3/year, at 0.8 load factor)[6]; and high, at \$3.5/gpd (\$3.2/m^3/year, at 0.8 load factor).

VC. The cost instability noted holds here, too. The values assumed are \$3.0 and \$4.0/gpd for low and high costs, respectively.

ED for Brackish Water. Local site conditions, brine disposal facilities, and the pretreatment predicated by feedwater quality may strongly affect ED plant investment costs. The number of ED stacks needed for a given desalting capacity is also strongly affected by the amount of salts to be removed from the feedwater, so that total investment depends on feedwater salinity. The low and high costs assumed are \$0.7 and \$1.2/gpd, respectively.

RO for Brackish Water. The same factors affecting investment for ED hold for RO. However, the total investment for RO is less affected by feedwater salinity. The low and high costs assumed are \$0.7 and \$1.1/gpd, respectively.

RO for Seawater. Low and high costs considered here are \$3.0 and \$5.0/gpd, respectively. For systems using power recovery, \$0.1/gpd was added to the investment cost.

---

[5] All investment costs are based on early 1975 prices.
[6] Load factor is defined as the ratio of the actual yearly plant production to the anticipated production if the plant operates continuously at the nominal rate.

Table 2 Summary of Desalting Plant Performance

Ref. no.	Process	Units	Capacity (m^3/day)	Location	Feed source	Feed, total salinity (10^3 ppm)	Years in operation[a]	Accumulated water production[b] (10^3 m^3)	Capacity factor for year[c] (1)	(2)	On-stream factor for year[d] (1)	(2)
1	MSF	1	3,800	St. Thomas	Seawater wells	30-46	5	1,440	77.0	72.0	42.0	65.0
2	MSF	1	9,450	St. Thomas	Seawater wells	30-46	3	6,100	78.0	78.5	77.0	85.0
3	MSF	1	9,900	Key West	Seawater wells	30-46	4	5,570	67.0	55.0	67.5	78.5
4	MSF	1	9,100	Nassau	Seawater wells	30-46	1/4	570	—	98.7	—	71.0
5	MSF	2	6,800	Eilat[e]	Seawater	46	9 & 4	10,970	97.2	93.2	76.0	69.0
6	MED (HTE)	1	4,000	Eilat[e]	Seawater	46	1	660	—	75.8	—	59.6
7	MED (VTE)	1	3,800	St. Croix	Seawater wells	36-42	3	2,220	85.0	64.0	78.0	88.0
8	VC	4	1,300	Ophira[e]	Seawater	46	5 & 2	450	—	79.7	—	67.8

9	ED	1	2,400	Mashabei Sade[e]	Brackish water	2.5	3	609	80.6	81.8	19.2	18.4
10	RO	3	176	Eilat[e]	Brackish water	2.5	1-2	38	74.7	104.0	55.0	72.0
11	RO	1	20	Di-Zahav[e]	Brackish water	2.5	2	8	103.0	100.0	71.9	84.1

[a] Years in operation: reference numbers 5, 6, 8-11, refer to the year ending March 31, 1975; reference numbers 1-3, 7 refer to the year ending December 31, 1971; reference number 4 refers to the three months September-Nobember, 1972.

[b] Accumulated water production refers to dates identical with those in note a above, except that reference numbers 1 and 3 refer to three years (1969-1971) only.

[c] Capacity factor is defined as the ratio of the total yearly production of all units at one site to the nominal hourly production multiplied by unit-hours in operation. Capacity factor for year 1: reference numbers 5, 6, 8-11 refer to the year ending March 31, 1974, reference numbers 1-3, 7 refer to the year ending December 31, 1970; Capacity factor for year 2; reference numbers 5, 6, 8-11 refer to the year ending March 31, 1975; reference numbers 1-3, 7 refer to the year ending December 31, 1971. Capacity factor for reference number 4 refers to the three months September-November, 1972.

[d] On-stream factor is defined as the number of yearly operating hours of all units at one site divided by 8,760 (hours per year), multiplied by the number of operating units. For the definition of years 1 and 2, see note c above.

[e] Sites in Israel.

Energy Requirements and the Equivalent Fuel Consumption and Costs for Small Plants (1-MGD Range)

MSF and MED. The value of heat input to these processes is 50 to 70 kcal/kg of product water. Assuming a mean value of 60 kcal/kg, the fuel consumption in a single-purpose plant amounts to 6 kg/m^3 of product water.[7] If the thermal efficiency of the low-pressure boiler is considered, and some allowance is made for process steam consumed by the ejectors for removal of noncondensables, the actual fuel consumption amounts to about 7 kg/m^3 of product. In addition to energy requirements in the form of thermal energy, the plant process pumps consume an average of 4 kWhr/m^3 of product, an equivalent fuel consumption of 1 kg for each cubic meter of product water.

VC. The total energy consumption of VC units is about 18 kWhr, an equivalent of 4.5 kg fuel/m^3 of product water.

ED. As already indicated, the direct current consumption for ED is proportional to the amount of salts to be removed. For optimum current densities, about 1.3 kWhr/m^3 of product is required for every 1,000 ppm TDS removed. In addition, about 0.5 to 0.8 kWhr/m^3 are required for pumping power. Thus, to reduce the salinity of a 2,500-ppm TDS feedwater to 500 ppm, and ED plant will consume about 3 kWhr for each cubic meter produced.

RO. The average energy consumption for present brackish-water technology amounts to about 2 kWhr/m^3. The predicted energy consumption for RO seawater desalting is about four times higher because of higher process pressures and lower extraction rates. About a quarter of this energy, however, could be recovered by applying water turbines. The equivalent fuel consump-

Table 3 Energy Consumption and Cost for Various Desalting Processes (1-MGD Range)

Process	Fuel consumption[a] (kg/m^3)	Cost, \$/m^3 at fuel price (\$/ton)		
		50	75	100
MSF and MED	8.0	0.40	0.60	0.80
VC	4.5	0.23	0.34	0.45
ED for brackish water	0.75	0.038	0.056	0.075
RO for brackish water	0.5	0.025	0.038	0.050
RO for seawater	2.0	0.10	0.15	0.20
RO for seawater with power recovery	1.5	0.075	0.113	0.150

[a]Assuming a lower heating value of 10,000 kcal (41,800/J) per kg of fuel oil.

---

[7] Assuming a lower heating value of 10,000 kcal (41,800 kJ)/kg of fuel oil.

Table 4 Basis for Computing Operating and Water Cost for 1-MGD Range Plants

Cost parameter	Process					
	Distillation		Membrane			
	MSF; MED; VC for seawater		RO and ED for brackish water		RO for seawater	
Interest rate (%)	10.0		10.0		10.0	
Insurance & taxes (%)	1.2		1.2		1.2	
Plant life (years)	30		30		30	
Plant load factor (%)	80		80		80	
Operating & maintenance staff at $12,500/man-year	12		8		10	
Maintenance (% of total investment)	3.0		3.0		3.0	
	Low	High	Low	High	Low	High
Feed treatment & chemical cleaning ($/m^3 product)	0.04	0.06	0.04	0.06	0.12	0.18
Membrane replacement ($/m^3 product)	–	–	ED 0.02 RO 0.07	ED 0.03 RO 0.10	0.28	0.40
Electrical power, all costs less fuel ($/kWhr)	0.015	0.025	0.015	0.025	0.015	0.025

tion for this process therefore amounts to 0.5 kg/m^3 for brackish-water desalting, and 2 or 1.5 kg/m^3 for seawater desalting, without or with energy recovery, respectively.

Table 3 summarizes the fuel consumption for the various processes detailed above and presents the costs resulting from fuel at low ($50/ton), medium ($75/ton), and high ($100/ton) prices.

Operating and Water Costs for Small Plants (1 MGD Range)

As for investment costs, local conditions may have some effect on operating costs. The basis for computing the operating and water costs is summarized in Table 4.

Water costs for the various processes resulting from the combined extremes for investment and operation and fuel cost are summarized in Table 5.

Investment Costs for Large Plants (100 MGD Range)

With present technology only two processes, multistage flash and multi-effect distillation, are practicable for large plants. The MED has two versions:

Table 5 Summary of Water Costs for Combined Extremes of Investment[a], and Fuel Costs 1-MGD Range ($/m³)

Process	Investment-operation-energy cost combinations			
	Low-Low	Low-High	High-Low	High-High
Distillation processes				
MSF and MED	0.97	1.37	1.17	1.57
VC	1.08	1.30	1.40	1.63
Membrane processes				
ED for brackish water	0.34	0.38	0.48	0.54
RO for brackish water	0.34	0.37	0.48	0.50
RO for seawater	1.14	1.24	1.66	1.76
RO for seawater with power recovery	1.10	1.17	1.61	1.68

[a] All investment costs are based on early 1975 prices.

vertical tube evaporators (VTE) and horizontal tube evaporators (HTE). Israel Desalination Engineering, Ltd., has developed a relatively low-temperature version of the HTE process, which is considered here, together with the more conventional, and as yet most-applied MSF process. For both processes, $1.5/gpd ($1.4/m³/year at a 0.8 load factor) to $2.0/gpd, seems a reasonable range for specific investment costs, with slight advantage to the HTE.

Energy Requirement and Costs for Large Plants (100 MGD Range)

The two processes were analyzed for a 600 MWe (megawatt electric) net 100 MGD light-water, nuclear dual-purpose plant. The energy requirements of the two processes, for an assumed performance ratio of 10 lb of product water per 1,000 Btu of steam, are as follows:

1. Process steam: 8.7×10^8 kcal/hr at 127°C for the MSF process and at 73°C for the MED process
2. Electrical power: 2.1 kWhr/m³ and 2.2 kWhr/m³ for the MSF and MED processes, respectively

The thermal output for a 600 MWe net power reactor is about 1,875 MWt (megawatt thermal); for a dual-purpose plant thermal outputs would need to be increased by about 600 and 330 MWt, respectively, if integrated with a MSF or MED plant. These values allow for increase of the gross electrical power output needed to supply the desalters. Incremental energy requirements for the MSF and MED processes, respectively, amount to 38 and 21 thermal kWhr/m³ of desalted water.

If the economic benefits of the combined power and water production are attributed to the latter, only incremental costs are allocated to the energy requirements of the desalting plants. Thus, in addition to the incremental nuclear fuel consumption, only the incremental investment for enhancing the thermal output of the nuclear reactor is allocated to the energy cost for desalting. The present nuclear fuel price for light-water reactors is about $0.8/thermal megawatt-hour. In this analysis a range of fuel prices—from the above to an assumed 100% increase—is considered.

A wide range is also assumed for estimating capital costs for nuclear reactor enhancement. At low value, the investment cost of the nuclear reactor (not including turbogenerator), for a 600 MWe power-only plant, was assumed to be $300 million. Incremental costs allocated to the energy required for water production was calculated by assuming a scale factor of 0.6. By applying the same economic ground rules as used for calculating the capital costs of the desalting plant, the resulting low value of total energy cost amounts to an approximate $2.35/thermal megawatt hour, or $0.685/million Btu, the high value being $1.37/million Btu.

Table 6 summarizes the incremental thermal energy requirement and cost for MSF and MED (HTE), in dual-purpose plants, at low and high fuel prices.

Operating and Water Costs for Large Plants (100 MGD Range)

The basis for computing for operating and water costs is as follows: The value of economic parameters used to calculate the capital cost (interest rate, plant life, and plant load factor) are the same as those taken for the 1-MGD-range plants, see Table 4); the operating and maintenance costs, less energy, are based on an early 1973 data compilation and estimate of the same plant types and sizes. The respective values are $0.060 and $0.045/m^3 for the MSF and MED processes.

Table 6 Incremental Thermal Energy Requirement and Cost for MSF and MED (HTE) 100-MGD Dual-Purpose Plants

Distillation process	Energy requirements (incremental)		Nuclear energy cost ($/m^3)	
	MWt	KWhr(t)/m^3	low ($0.685/10^6 Btu)	high ($1.37/10^6 Btu)
MSF	600	38	0.09	0.18
MED (HTE)	330	21	0.05	0.10

A summary of the resulting water costs for the two plant types at the extreme ranges of investment and nuclear energy costs is given in Table 7, the effect of varying nuclear energy costs is illustrated in Fig. 13.

The art of large evaporator design, construction, and operation is as yet a subject treated only in engineering and technological studies. It may be reasonably assumed that the advent of large, seawater, evaporative-type systems will bring about a substantial improvement in many if not all of the critical system components, such as construction materials, plant geometry and heat-transfer rates, feedwater pretreatment, and construction and production methods. The cost figures quoted in Table 7 can at best give a notion as to what a water economy may encounter if and when it becomes imperative to introduce a large desalting plant into the system.

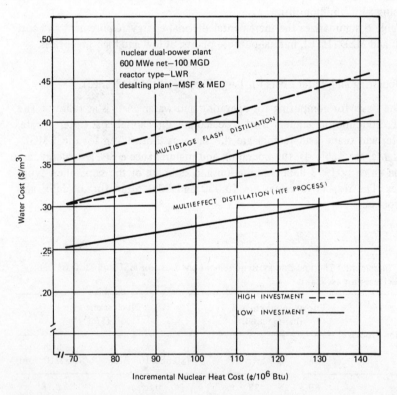

Fig. 13 Water costs versus nuclear heat costs. All investment costs are based on early 1975 prices.

Table 7 Summary of Water Costs for Possible Combination of Investments and Nuclear Energy Costs for a 600-MWe, 100-MGD Nuclear Dual-Purpose Plants

Distillation process	Investment-energy cost combinations ($/m^3)			
	Low-Low	Low-High	High-Low	High-High
MSF	0.31	0.40	0.36	0.45
MED (HTE)	0.26	0.31	0.31	0.36

References

Alexander, L. G., and Hoffman, H. W., (1971). Performance characteristics of advanced evaporation tubes for long-tube vertical evaporators. Oak Ridge National Laboratory, ORNL-TM-2951.

Glueckstern, P., Arad, N., Greenberger, M., Kantor, Y., Streifler, I. E. (1972). Study of nuclear dual-purpose plants suitable for non-base load applications. Mekorot Water Co. Ltd., Tel-Aviv, Publication 231-102B, International Atomic Energy Agency, Vienna, Microfiche IAEA-R-700-F.

Glueckstern, P., Steingarten, A., Bar-Peled, A., and Cohen, U. (1975). Operating data of Israeli desalting plants in 1974. Mekorot Water Co., Ltd., Tel-Aviv.

Hornburg, C. D., Bailie, R. E., Morin, O. J., and Suratt, W. B. (1973). Commercial desalting plant data and analysis—volume 1. Summary and comparison. National Technical Information Service, U.S. Department of Commerce, PB-226809.

Newton, E. H., Birkett, J. D., and Ketteringham, J. M. DSS Engineers, Inc. (1972). Survey of materials behavior in large desalting plants around the world. Arthur D. Little, Inc., Report to the Office of Saline Water.

Rapier, P. M., Rowe, W. H., Ko, A. C., DeReinzo, P. P., and Gitterman, H., Burns and Roe, Inc. (1972). Analysis and summary report of operation Guatanamo Naval Base desalination facility. Office of Saline Water Research and Development Progress Report No. 769.

15

The Strategy of Water Resource Management and Development in the Arid Zone

Aaron Wiener

Tahal Consulting Engineers Ltd., Tel Aviv

Some Definitions

Before entering upon the subject matter proper, the meaning of a few terms, as used in this chapter, will be defined.

> Arid zone: Emphasis will be on actual or anticipated water resources scarcity, rather than on climatic criteria.
> Phase space: A space in all the dimensions that are significant for planning.
> Planning space: The phase space within a specific geographic region.
> Geometry: Distribution within the phase space, of resources, demand, and pollution.
> Development: Interventions aimed at upgrading the resources, pollution, and demand geometries with a view to achieving a set of predetermined objectives.
> Management: The decision-making process related to the operation of the overall system, comprising the resources, demand, and pollution geometries, and their modification by development.
> Decision space: Space comprising only decision environments (i.e., factors influencing the decision-making process) and decision streams.
> Policy: Incentives and disincentives aiming at modifying part of a decision space.

Planning and management of a resources system represent interventions aimed at establishing a conscious and reasoned allocation of resources that will lead to the achievement of predetermined objectives, within the limitations of specified constraints. In a market-oriented economy, planning will not evolve spontaneously and the pattern of resources allocation will be the aggregate outcome of individual decisions by numerous decision makers. Only when scarcity becomes imminent will planning be resorted to.

Water resources constitute one of the preconditions for the existence of man, and one of the most important raw materials for his economic activities. This applies especially to arid countries, where rainfall is insufficient for the growing of crops. Because of their supreme importance, water resources generally become the subject of development and management planning some time before water scarcity becomes acute. The institutional frameworks of planning, and the phase space correlated with these frameworks, will, as a rule, be guided by purely politico-administrative considerations, which bear no relation whatsoever to the water resources geometry. The discrepancy between the administratively defined phase space of planning and the planning space that would best fit the water supply and pollution geometry will constitute a serious barrier on the road to optimal solutions. With increasing polarization of demand, this discrepancy will bring about inefficiencies that will press, in turn, for the phase space of planning to expand so that it will embrace appropriate sectors of the water resources demand and pollution geometries. In time, the growing imbalance between these geometries will reach a point where the merging of a number of phase spaces into regional spaces will become necessary, and once these imbalances become sufficiently serious, a national planning space may have to be adopted.

The Phase Space and Its Dimensions

The comprehensiveness of the phase space of planning (i.e., the number and type of its dimensions) and its physical extension (i.e., geographic coverage and time horizon) should be limited to the minimum necessary to obtain reasonably effective solutions and to assess the principal, side, and spilloever effects. This limitation of the phase space is dictated by several reasons: the prevailing fragmentation in the politico-administrative decision process, the stochastic nature of many of the variables included in the phase space, the numerous uncertainties inherent in many of the functions involved, insufficient information, and economic considerations. However, boundaries of the phase space of planning should not be considered permanent; the comprehensiveness and extension of the phase space can and should be adapted to changes in the situation and to the anticipated rate of these changes. Thus, with increasing scarcity of water, the phase space will have to become both more comprehensive and more extensive. Since in arid regions scarcity tends to increase with time, the phase space of planning will have to be an *expanding* phase space.

As long as the emphasis remains confined to upgrading and otherwise manipulating the physical aspects of the system, the dimensions of the phase space (embracing resources, demand, and pollution geometries) consist of: the two dimensions of the plane (geographic distribution), the third dimension of space (potential energy), time, flow, and the qualitative dimensions (subdivided at least into biological and physicochemical quality dimensions). As water becomes scarcer, emphasis will have to shift to other dimensions—manipulation of the resource base itself into which water enters as a production input, manipulation of demand, water allocation, and similar considerations. This implies an expansion of phase space to include the principal dimensions of the economic, institutional, and political universe. When this stage is reached, the problem of subdivision of the phase space will arise and hierarchic structure of the phase space may have to be adopted in order to facilitate analysis and to conserve resolution.

Implications of Aridity

Aridity brings to the fore a number of issues that in regions with ample water resources may for a long time remain of secondary significance only. The most important among these issues are as follows:

1. With increasing scarcity, the need for an efficient water allocation process becomes more pressing, and one can no longer afford to tolerate allocation inefficiencies carried over from periods when scarcity was not a governing factor.
2. As the rate of water utilization in arid or semiarid climates increases, the problem of maintaining quality standards becomes increasingly pivotal and more difficult to solve. The high rate of evaporation that characterizes such regions, combined with intensive utilization and the corollary pollution (relative to available flows that could act as dilutants or flushing agents), will make it increasingly difficult for water resource management to achieve qualitative steady states along with quantitative ones.
3. The necessity to maintain the quality of the water resources and to improve the socioeconomic efficiency of their allocation calls for the use of a more comprehensive phase space in planning—one that will, in addition to having resource demand and pollution geometries, also have the relevant economic, institutional, and political dimensions.
4. Since there should be a homologous relationship between comprehensiveness and extension of the phase space and the structure of the supporting organizational framework, organizational structure will have to be adapted to suit the expanding phase space. The same applies to the legislative underpinning of water resources management.

Orthodox Planning

In orthodox planning, development and management are considered separately. The limelight is predominantly on development, whereas resource management is usually an afterthought, taken up only when the development phase is completed. Development concentrates mainly on increasing the availability of water to satisfy an exogenous demand function. Development planning, then, becomes a definition of the interventions that have to be applied to the water resources geometry, and qualitative deterioration, in turn, are considered merely as undesirable side effects to which only corrective measures have to be applied in order to keep deterioration within tolerable limits.

Orthodox planning operates with static inventories: Its point of departure is the status quo of water resources geometry; its aim is to reach a target status dictated by demand forecasts; the transitional states of the resource system and the opportunities that such transients offer are ignored. This applies, of course, mainly to groundwater sources that can often be shown to hold considerable reserves of water, available on a one-time basis during the transitional states. Orthodox planning usually treats groundwater and surface flows as separate systems and does not attempt to utilize the comparative advantages of both and the synergetic effect that can be achieved by an integrated use that combines surface storage of water with storage in a groundwater formation.

The inefficiency of water resources allocation resulting from the assumption of an exogenous demand function is often compounded by unrealistic water rates quite unrelated to real costs. This distortion of water rates counteracts any attempt to rationalize the allocation and use of water resources, to compensate for increasing scarcity. Inefficient allocation will continue until the resources approach exhaustion and administrative rationing measures have to be imposed. The artificial removal of all self-regulative mechanisms will only hasten the exhaustion of the resources, and when the moment of truth arrives, the planner will be faced with the dilemma of either changing the rate system and introducing regulative market mechanisms or supplementing the dwindling water resources by inordinately expensive water. Neither of the two choices are, in fact, feasible. Any attempt at introducing regulative market mechanisms will meet with violent resistance by water users who have become accustomed to nominal pricing scales. On the other hand, the introduction of high-cost water without first increasing use efficiency will be self-defeating, for inefficient use is incompatible with high-cost water.

In the long run the most damaging element of orthodox planning may prove to be the separation of the treatment of the pollution geometry from that of the water resources geometry. By failing to include the social cost of pollution early in the planning and evaluation process of water resources development, water users will tend to consider the environment in general, and more specifically the downstream range of the water resources system, as a medium for waste disposal. The counterproductive effects of such a "tragedy of the Commons" may, in some cases, more than offset the direct benefits achieved.

With time, public legislation will assume the role of the "warden" of the environment in general and especially of water resources; it will establish standards for the quality of pollution streams, and impose fines proportional to the degree of pollution. Such an enforced "internalization" of so far neglected social costs may motivate the planner of waste disposal and treatment facilities to adopt pollution control measures that at best allow for the social costs of pollution streams; it cannot, however, take the place of a comprehensive water resources planning process that comprises pollution geometry.

The Integrative Planning Process

As the shortage of water becomes more evident and pollution streams more objectionable, the phase space of planning will have to expand until it will encompass also, in addition to resource geometry, demand geometry and pollution geometry, and extend into the political, institutional, and social dimensions. The role of demand will, thereby, shift from that of an exogenous variable, taken over from the national or the sectorial national plan, to that of an endogenous variable, amenable to direct manipulation by administrative measures or, preferably, indirect manipulation by economic policies, designed as part of a process of integrative planning.

In upgrading the water resources geometry, underground and surface waters will be considered as two interdependent resource phases that have to be managed in a way that will lead to the best combination of objectives achieved. Integrative planning implies that the development of the system—permanent physical upgrading of the resources system through the construction of appropriate facilities—and its management, will no longer be treated as separate concepts, and that development planning will be based on the selection of the most attractive management alternatives of the system.

Simulations will be conducted to explore alternative management patterns and demonstrate the potential operational uses of transition stages of water resources, and especially of the groundwater phase for lengthy interim periods. This will also entail abandoning static inventory concepts such as the "steady-state safe yields." Instead of this, alternative resources system management patterns should be investigated for their compatibility with system objectives and with the specifications laid down for the target status, that is, the quantitative and qualitative state of the system at the time horizon of planning and the management it implies beyond the time horizon. Thus, compliance of the system with planning objectives up to the time horizon—while conserving its integrity for use by future generations—will be the criterion for the eligibility of a program, while preference ranking of alternative eligible programs will be established in accordance with predetermined economic or other criteria.

Possibly the most important implication of the use of an expanded phase space will be the highlighting of the qualitative features of alternative management patterns of water resource geometry, and especially of its groundwater phase. Every alternative management pattern aimed at certain program objectives will imply a change in water quality (biological or, more important, physicochemical change). This change can be the result of a shift in the natural balance between pollution streams and fresh water flows, brought about by the exploitation of the resource; it may result from a change in the flow patterns of underground water bodies of different qualities; it may be a side effect of water use (sewage, drainage); it may be caused by the out-of-basin import of waters containing mineral or other pollutants; or it may be the side effect of an economic activity of man not related directly to water resources (soil pollution). In arid regions, because of the relative scarcity of water and the ubiquitousness of mineral pollutants, quality deterioration will proceed more rapidly, affecting groundwater basins, or the middle or lower reaches of rivers. Since the various uses of water are subject to restrictions corresponding to their tolerances to pollutants, water resources may, in time, become unusable for certain purposes.

It follows that integrative planning has to take into account qualitative changes that result from alternative management patterns, along with quantitative ones. Every management pattern will imply a specific sequence of flows and quality changes, and every quality change will result in a change of the economics of water use, a change that will become especially conspicuous when quality levels approach or exceed permitted tolerance levels for an important water use. Taking appropriate management measures in good time may make it possible to postpone the date at which these quality tolerance levels will be exceeded. Such anticipative quality management measures may, inter alia, include the following; blending of waters of different qualities; dilution of substandard water with higher-quality water, storage of water of high mineral content at the bottom of a groundwater formation and withdrawal in its stead of higher-quality water from the upper part of the formation, and so on. Management measures for quality are often slow in taking effect and may, therefore, have to be undertaken long before the anticipated situation becomes critical.

Since the imbalances between pollution streams and fresh water flows usually make for progressive pollution and a continuous increase in the concentrations of pollutants in water, the quality of water resources tends to gravitate toward nonsteady states. Some of these qualitative changes are subject to inertial delays and may, therefore, escape detection for some time, but their long-run effect may become ruinous, unless cumulative deterioration of the source is prevented in time or—and this applies to the majority of groundwater management policies—delayed by appropriate management measures.

The selection of the development plan should be based on a comparison of the relation between the respective quantitative and qualitative benefits (or regrets) and costs of the various development and management alternatives; thereby, the conservation of reasonable quality standards of water resources will be built into the original development plan, instead of being treated only as an afterthought, and a satisfactory management pattern achieved.

The incorporation into the planning framework of long-term quantitative and qualitative changes in the resources system is, of course, conditional on an extension of the time horizon and implies an extension of planning considerations to the conservation of the quantity and quality of water resources for continued use by future generations.

Once conservation and the maintenance of acceptable steady states (or disequilibrium states made acceptable by continuous management intervention) are accepted, there is reason to doubt the value of discounting procedures that reduce the influence of the benefits and regrets assumed to accrue to future generations from decisions of the present generation. In this context, discounting would be unjustified and the maintenance of an acceptable target status of the resources system should be considered as a planning constraint[1].

The adoption of such a far-sighted approach, incorporating joint consideration of the water resources and pollution stream geometries, will, in many cases, have a decisive influence on the shaping of development plans. For example, where major agricultural and nonagricultural water users are located above an intensively utilized groundwater formation and drain into it, or where return flows from irrigation add a significant mineral load to a surface stream, the concurrent review of potential long-term quantitative and qualitative deterioration implied in alternative development schemes might lead to the relocation of development areas, to a different scheduling of development phases, or, in extreme cases, to a complete remodeling of the development scheme.

The responses of quantitative and qualitative characteristics of the water resources system to development interventions will often prove to be a slow process; in the long transition periods until new steady (or unsteady) states come about, transients in the form of one-time stocks of groundwater will become available. The proposed dynamic planning process will highlight the opportunities and potential benefits offered by these transients, especially in resources systems in which groundwater plays a significant role. The proper utilization of these transient groundwater stocks should represent an important planning consideration and prevent their loss to base levels from regions of anticipated scarcity during such transitory periods. In some cases the judicious use of tran-

---

[1] Futures, Vol. 19, No. 5, October 1977, pp. 383-403.

sients may create the opportunity to initiate and predevelop the demand and to improve the effectiveness of the production process in irrigated agriculture, well ahead of the major investment in large and undivisible engineering works. This can greatly reduce the costs of gestation and maturation that characterize many large-scale irrigation projects.

Hierarchic Structuring of the Planning Space

The expansion of the phase space into the institutional, social, and political dimensions will, of course, imply the incorporation of subspaces of low resolution and quantifiability. The resulting expanded comprehensive phase space will thus become extremely nonhomogeneous, since it will consist of well-structured subspaces of relatively high quantifiability and resolution, along with relatively ill-structured subspaces of low quantifiability and resolution.

The analysis of a nonhomogeneous phase space can best be effected by a hierarchical structuring of the overall phase space into more homogeneous subspaces. In such a hierarchically structured space, every subspace can be analyzed separately, the well-structured ones by a fully quantified analysis, the ill-structured ones by a more ordinal type of analysis, whereas the interaction among subspaces can be the subject of a separate low-resolution analysis that will produce the parametric values and constraints needed for the analyses of the separate subspaces. When analyzing a specific subspace, some variables of the other subspaces will, of course, serve as exogenous variables for the subspace under analysis. To arrive at quasi-optimal solutions, the analysis of the overall space and its constituent subspaces may have to be iterated.

Within the most general phase space, the water resources subspace will have the highest degree of resolution and quantifiability. The pollution subspace will have less resolution, but will still be moderately well structured and quantifiable. The demand subspace, which will comprise numerous production processes and depend on various macroeconomic, institutional, and political factors, will have the least resolution. Therefore, the water resources and pollution subspaces can often be merged into a single relatively well-structured subspace, while the demand space will have to be analyzed separately.

Difficulties Inherent in Analysis

Comprehensive analysis of water resources is subject to a number of unavoidable difficulties that must be carefully considered in the course of analytical procedures. Some of the more common difficulties encountered are reviewed in the following.

The Stochastic Nature of Water Resources Variables

Some water resource variables can, of course, be expressed only in terms of probability distributions, derived, faute de mieux, from past records. This, in turn, implies the assumption (unwarranted according to some meterologists) that future probability distributions of water resources variables will be similar in shape and parameters to those of the past. It would be uneconomical to design solutions applicable to all climatic conditions. A decision will, therefore, have to be made, stipulating the frequency for which the supply may fall below design values. Where such a shortfall would cause unwarranted complications or damage, one would have to design in addition to the "design solution" resulting from the selected design frequency of shortfalls, additional interventions ("contingency solutions") to be undertaken only if such a shortfall is considered imminent. In this case the state of the system would also have to be specified that would give such an intervention the green light, taking into account lead time for construction. Because of the stochastic nature of water resources variables, regulative storage will be needed to upgrade the actual availability of flows, and sometimes also the quality of water in accordance with demand. Often it will be found uneconomical or even unfeasible to provide sufficient surface storage facilities to ensure a safe design flow (with predetermined frequencies for shortfalls). In such cases, conjunctive utilization of water resources, comprising both the surface and the underground phase (where there is a sufficiently large groundwater formation) can greatly increase the regulative capacity of the system and thereby its dependable yield. Given a total storage capacity (surface and underground), the annual yield of a system will depend on the degree of regulation that we wish to impose upon the system: If we insist on a constant annual supply under all flow conditions (or, even more so, if we specify a larger supply in dry years to make up for the deficiency in rainfall), the "safe sustainable" supply will be smaller than if we accept a reduced supply in dry years.

Uncertainty Regarding Economic Parameters

Fluctuations of national and sectorial economic parameters exert a decisive influence on water resource development planning. Variations in macroeconomic parameters influence the rate of industrial expansion and housing construction and the corresponding rate of growth in water demand. Even more important are the extreme fluctuations in the prices of agricultural commodities for which water is one of the principal production inputs in arid countries. Whereas in the 1960s and early 1970s these fluctuations remained within a reasonable range, they have become more extreme and erratic recently. This applies to the agricultural sector as a whole but particularly to some specific crops. Prices of some crops in the international market are extremely volatile because the volume of international trade represents only a small percentage of total production (in the case of rice, for example, about 5%), and small variations in production are, therefore, magnified into extremely large variations in supplies available for export.

During the last few years, a new factor has entered planning considerations connected with water resource development for irrigation, namely, food security. Opinions are divided as to whether a critical food shortage may develop in the near future as a consequence of an unfavorable regional climatic cycle, but a consensus was reached at the 1974 World Food Conference in Rome that such a contingency is sufficiently probable to warrant the establishment of worldwide food stocks which could be resorted to in order to bridge a critical shortfall of production. Even if such food stocks were established, national responses to a really serious food shortage on a global scale might become panicky, and it might prove impossible to convince national legislators to participate in mutual aid operations. Therefore, food security might become an issue that will have to be considered as a constraint transcending purely economic considerations.

Uncertainty Regarding Future Production Technologies

Uncertainty regarding the nature of future production technologies, both in industry and in agriculture, and uncertainty regarding the type of plumbing that will be used for new housing schemes, has an increasingly important influence on planning the development of water resources. In agriculture (and in some industries), water is an important production input in the resources base, and it can, to a certain degree, be replaced by other inputs or by appropriate investments, such as investments for control of the crop environment. In situations of water scarcity in which the marginal cost of water is increasing as a consequence of resorting to resources requiring higher capital investments and operational costs, it will be helpful to review the resources base of agricultural production with a view to reducing (within economic reason) the water input. Since agricultural production techniques are now entering a phase of intensive innovation, it is difficult to assess the most economical water input over the lifetime of a project. This difficulty will, of course, increase with the extension of the time horizon.

Uncertainty Regarding the Effects of Pollutants and Future Quality Standards

Industry and agriculture use more and more substances that may directly or indirectly reach surface or underground water resources. The long-term cumulative effect on human health of many of these substances is only imperfectly known, and every few months a new pollution scare causes commotion in the professional world and creates pressures for a revision of quality standards. Pollution, as a rule, tends to be more acute in arid countries, where dilution flows are rarely available and where groundwater utilization is liable to create quasi-closed basins in which pollution accumulates. We may assume that in the future some of the existing pollutants, and some of those that will be added in the course of time, will, by themselves, or, in combination with other substances, prove to be health risks at concentrations that are now considered tolerable. As a consequence, we have to assume that quality standards will, in time, become stricter. Prevailing water quality standards cannot, therefore, be considered as invariants, and development plans should have sufficient built-in flexibility to permit future compliance with unforeseeable additional restrictions regarding water quality. Because of our ignorance of the potential chronic health effects of existing and future pollutants, and in order to reduce the cost of additional treatment of water, we should, generally, aim to maintain, within economic reason, the natural quality of water resources and to prevent their pollution. Considerations of this nature may lead us to adopt, in the long run, a development scheme that would provide for selective uses of water, reserving the highest quality water for human and other low-tolerance consumption and allocating partly polluted waters and reclaimed waste waters selectively to less sensitive uses.

Conclusions

The stochastic nature of the many important variables of the water resources planning space and the increasing number of uncertainties that characterize them call for the development of appropriate planning strategies. To ascertain the influence of the various uncertainties, sensitivity tests will have to be applied and ranges of outcomes reviewed when making the final selection between alternative solutions.

Planning under conditions of uncertainty will require the incorporation of a considerable measure of flexibility into the project outline, so as to keep open as many options as possible for the future, while meeting the immediate needs of the project area. The prevalence of uncertainty implies that flexibility, subdivisibility, and phasing may have a very substantial economic value which should be allowed for in evaluations of alternative solutions. Ex-post-type analyses indicate that the planner rarely, if ever, assesses the extent of uncertainty correctly and that the error is, in the great majority of cases, an underestimate. This empirical fact ought to justify a greater emphasis on flexibility than has been customary in the past.

Planning and the Dispersion of Decision-Making Processes in Agriculture and Industry

Planning is usually confined to a definition of the intervention of the public sector into ongoing processes. However, the majority of the decisions regarding the utilization of water and the discharge of wastes are made by individuals whose interests are often at variance with the public interest. For the individual the water resources system is a "commons"[2] that is, a public domain without a "warden." The "commons" is an economic situation in which Adam Smith's "invisible hand" does not operate, that is, in which the striving of all individuals to achieve maximum personal benefits does not lead to a satisfactory resources allocation.

Let us take a pollution problem as an illustration. For the farmer the addition of nitrogen fertilizers will, up to a point, increase crop yields. He will, therefore, tend to increase fertilizer applications as long as (considering the risks involved) the value of incremental yields will exceed that of the additional dose of fertilizer. Intensive fertilization, however, will often entail a decrease in fertilizer uptake and, as a consequence, an increase in surplus fertilizers that will, in time, be transported (by irrigation and rainwater) to the groundwater formation and increasingly pollute it. Since the cost of the "externality" pollution is borne by the public sector, it will not enter the farmer's optimization calculations; however, as seen from the social point of view, the allocation of resources resulting from the farmer's optimization will not be an optimal one.

As long as resources are plentiful and the "metabolism" of water resources systems is rapid, these problems will rarely become acute. However, in arid countries, and especially in arid regions relying on groundwater formations located in areas of intensive economic activity, the discrepancies between the optima of the individual decision makers and those related to public welfare will rapidly increase and will require action by the public sector. It will, therefore, become necessary to include in planning not only those public interventions that are connected with the "hardware" aspects of the project, but also such interventions in decision space (controlling the dispersed decision processes of individual water users) as might be necessary to minimize the discrepancy between private and public optima. Where the rate structure of water does not reflect the true cost to the economy (and this is the case in the majority of irrigation uses), similar discrepancies will also affect quantitative aspects.

Modification of the decision space controlling a plurality of individual decisions can be effected in a number of ways, from mandatory legislation, through cooperative regional resources management organizations, to financial

---

[2] Garett Hardin, The tragedy of the commons, Nature, December 13, 1968.

incentives and disincentives. Because of the dispersal of the decision processes concerned, financial policies should be preferred, where applicable, aiming at internalization of the social costs of the decisions of the individual decision makers. The spelling out of such policies is, however, hampered by considerable theoretical and practical difficulties. Internalization of "externalities" related to water will often be resisted by both the water user and the macroeconomic planner: The former will condemn it because it increases his costs, the latter, because of its possible inflationary impact.

Institutional Aspects

A relation of homology ought to exist between the structure of the planning space and that of institutional organization. To ensure this homology, all the issues of a project or program that require public intervention should be placed under a single administrative, or, at least, coordinative, authority. The same applies to geographic coverage.

Such homology is a precondition for ensuring the synchronization and proper sequence of the numerous interventions that enter into development programs. This is especially true in the less developed countries, in which interventions cannot be confined to the customary project hardware, but have to extend to technological, social and institutional transformations. In reality we find, in most countries (and especially in less developed countries), a fragmentation of institutions and decision processes that will make synchronization of project intervention all but impossible.

If we wish planning to be more than an intellectual exercise, we have to ensure that this relation of homology will be maintained when the planning space has to expand and to become more comprehensive. Expansion of the planning space requires a parallel and concurrent expansion of institutional boundaries, manifested in the concurrent merging of correlated institutional decision processes under one authority, or at least, under effective and authoritative coordination. The absence of sufficiently wide (geographically speaking) and sufficiently comprehensive frameworks will often delay or prevent the drawing up of badly needed comprehensive development plans.

Institutions—or at least those structuring the public sector—are, by their very nature, averse to change and liable to resist change. The modification of the institutional dimension will, therefore, have to become an essential dimension of the planning space. To maintain "institutional feasibility," planning will have to take this initial resistance into account and make provisions for appropriate countermeasures. The intervention may take the form of merging a number of administrative district institutions into a single water resources development and management institution, or merging the fragmented authority over the various elements of a water-related, production process (e.g., agricultural extension, local institution building, incentive policies in irrigated farming), and investing one ministry, or one coordinative body, or one project organization with this authority.

The detailed development and management planning we have in mind here will be confined to regional or, at best, to sectorial planning. All sectors of the economy are, however, to some degree interconnected and interacting. To compensate for the neglect of intersectorial interaction implied in sectorial planning, we shall have to resort to intersectorial reconciliation. This coordinative planning operation will first define the type of services and inputs the water resources sector will require from other sectors and then determine to what extent such services or inputs can, in fact, be made available, or, alternatively, how the water program will have to be modified in order to harmonize it with development programs of other sectors.

The expansion of institutional frameworks and the increasing comprehensiveness of their operation will, of course, require a revision of the pertinent laws and regulations. This legal adaptation will be in the direction of increasing and centralizing public authority and control. Considerable resistance will have to be overcome before legal adaptation will become politically feasible, and, for this reason, the legal underpinning of the development process will also have to become one of the dimensions of comprehensive development planning.

Extension of the legislative framework will necessitate a parallel expansion of the means of the collection and generation of appropriate information and the setting up of an information and monitoring system designed specifically to serve the needs of the administration of water legislation.

Conclusions

To ensure the availability of water when and where it is needed, and to safeguard its quality in an era of rapid demographic and economic expansion, water resource management will be needed even in humid regions having an abundant supply of water. In arid regions, water resource management is indispensable if we wish to ensure an adequate supply for the most important uses and maintain the flow and quality of the water source for future generations.

In such regions, problems of biological and physicochemical pollution are liable to become acute at an early stage, and some of these deterioration processes may prove to be irreversible, or reversible only at high cost and after a long delay. If we wish to prevent deterioration of the resource, early action will be required well ahead of the acute pollution stage.

The selection of a sufficiently comprehensive planning space is, therefore, a precondition for the design of an appropriate management policy of water flow and quality in arid regions. Such a planning space should comprise, in addition to all the dimensions relevant to quantitative management, dimensions related to water quality, and it should extend over a sufficiently large geographic area. As scarcity becomes more acute, the planning space will have to be made more comprehensive and its area of coverage extended.

When scarcity reaches the point where it becomes unfeasible to meet the potential demands of all use sectors, the resources base of the principal production processes into which water enters as an important raw material will have to be included in the planning space, and demand will thereby become an endogenous variable in the water resource planning space. Institutional structure and legal administration will have to be adapted to the needs of water resource planning.

The expansion of the planning space (in the number of dimensions and in geographic and time coverage) is a reflection of the universal necessity to broaden the planning approach from consideration of ad hoc technological solutions, to the extension of planning to "externalities" (i.e., side effects on issues or sectors not included in the planning space) and, ultimately, to systemic planning. For economic reasons, and in view of the fragmentation of the politico-bureaucratic institutional framework, systemic planning might, in most cases, have to be limited to sectorial planning.

Arid regions will have to adopt systemic planning well before humid regions will, because the penalty for failing to do so will be much heavier.

16

The Future of Arid Lands

Allen V. Kneese and Jennifer E. Zamora

*University of New Mexico, Albuquerque, New Mexico
and Resources for the Future, Washington, D.C.*

Introduction

For many of us who have spent major parts of our lives in them, the arid lands have an appeal that is naggingly there no matter how long or far we may have been away from them. At the same time we are aware that even though they are the sites where civilization was spawned they present many severe problems for the populations that currently inhabit them. Moreover, even those persons with a rather casual knowledge of these lands have come to realize that for various reasons they are currently suffering enormous stresses and strains, alas, in many instances not for the first time in their history.

In this chapter we try to provide a broad perspective on the situation of the world's arid lands, the trends affecting them, and possible future developments. The reader will find that the tone of this chapter is rather pessimistic. This is not congenial to us, and we hope that events will prove us wrong in those instances where we see strong grounds for concern.

The first substantive section of this chapter is a modest statistical analysis of some published data. It is aimed at exploring two major hypotheses. The first is that climate as such, when other influences are accounted for, tends to affect per capita income and that aridity, in particular, affects per capita income negatively. The second is that arid regions that are hinterlands of more humid regions are in quite a different situation economically than those that are not. Other hypotheses concerning the influence on income of the age structure of the population, education levels, and so on, are also tested. Our main interest in these, however, is to assure that we have identified them and adjusted for their effect so that the influence of the factors of primary interest can be identified. We then go on to explore further the arid hinterlands type of situation and to a (necessarily) more qualitative and anecdotal discussion of other tendencies influencing the arid lands and their economic development.

Statistical Analysis

Data Collection and Procedures

Per capita income data for 118 countries were obtained for the year 1963 from U.N. Statistical Office sources.[1] The sample excludes Communist countries because, although data were available from the United Nations and other sources, it was felt that the possible effects of climate differences could be obscured or distorted in centrally planned economies.

1963 was chosen because it was the year for which the United Nations had per capita income figures for the largest number of countries. 1963 data were also gathered from U.N. sources for such socioeconomic variables as population, rate of population increase, age distribution of the population, life expectancy at birth, education levels, and urban population. Other economic variables for which data were collected included agricultural land as a percentage of total land area, and level of foreign aid received.

A climate classification was devised by collapsing Trewartha's classification into six major types: desert, steppe, tropical, mediterranean, humid subtropical, and humid continental. Each country in the sample was classified into one of the six types according to its predominant climate type. A second more aggregative classification of arid versus nonarid countries was made in which the countries classified as either desert or steppe were reclassified as arid and the countries in the other four climate types were designated as nonarid. Finally, arid lands as a percentage of total land area was estimated for each country, again using desert and steppe climate areas together as the definition of arid lands. A subset of the original sample was then created, composed of each country that had some proportion of arid land, however small, within its borders. The arid lands subset had 53 countries.

A simple linear regression model was used to estimate several equations formulated to determine if differences in economic well-being among countries, as denoted by the level of per capita income, can be attributed in part to climate differences in general and to aridity in particular. The equations were also designed to attempt to eliminate the influence of some of the variables that one would expect to have a significant effect on per capita incomes. In the following

---

[1] All data sources are listed in the Appendix to this chapter.

section the hypotheses and equations being tested are set out in more detail and the results of the statistical anaysis are presented.

Results of Statistical Analysis

Climatic Hypotheses

It was hypothesized that there would be significant differences in per capita income among groups of countries as a result of climate differences. To test this hypothesis two regression equations were estimated. The first was used to estimate the effect on per capita income of seven socioeconomic variables:

$$P_i = \beta_0 + \beta_1 R_i + \beta_2 PF_i + \beta_3 I_i + \beta_4 FA_i + \beta_5 LE_i + \beta_6 AL_i + \beta_7 PL_i \tag{1}$$

where
P   = per capita income
R   = rate of population increase
PF = percentage of population under age 15
I   = illiteracy rate of population over age 15
FA = foreign aid received
LE = life expectancy at birth
AL = arable land as a percentage of total land area
PL = pasture land as a percentage of total land area
i   = number of countries (118)

The second equation was estimated including dummy variables for the six major climate types used:

$$P_i = \gamma_0 + \beta_1 R_i + \beta_2 PF_i + \beta_3 I_i + \beta_4 FA_i + \beta_5 LE_i + \beta_6 AL_i + \beta_7 PL_i + \gamma_1 D_1 \\ + \gamma_2 D_2 + \gamma_3 D_3 + \gamma_4 D_4 + \gamma_5 D_5 \tag{2}$$

The two equations were estimated, the variables whose coefficients were not significant at the 0.90 level of confidence were discarded, and the equations were estimated again using only the significant independent variables. An F-statistic was calculated to test the null hypothesis that there is no significant difference between the intercept of Eq. (1) and the intercepts of the six lines represented in Eq. (2).

Climate differences do have a significant impact on per capita income since the results of the estimate and F-test given in Table 1 allow the null hypothesis to be rejected. These influences, however, are not as strong an explantory factor as might have been expected, as indicated by the low R^2 values obtained for each equation. However, this may be accounted for by problems of data comparability over such a wide sample of nations. The fact that a significant

Table 1 Climate and Aridity Effects on per Capita Income: Results of Estimating Regression Equations

Variable	Equ. (1)	Eq. (2)	Eq. (3)	Eq. (4)	Eq. (5)	Eq. (6)	Eq. (7)	Eq. (8)	Eq. (9)
Dependent Constant	p = 685.54 (2.325)[a]	p = 223.53 (0.567)	p = 1939.5 (9.585)	p = 1850.1 (9.114)	p = 1837.2 (9.089)	p = 370.19 (1.359)	p = 464.35 (1.623)	p = 315.72 (1.192)	p = 472.56 (1.735)
PF	− 23.999 (5.303)	− 20.625 (4.1941)	− 39.33 (7.103)	− 33.273 (5.33)	− 32.834 (5.327)	− 13.933 (2.971)	− 14.518 (3.079)	− 13.318 (2.929)	− 14.485 (3.152)
LE	12.303 (3.693)	9.243 (2.373)				9.045 (2.873)	8.743 (2.77)	9.384 (3.077)	8.831 (2.867)
AL	− 0.0026 (1.45)	− 0.0017 (0.996)	− 0.004 (1.631)	0.0038 (1.5682)	− 0.0044 (1.81)	− 0.0023 (1.73)	− 0.0025 (1.904)	− 0.0034 (2.441)	− 0.0029 (2.169)
FA						0.367 (1.381)	0.412 (0.153)	0.345 (1.339)	0.434 (1.653)
D1		374.54							
D2		362.66							
D3		470.99							
D4		761.43							
D5		352.04							
PA				− 2.995 (1.948)			− 0.994 (1.065)		

	(1)	(2)	(3)	(4)	(5)	(6)	(7)	(8)	(9)
DA					−278.41				
HT1								227.13	
HT2									−122.54
R^2	0.3724	0.4504	0.4653	0.4976	0.5043	0.3147	0.3309	0.3719	0.3585
SSR	21,420,000	18,760,000	12,400,000	11,650,000	11,500,000	28,810,000	28,130,000	26,400,000	26,970,000
DF	114	109	60	59	59	48	47	47	47

F-test for Eqs. (1) and (2) = 3.091[b]
F-test for Eqs. (3) and (4) = 3.798[c]
F-test for Eqs. (3) and (5) = 4.617[b]
F-test for Eqs. (6) and (7) = 1.136[d]
F-test for Eqs. (6) and (8) = 4.291[b]
F-test for Eqs. (6) and (9) = 3.207[c]

[a] Absolute values of t-values are given in parentheses below the coefficients t ⩾ 1.289 significance at the 90% confidence level.
[b] Significant at 95% confidence level.
[c] Significant at 90% confidence level.
[d] Nonsignificant.

effect can be attributed to climate differences is interesting. Another observation which can be made by examining Table 1 is that although percentage of arable land (AL) was a significant explanatory variable in Eq. (1), its effect is greatly reduced in Eq. (2) when climate differences are taken into account. This suggests that this variable was actually masking climatic effects in the first equation.

Socioeconomic Variables

The results in Table 1 also provide a test for other hypotheses regarding the effect of several socioeconomic variables. We report these and give the rationale for using the ones we did but reiterate that our primary interest is in the climatic variables. We must also caution that the data on the socioeconomic variables are often of very poor quality and are sometimes highly intercorrelated. We hope, however, that their influence as a group of variables will have been adequately accounted for in the equations estimated.

Population Change. The hypothesis that rapid rates of population growth tend to be associated with low per capita income is not borne out statistically in this analysis. However, the correlation between the rate of population growth and other independent variables such as percentage of population under age 15 (PF) was relatively high (0.52). (The correlation matrix for the independent variables for which data were gathered is given at the end of the Appendix to this chapter). Since (PF) was a significant explanatory variable in the first estimation of Eq. (1), it was felt that it probably included the population growth factor and, therefore, R was eliminated.

Population Age. A rapidly increasing population will usually be a young population, which implies that a relatively small proportion of the population will be involved in the labor force. This would be expected to have a depressing effect on per capita income, regardless of whether the country is predominantly arid or humid. This conjecture is supported by the statistical analysis; percentage of population under age 15 (PF) is a significant variable in both equations and, as explained above, most likely accounts for some of the variation expected as a result of differences in the rate of population growth.

Level of Education. It was hypothesized that low levels of education in a country would be related to low levels of per capita income without reference to climatic conditions. Although the analysis did not indicate education level, as measured by the rate of illiteracy, as a significant variable, this may be due to the fact that illiteracy may not be the proper measure of education level to use. Perhaps some other measure, such as median years of school completed for various age segments of the population, would produce better results; however, these data were not available for enough countries in the sample to permit this supposition to be tested.

Health. It was hypothesized that the health of a country's citizens as measured by life expectancy at birth would be positively related to per capita income. This conjecture was supported by the analysis as demonstrated in Table 1. Life expectancy is a significant explanatory variable in both equations, although in Eq. (2) its t-value is somewhat lower as a result of the inclusion of direct climatic effects.

Arable Land. The percentage of arable land (AL) in a country can be taken as a measure of potential agricultural productivity. We expected this variable to be positively related to per capita income, but, in all of the equations we estimated, this variable, whether significant or not, showed a negative relationship to per capita income. After some reconsideration we realized that although agriculture is an important industry in most countries, it is also one of the lowest in value added in production per value of productive inputs. The negative coefficients obtained in Eqs. (1) and (2) imply that, in general, an increase in arable land will actually have a depressing effect on per capita income. In other words, the resources required to bring more land under cultivation are greater than the addition to national income from the increased agricultural production. This situation may be aggravated by the heavy subsidization of agriculture in many countries.

The results in Table 1 show that the initial significance of the arable land variable in Eq. (1) is actually in response to climatic differences. When these differences are accounted for in Eq. (2), the general effect of (AL) is decreased although still negative. This implies that the significance in Eq.(1) was due to larger negative effects of AL for climate extremes.

The other variables in the first formulation of Eqs. (1) and (2) were nonsignificant and/or highly correlated with some other independent variables and were, therefore, eliminated from the analysis.

Aridity Hypotheses

Two additional equations were formulated to determine if aridity, in and of itself, has a significant influence on per capita income. Equation (4) is designed to test the hypothesis that the percentage of arid land area in a country is inversely related to per capita income, whereas Eq. (5) is designed to test for significant differences in per capita income between predominantly arid and nonarid countries. We felt that perhaps the full sample should not be used to test the impact of aridity on per capita income level because predominantly tropical countries also had very low per capita incomes and it was expected that this would obscure the aridity effects. Therefore, a subsample of 62 countries was constructed that excluded tropical countries. The estimates of Eqs. (3), (4), and (5) in Table 1 give the results of testing the aridity hypotheses with this "nontropical" sample. Equation (3) is the hypothesis equation without a climate

variable, whereas Eqs. (4) and (5) include the two aridity variables:
PA   = percentage of arid lands in a country
DA   = a dummy variable denoting countries with 50% or more arid land area

The results of the estimation show that when tropical countries are excluded from the sample, the aridity hypotheses are supported. The t-value for percentage of arid land (PA) in Eq. (4) was significant at the 0.95 level and produced a substantial increase in the R^2 value. The F-statistics for comparing Eqs. (3) and (4) and Eqs. (3) and (5) were significant, indicating that conditions of aridity do have a significant impact on per capita income differences between countries.

Arid Hinterlands Hypotheses

The main hypothesis here was that those arid lands that are hinterlands of a more humid region, that is, located in a nation in which less than 40% of the land area is arid, will have a higher per capita income than those not so located. We want to test the notion that arid hinterlands can serve a number of useful purposes for a prosperous humid region and thereby enjoy relatively high per capita incomes, but that when nearly the entire country is arid there will not be a basis for this higher level of income.

To test this and other hypotheses, a set of equations was formulated and estimated using the original variables with the arid countries subsample of 53 countries. The equations were estimated, the nonsignificant variables discarded, and the final equations subjected to further statistical testing. The variables in these equations have all been identified previously with the exception of (HT_1), which is a dummy variable for countries with arid hinterlands defined as countries having between 25 and 40% arid lands, and (HT_2), which is a dummy variable for those countries having 75% or more arid land area.

The results shown in Table 1 support the hypothesis concerning arid hinterlands. An F-statistic calculated to compare Eqs. (6) and (8) was significant at the 0.95 level of confidence, which indicates that among countries with arid lands those that have an arid hinterland as defined previously also have significantly higher per capita incomes than those without.

The hypothesis that, among arid countries, the percentage of arid land area in a country would negatively influence per capita income is borne out to some extent. The coefficient for the (PA) variable is negative as predicted, although the t-value for the variable and the F-test between Eqs. (6) and (7) are nonsignificant. This appears to be at odds with the results in Table 1, but it must be recalled that no attempt was made to control for the influences of tropical countries in this sample.

Another test for aridity effects among arid countries is a comparison between Eqs. (6) and (9). The dummy variable (HT_2) was added to test the

The Future of Arid Lands

hypothesis that countries which are predominantly arid (75% or more) have lower per capita incomes, and the results support this contention.

General Results

The general result of the statistical analysis presented in this chapter supports our basic hypotheses about the effects of climate and arid climates in particular on the level of per capita incomes among countries. Climate differences in general are a significant explanatory factor of the variation in per capita incomes among countries. The effect of aridity on per capita income is generally negative; the more arid land a country has, the more likely it is to have a low per capita income. However, among countries with arid lands there is some relatively low proportion of arid land area that enhances the economic condition of these countries in relation to other arid countries. The next section of this chapter deals with a more detailed explanation of the concept and significance of "arid hinterlands."

Arid Hinterlands

The statistical analysis just presented suggests that the presence of an arid hinterland within a more humid nation does not have a depressing effect on per capita income, indeed may even increase it. Here we discuss the nature and uses of hinterlands in a little more detail.

One may regard arid hinterlands as being made up of essentially two types: first, those that are de facto hinterlands of other nations even though they are not politically part of them, and second, those situations where the hinterland function can be served effectively only if in fact the arid land is located within the more humid political jurisdiction.[2] The first category includes activities such as the following.

Recreation. A good instance of this is the North African beaches heavily used by European tourists. The excellent beaches with warm and dry climates of many coastal arid areas are an attraction to visitors, and political boundaries may have little influence on this attractiveness. Spain, vis-a-vis the rest of Europe, is an excellent illustration.

Winter Crops. Since most arid areas, aside from the polar ones, are located near the equator, they tend to have relatively warm climates. In many

---

[2] In the former case arid hinterlands may coincide with arid countries and the statistical analysis of the previous section is not applicable.

cases this means that high-value winter crops can be grown. The supplying of fruits and vegetables from several Middle Eastern countries to Northern Europe is a case in point.

Archeological Interest. Since (as we discuss below) the world's major civilizations and cultures arose in more arid climates, and because cultural artifacts persist much longer in arid than in humid climates, many arid areas are of archaeological interest. Again, this is true of the Middle East and China, and of Latin America and, to a somewhat lesser degree, the southwestern parts of the United States.

Astronomical Observations The clear atmosphere of many arid areas has made them outstanding for purposes of astronomical observation. The presence of observatories in arid mountain regions in various parts of the world attests to this. Because of the lack of conflict with other land uses, installations that are not particularly dependent on aridity as such are frequently located in arid areas (e.g., large radio telescopes).

This discussion is by no means meant to imply that the indicated activities are not also heavily conducted in areas where arid lands are hinterlands of more humid areas in the same political jurisdiction. The United States provides an excellent case illustrating this whole range of activities in a single political jurisdiction containing arid hinterlands. The only point is that these particular activities do not appear to be so highly dependent on location in the same political jurisdiction as some others discussed below. Except for the first two, recreation and winter crops, these activities are unlikely to have a large effect on per capita income.

It is possible, however, to think of a number of man's activities where having the location of both the arid and humid regions in the same political jurisdiction is very important.

Military Installations. Because of considerations of both weather and remoteness, many major military installations throughout the world are located in arid regions. This is particularly so of those requiring aircraft operations. In addition, availability of land and relative ease of securing installations have promoted their development in arid regions. Associated research activities are often large enough to be of economic significance. Installations of this nature are a major source of income in a number of arid regions, perhaps most strikingly in the western United States.

Residential Choice. For well-known reasons some arid areas are regarded as highly desirable places of residence. Although there are many instances of persons establishing residences in political jurisdictions outside the one in which they hold citizenship, for example, many people from the United States retire in Mexico, and from other parts of Europe in Spain, there are still in general fairly strong reasons for selecting a place of residence inside one's own country. These have to do with language, customs, legal rights, taxation, and a variety of other considerations. Perhaps the outstanding example of choice of arid areas as a place of residence within a hinterlands type of situation, is, once again, in the United States. People have moved in large numbers to southern California and are doing so increasingly to other states in the American Southwest. Over and above this, an enormous amount of land sales have occurred. Should all the plots that have been sold be occupied, the population of the area would be multiplied.

Footloose Industry. A number of modern industries are relatively free of direct ties to raw materials supplying locations or to markets. This is particularly true of industries in the electronics field, for example. To a considerable extent, these footloose industries locate in areas that are regarded as environmentally desirable, including arid areas in the United States and other countries. The possibility of living in a desirable environment has become increasingly important to many persons, and industries located in such environments may have a strong recruiting advantage for highly trained personnel.

The presence of arid lands within a more humid nation, may, for these reasons, be of substantial benefit to the nation as a whole. Since some of the benefits from using the arid lands have the nature of public goods type environmental assets, not all of the beneficial effects are recorded in per capita income. One may expect that as income continues to grow in a number of the more developed countries in the world, the desirability of arid lands as hinterlands will grow as well and in many regions this will be a source of sustenance and exert an upward influence on per capita income. Our statistical analysis tends to indicate, however, that aridity *as such* is a substantially negative factor in per capita income. Indeed, as one looks at developments in the arid countries, one can see a number of long-term, relatively irreversible tendencies that will work further in the direction of hindering their economic development.

Tendencies That Will Adversely Affect the Future of the Arid Lands

Desertification

Grazing and the search for fuels and additional farmland are denuding vast areas of North Africa, the Middle East, and the Indian subcontinent. Deforestation in India and Pakistan is forcing the use of dung for fuel and is depriving land of a natural source of fertility.[3] In papers prepared for the U.N. Conference on the Human Environment,[4] many countries documented this phenomenon of

---

[3] Lester Brown, *Population and Affluence, Growing Pressure on World Food Resources*, Population Reference Bureau, Washington, D.C., 1973, p. 11.
[4] *The Human Environment*, Vol. 2, Summaries of National Reports on Environmental

increasing desertification of semiarid and steppe areas. Israel, in its country report, pointed out that much of the Negev is a man-made desert, the product of ancient intensive grazing and wood gathering, which finally denuded the land and caused subsequent wind erosion, precluding the possibility of natural regeneration. It was pointed out that to return this land to any kind of useful status would require an enormous investment if it is possible at all. In addition, the modern return to this land has increased the tendency for erosion by disturbing the equilibrium established over past centuries of very slight human activity.[5] Iran's foreage-producing areas, which have also been grazed by domestic livestock since ancient times, are to a large extent rather marginal in quality. Iran's country paper reported that population growth and the rise of the standard of living there (which may or may not be sustainable when the oil runs out) have increased the demand on livestock production, while the amount of foreage space has been reduced, the number of animals has doubled in 20 years and is clearly in excess of the feed potential.[6] The forested areas of Iran have been greatly diminished in the last 50 years. Main factors have been charcoaling, shifting cultivation, and overgrazing, especially by goats. There are now some efforts to bring the situation under control. The country paper of Morocco reported that the soil is becoming degraded there in an alarming fashion. Mountain areas are highly populated and land clearing is pushed insistently, often without the most elementary precautions, leading to the loss of arable soil to erosion and floods. The livestock population is large; overgrazing and pasture degradation causes erosion and converts arid zones into desert. The forest area in Morocco has also been greatly reduced. Only about 5 million hectares remain, and these are continually subjected to invasion and depletion by agriculture.[7] Nigeria reports that soil depletion and erosion are continuing apace. The Enugu region is said to provide a striking example of gully erosion. Population pressure has reduced the former fallow periods of from 10 to 15 years to 1 to 2 years, resulting in serious soil degradation. Senegal reports that the activities of man are compounded by unfavorable climatic conditions in the intertropical areas. Charcoal is produced at an enormous rate, and annual brush fires destroy natural growth and wildlife. Thousands of hectares have been reduced to sand by overgrazing and subsequent abandonment.[8] These are only a few of the many examples that one can draw from the U.N. Reports, showing that the process of desertification is going on at a rapid rate and has been stemmed in only a few locations, if in fact, anywhere.

Problems 1972, Woodrow Wilson International Center for Scholars, Washington, D.C., March 1972.
[5] Ibid., p. 49
[6] Ibid., p. 43.
[7] Ibid., p. 63.
[8] Ibid., p. 79.

The Future of Arid Lands

The Deterioration of Mountain Environments

Nearly all the world's high mountain environments adjacent to arid areas are suffering rapid deterioration.[9] The causes are very similar to those affecting the arid lands themselves—primarily population pressure leading to efforts to extract more production from these lands than they can sustain with the technologies being applied. The results are erosion, floods, siltation of reservoirs, and migration of mountain people to already poverty-stricken cities in search of work. In these ways millions of people residing in adjacent lower lands are affected. UNESCO has reported accelerating damage to the basic life support systems today in practically every mountainous region of Asia, Africa, and Latin America.

Eckholm has stated the situation well:

> When the environment starts to deteriorate on steep mountain slopes, it deteriorates quickly—far more so than on gentler slopes and on plains. And the damage is far more likely to be irreversible. The mountain regions are not only poor in economic terms; many areas are rapidly losing any chance of ever prospering as their thin natural resource base is washed away. Degenerating economic and ecological conditions in the mountains, in turn, often push waves of migrants into the lowlands, leaving behind an aged, dispirited population incapable of reversing the negative spiral.[10]

General Environmental Disruption

The environmental media—atmosphere, water, and the land—are usually very sensitive to human interventions in arid areas. Because of frequency of inversions, air pollution easily becomes a serious problem. Severe air pollution already exists in most of the world's large arid-zone cities, where growth in population and economic activity continues apace.

Salt

The salinity of several of the great river systems that feed the agriculture and other activities of the arid zones has increased markedly in recent decades and will continue to do so. An important case is the Colorado River in the

---

[9] See Eric P. Eckholm, The deterioration of mountain environment, Science, September 5, 1975.
[10] Ibid., p. 764.

United States, where salinity has been on the increase for some time in the lower reaches, which also are the most productive for agriculture. Further upstream, development threatens to aggravate this problem considerably in the next two decades. A similar, but perhaps less well-known situation exists in the River Murray in Australia. This river deteriorates progressively. In the upper reaches it has less than 30 mg/liter filterable residue but at Waiderin in South Australia it exceeds 600 mg/liter for much of the irrigation season during low flow periods.[11] Numerous other instances could be cited.

There is every reason to suppose that the water resources of arid lands will become increasingly saline. The manifest pressures of scarcity of food supply plus the availability of large amounts of capital in some arid countries may be expected to give rise to intense efforts to increase irrigation and in consequence to permit less and less fresh water to drain into the seas. This will be pushed by the introduction of crop varieties that both need and respond effectively to irrigation. Less salt will tend to be "flushed" from the continents through surface and subsurface channels, and the salinity of waters in arid areas accordingly will tend to increase. In Israel, for example, the coastal plain has been developed to the point of a "closed" water system, and little flushing to the sea takes place any longer. Since a high level of salt is the deadly enemy of many crops, this tendency, which is already so evident in areas developed for intensive agriculture, is distressing.

Fortunately some technological "helps," if not solutions, exist. In some instances it is possible to control relatively concentrated natural sources of salinity which contribute to the problem. For instance, it was possible to divert salt springs flowing into the Kinneret, the main reservoir in Israel's water system, and thereby reduce the delivering of salt to the lake by some 50%. Presently there are plans to try to control several major sources of salt loadings in the Colorado River system in the United States.

Furthermore, it is possible to reduce, or sometimes reverse, the adverse effects of salinity on crop production by selecting and breeding salt-tolerant species of crops and by modifying the technology of delivering water to the plants. For example, it appears from recent experiments conducted in several countries that drip systems of irrigation are substantially more amenable to the use of saline water than the conventional flooding and sprinkler systems.

Unfortunately, these methods for dealing with salinity normally require heavy, although perhaps cost-effective, investments. Moreover, adaptation to salinity rather than control of it greatly reduce the flexibility with which the affected water can be used.

---

[11] *Water for the Human Environment*, Proceedings of the First World Congress on Water Resources, Vol. 2, Country Reports, published by the International Water Resources Association, Chicago, 1973.

Groundwater Problems

Groundwater overdraft and quality deterioration are very substantial in many areas. Perhaps the most spectacular incidence of groundwater overdraft is found in the southwestern United States, where many millions of acres of irrigated land are supported by wells drawing upon limited amounts of fossil groundwater. This resource is diminishing and is doing so irreversibly. Another serious instance of groundwater overdraft is in the Cuyo region of Argentina. This region is one of the world's major producers of wine grapes and is wholly dependent on a depleting irrigation water supply for its economic existence.[12] Other problems associated with groundwater overdraft are subsidence in some urban areas and, in the coastal zones, intrusion of salty water into the groundwater aquifers. In Israel, for example, the coastal aquifer has been pumped much in excess of natural replenishment, especially in the 1950s and 1960s. Water levels have been lowered substantially. The fresh water-seawater interface, the location of which was originally seaward of the coastline, has intruded inland. In the Tel Aviv area, all city wells located within 2 to 3 km of the coast had to be abandoned in the late 1950s because of seawater intrusion. In some areas, especially north of Tel Aviv, seawater intrusion is still continuing. Intrusion of salty water into an aquifer is again a change that is reversible only to a limited extent.[13] Water quality in aquifers may also be otherwise affected by a variety of human activities, including in some instances the purposeful discharge of waste materials. In heavily irrigated areas using applications of nitrate fertilizers, nitrates may penetrate the aquifer. Once, again, to cite the case of Israel, a rather steep ascent of the nitrate content in the groundwater below the most populated areas of the coastal plain has been found. Causes are both fertilizer and sewage water infiltration. Some wells have passed the limit of allowable nitrate concentration for drinking water and have had to be used for agricultural purposes only, or closed entirely.[14] As the Green Revolution penetrates to additional areas, groundwater problems associated with the heavy application of fertilizers and pesticides can be expected to increase.

---

[12] A detailed examination of this case is found in Kenneth P. Frederick, *Water Management and Agricultural Development*, John Hopkins University Press, Baltimore, 1975.
[13] *Water for the Human Environment*, Vol. 2, p. 226.
[14] Nitrate levels of more than 100 ppm are frequently found in wells in agricultural parts of the country. The drinking water limit in the United States is 50 ppm. See Ralph Mitchell, Environmental deterioration, *Selected Papers on the Environment in Israel*, No. 3, Jerusalem, 1975.

Siltation of Scarce Reservoir Sites

Reservoir sites for water storage for various uses are a scarce resource. In many river basins, large-scale development of these sites has already taken place and they are in the process of being destroyed by siltation at a greater or lesser rate. Brown[15] cites the instance of a large irrigation reservoir built in Pakistan, which was originally expected to have a lifetime of 100 years but is now expected to last only half that long because of the siltation problems. Eckholm also reports many instances of reservoirs silting up with startling rapidity (in as little as 7 years) because of the deterioration of mountain environments discussed earlier.[16] If one looks at this matter in a larger time scale, it may well be that suitable reservoir sites in the world will have been used up and many will have become useless in the next century or so.

Decreasing Quality of the Marine Environment

Coastal arid regions are frequently highly dependent on the marine environment; most obviously perhaps for fish as a source of protein-short lands, but also on the magnificent beaches that are often found in arid areas and are a salable resource for recreation. Fishing has been greatly disturbed in many areas because of oil pollution in the rich coastal waters and, in the case of the eastern Mediterranean, the effect of the Aswan Dam on the amount of nutrients in the coastal waters. In addition, enormous spans of beach have been polluted by oil and other types of waste materials, greatly reducing their wholesomeness and attractiveness. Once again, this situation is richly reflected in the country reports for the U.N. Environmental Conference. Morocco reports, for example, that the major source of coastal water pollution is urban sewage. Existing services are overaged and inadequate and drain without any treatment into the sea and cause pollution of beaches.[17] Israel reports that oil pollution is visible along all its beaches, and that it is not yet clear what percentage is due to spill at terminals and ports and what to general contamination of the Mediterranean. The Mediterranean, it might be noted, has perhaps the most valuable beaches in the entire world. There is little or no effort made to protect them, and they are universally becoming seriously contaminated. Kuwait reports that marine pollution is an outstanding environmental problem, damaging coastal recreation areas and adversely affecting the fishing industry which supplies the local market and supports specialized industries. The sources of marine pollution are oil, oily

---

[15] Ibid., p. 12
[16] Ibid.
[17] Ibid., p. 63.

The Future of Arid Lands

mixtures from ships and tankers, industrial effluents, and raw sewage discharged to the sea.[18] One of the most extraordinary cases of beach pollution is found in Spain. An enormous number of large hotels have been built along the Spanish Mediterranean since World War II. All of them discharge raw sewage to the ocean close to the beaches. In fact, we do not know of more than a few waste water treatment facilities along the entire Mediterranean coast.

Declining Quality and Quantity of Wildlife Populations

Population pressures that, as noted above, are contributing to desertification of vast areas and the massive deterioration of mountain environments are also leading to the diminution of numbers and diversity of wildlife. In many cases extinction is near at hand. Pressures to increase agricultural productivity are leading to heavy applications of pesticides, frequently of persistant varieties. The combination can be devastating to wildlife. Biocides are likely to become increasingly important as the Green Revolution spreads. The most diverse and many would say the most interesting wildlife populations in the world exist in Africa. In addition to whatever inherent interest their existence may have, they have also become an important source of income for several African countries. The country report for the U.N. Conference on Environment from Kenya is therefore of particular interest. It indicated that the main environmental problems connected with wildlife are conflicts in land use and threats directly to the wildlife resource and/or to its habitat. The report notes that these overlapping problems together are liable to lead to a general deterioration of the natural environment. The serious threat that extending cultivation in marginal areas poses for wildlife populations is especially insidious when it occurs in areas peripheral to sanctuaries or across traditional migration routes. Population pressures make settlement within game reserves an increasingly pressing issue, requiring early resolution by the government if the original purpose of these areas is to be preserved. The disruption to wildlife in these cases is out of all proportion to the area settled and is the present major cause of wildlife elimination.[19] Judging by reports by other arid and semiarid countries, the situation in Kenya can by no means be regarded as the worst one.

The Viability of Nomadic Cultures

Through ages of evolution, nomadic cultures have arisen in most arid areas.

---

[18] Ibid., p. 58.
[19] Ibid., p. 57.

Their willingness to follow opportunity wherever it leads is the reason why some production useful for human beings is possible from many arid areas. The nomadic peoples are those that today are furthest removed from modern life in countries in which they exist. The Bedouin in Israel and the Navajo in the southwestern United States are enormously removed from the contemporary cultures within which these groups exist as human enclaves. Whether or not such societies are viable in these circumstances and can continue to make their far-flung harvest from the dry lands of the world is very much an open question. The ethics of the policies of dominant cultures with respect to these people is a matter of debate in many countries. If the nomadic way of life could persist, it is possible that modern science could be helpful in making it more productive. The camel, for example, is the most highly adapted to arid conditions of the large mammals. If it could be bred in such a manner as to be more generally useful, say for fiber and meat, its ability to range far from watering places might greatly increase the productivity of the desert lands ranged by the nomads. Similar results might possibly be achieved by increasing the number of watering places. But recent experience in the Sahel region of central Africa suggests that this approach would have to be used with great caution. A case (*Science Magazine*) has been made that a tube-well program to produce watering places, which was conducted under French auspices, made an important contribution to the desertification of much of the Sahel region in recent times. By permitting a great increase in the number of cattle and changing traditional migratory patterns, the program is said to have increased pressure on the land and possibly made as much of a contribution to the recent troubles in the Sahel as did the drought.[20] It seems likely, however, that the nomadic cultures will find it impossible to exist for many generations more if they are in contact with technologically more sophisticated societies, which in almost all cases they are. Their disappearance would tend to have a depressing effect on production from the arid regions.

Future Development

The previous section has indicated that there are a number of significant adverse tendencies insofar as the future of the arid lands is concerned. Most of these are not new, but some of them have intensified in the last few decades, particularly those related to population pressures. On the other hand, there has been a more or less continuous, if slow, increase in per capita income in most of these countries over a rather long period of time. This seems to have been the result of a

---

[20] Sahelian drought: No victory for Western aid, Science, July 1974.

number of influences, including some temporary ones such as the large-scale exploitation of mineral resources, especially energy, in some of them and an increased agricultural productivity resulting from the application of higher-technology methods in agriculture in others. A certain amount of industrialization has also been carried out in a number of arid countries. It was pointed out earlier that if per capita incomes continue to rise in the more humid regions, the hinterland influences on the arid regions may be expected to accelerate. However, the adverse developments in the environmental resource situation that we have indicated above and that it appears are nowhere near being brought to a halt, could have a severely dampening effect in the absence of some sort of bonanza resulting from mineral discoveries, and to counter depressing influences, the future of living standards in most arid regions appears to be heavily dependent on the improved use of scarce water resources, especially in agriculture.

Large-scale water transfers and the desalting of seawater, which are often looked to as ways of "making the deserts bloom," do not appear to have any decisive potential in this direction. Enormous schemes for transferring water have been proposed in several countries. There is what is referred to as the North American water and power alliance. There is a proposal for a large-scale divergence of waters from the Mississippi River to the high plains of Texas. The Soviet Union has put forward a scheme for reversing the flow of the Siberian rivers and irrigating millions of acres in central Asia.[21] All of these proposals are subject to serious criticisms on ecological and economic grounds, and they appear not to be in the picture, at least for a longer period into the future; and in any case they would not help the poorer countries. Desalting and delivery of water to points of use are likely to be too expensive in the foreseeable future to be justified for agriculture or to be justified for industrial or municipal uses if water supplies can be acquired from agriculture. Estimates of the cost of desalted water in the near term seem to converge to around 5.3 to 6.6 cents (1970 prices) per cubic meter. Even this near-term figure is speculative, and is reached only in conjunction with production of electric power with a large-scale nuclear plant. There is no reason, therefore, to expect that in the near future desalting offers much likelihood of substantially improving the economies of the less developed arid countries through expanded agriculture, unless it is undertaken and supported as a gift. But were this to be done it is likely that a gift of such magnitude would contribute far more to the recipients' economic welfare if it were made in some other form.[22]

---

[21] Our information on the Soviet scheme comes from personal discussion with Igor Belyaev.
[22] For additional discussion see Nathaniel Wollman, Economics of Land and Water Use. In Harold E. Dregne (ed.), *Arid Lands in Transition*, Pub. No. 90, American Association for the Advancement of Science, Washington, D.C., 1970.

It appears, therefore, that future gains in agricultural production must be based on improved use of existing water resources. The literature can leave little doubt that there is a very large scope for such improvements. Crosson's study of Chilean agriculture showed that a 30% increase in investment in irrigation works had little effect on total output, whereas an equivalent expenditure on fertilizer and pesticides would have yielded substantial results. He found that agricultural labor was not educated adequately enough to make efficient use of modern inputs and that irrigated land was not a limiting factor but that other inputs were.[23]

For most of the countries in the arid zones of the world, the scattered and incomplete evidence that is available indicates that output could be doubled or tripled with the same water base if more contemporary farming methods were adopted, including associated heavy capital and operating costs. Gains of this magnitude appear to be possible even if the "miracle grains" are neglected. But, as indicated further below, tradition, lack of education, costs, and environmental effects make rapid introduction of such methods on a worldwide scale unlikely.

Also, the widespread crop failures of the early 1970s served to remind us that the Green Revolution has thus far not made us disaster-proof. Actually, the Green Revolution has been much more limited in its impact than many people suppose. This is to a considerable measure due to the limited area it has actually effected. High-yielding varieties of rice have thus far been semidwarf and suitable for shallow-water regions comprising only 25 to 30% of the world's total rice lands. Furthermore, such dwarf varieties are particularly sensitive to flooding. The International Rice Research Institute in the Philippines is working on new varieties suitable for deeper-water conditions, and researchers there believe that rice production could be doubled throughout the world in the next 15 years. One promising line of development is work on varieties that elongate and thereby survive floods. The Institute is also working on drought- and insect-resistant varieties and adaptability to saline and nutrient-poor soil. The importance of progress in these areas is underlined by the fact that about one-third of mankind, or roundly 1,300,000,000 people, depend on rice for more than one-half their food.[24]

One must recall, however, that achieving the enormous gains in production that the Rice Institute foresees as necessary and possible will require gigantic investments in capital, education, and operating costs. This is especially so since the increase in energy prices has raised the cost of fertilizer, pesticides, and mechanical power inputs to agriculture. Whether or not it would be possible to make these investments quickly enough to avert disaster in the arid regions

---

[23] See Pierre R. Crosson *Agricultural Development and Productivity*, Johns Hopkins University Press, 1970.
[24] David Spurgeon, Up-dating the Green Revolution, Nature *254*, April 24, 1975.

The Future of Arid Lands

experiencing rapid population growth is a very open question, it seems to us. Furthermore, the environmental impact of applying the vast amounts of fertilizer and pesticides that would be needed are still of somewhat unknown dimension but will no doubt contribute to some of the deterioration tendencies discussed earlier. It is perhaps unlikely that contamination by either will pose any acute threat to human life, but they could deal the death blow to many other life forms and lead to more rapid eutrification of major water bodies. Given the race between population growth and food production, such considerations are likely to receive scant attention except perhaps in some cases, such as parts of Africa, where wildlife is a major attraction and a substantial source of income.

In recent years the large exploitation of mineral resources, especially petroleum, has taken on major importance in the development plans of several arid countries. Since minerals are exhausting resources, their boosting effect on the economy is necessarily more or less temporary. In the case of petroleum income the effect will probably last less than half a century. Coal and nonfuel minerals may last somewhat longer. Many nations obtaining this type of income (including the Navajo nation in the arid hinterland of the United States as well as some Middle Eastern countries) are attempting to use the bonanza to provide a longer-term economic base for their countries. To do this successfully will require imagination and planning of a high order. In mineral-rich countries where irrigation is important, there are fairly straightforward opportunities for increasing agricultural efficiency through education and technological improvement. Industrialization is another matter. Renewable and other nonrenewable resources are often scarce; more often than not the location of arid lands (because they were not previously the locus of large-scale development) is poor with respect to markets and existing transportation routes. In some cases the recreational hinterland status of an arid nation may offer attractive opportunities for investment and development. But no instances in which this type of development has by itself provided a basis for generally high per capita income come to mind. High mineral incomes and large foreign exchange earnings do provide a unique opportunity for "a giant step forward," but how to make the step successfully is usually a very difficult question.

Epilogue

It seems rather anomalous that most of the arid regions of the world appear to be facing such a severely difficult future when they were also the areas where civilization first arose. What this suggests, however, is that such regions are fertile ground for the development of civilization and its (necessarily) rather sophisticated economic base, but that the level and magnitude of development that can be attained is somehow inherently limited. If this is true, it would follow that, up to a point, aridity, with some external source of river or spring water, usually from mountain ranges, is an advantage, but that the advantage is circumscribed. The favorable growing conditions typical of many arid regions, including long warm seasons and relative immunity of crops from natural disasters, mean that a surplus of food can be grown, sometimes year-round and without prodigous labor, if the population is small.[25] At the same time conditions for plant growth are not so favorable, as in the tropics or the coast of California, for example, that essentially no cultivation is needed. The Southern California Indians could go naked all year and pick berries and never develop any high order of culture. The forests of the north harbored animals to hunt, and the combination of weather and forest made primitive methods of cultivation rather unrewarding there. The need to cultivate in arid lands, and to move water, was a natural incentive to take an interest in mechanical things and physical principles. Because irrigation societies are necessarily settled societies of some size, there is a need to develop both permanent structures and enduring institutions. This reinforces the stimulus to develop engineering and naturally arouses an interest in human relationships and such matters as ethics and justice. Early arid lands civilizations have also usually been strong in astronomy and mathematics. It does not stretch the imagination to think that the clear skies and angularity of countryside and structures in arid lands have something to do with this.

Societies of gatherers, hunters, and nomads (other early forms for organizing economic activity) seem to lack one or more of the ingredients for generating a high level of civilization. Once the basis for civilization in mathematics, science, the arts, and human organizations has been laid, it does, however, seem to have high transfer value and to be able to achieve a self-generating dynamic of it own. Given the enormous flows of renewable resources and the more or less continuous development of technology (with a giant step up during the industrial revolution), the more humid lands have been able to develop their economies and their (to a large extent derivative) civilizations to unprecedented scales. Aridity has, however, held the development of the dry lands on a much tighter leash. At the same time, especially in the most recent times, population has grown enormously. The combination, plus many other complicating factors including those discussed in this chapter, has prevented most of the truly arid nations from rising up out of poverty. In some cases the question seems to be not population control but the much more difficult one of *reducing* the population to get it into a sustainable balance with available resources.

---

[25] This, of course, does not apply to nonirrigated semiarid areas, which contain probably the highest-risk agriculture in the entire world.

Acknowledgments

We wish to thank Nathaniel Wollman, Dean of the College of Arts and Sciences, The University of New Mexico, for many helpful comments on this chapter. Dr. Kneese also wishes to thank the International Institute for Applied Systems Analysis in Laxenburg, Austria, for support during the summer of 1975, part of which was directed toward the production of this chapter.

Appendix

Full Sample—118 Countries

Africa
- Algeria
- Angola
- Burundi
- Cameroon
- Central African Republic
- Chad
- Congo (Brazzaville)
- Congo, Democratic Republic of the
- Dahomey
- Ethiopia
- Gabon
- Gambia
- Ghana
- Guinea
- Ivory Coast
- Kenya
- Liberia
- Libya
- Malagasy Republic
- Malawi
- Mali
- Mauritania
- Morocco
- Mozambique
- Niger
- Nigeria
- Senegal
- Sierra Leone
- Somalia
- South Africa
- Southern Rhodesia
- Sudan
- Tanzania, United Republic of
- Togo
- Tunisia
- Uganda
- United Arab Republic
- Upper Volta
- Zambia

North America
- Canada
- United States

Central and South America
- Argentina
- Barbados
- Bolivia
- Brazil
- British Honduras
- Chile
- Columbia
- Costa Rica
- Dominican Republic
- Ecuador
- El Salvador

Guatemala
Guyana
Haiti
Honduras
Jamaica
Mexico
Netherlands Antilles
Nicaragua
Panama
Paraguay
Peru
Surinam
Trinidad & Tobago
Uruguay
Venezuela

Asia
Afghanistan
Burma
Cambodia
Ceylon
China (Taiwan)
Hong Kong
India
Indonesia
Iran
Japan
Korea, Republic of
Laos
Malaysia
Nepal
Pakistan
Philippines
Singapore
Thailand
Vietnam, Republic of

Middle East
Iraq
Israel
Jordan
Kuwait
Lebanon
Saudi Arabia
Southern Yemen
Syria

Europe
Austria
Belgium
Cyprus
Denmark
Finland
France
Germany, Federal Republic of
Greece
Iceland
Ireland
Italy
Luxembourg
Malta
Netherlands
Norway
Portugal
Spain
Sweden
Switzerland
Turkey
United Kingdom

Oceania
Australia
Fiji
New Zealand

Arid Sample—53 Countries

Africa
- Algeria
- Angola
- Chad
- Congo (Brazzaville)
- Dahomey
- Ethiopia
- Gambia
- Ghana
- Kenya
- Libya
- Malagasy Republic
- Mali
- Mauritania
- Morocco
- Mozambique
- Niger
- Nigeria
- Senegal
- Somalia
- South Africa
- Southern Rhodesia
- Sudan
- Tanzania
- Togo
- Tunisia
- United Arab Republic
- Upper Volta
- Zambia

North America
- Canada
- United States

Central and South America
- Argentina
- Brazil
- Chile
- Ecuador
- Mexico
- Paraguay
- Peru
- Venezuela

Asia
- Afghanistan
- India
- Iran
- Pakistan

Middle East
- Iraq
- Israel
- Jordan
- Kuwait
- Saudi Arabia
- Southern Yemen
- Syria

Europe
- Portugal
- Spain
- Turkey

Oceania
- Australia

Nontropical Sample—63 Countries

Africa
- Algeria
- Chad
- Ethiopia
- Kenya
- Libya
- Mali
- Mauritania
- Morocco
- Niger
- Senegal
- Somalia
- Southern Rhodesia
- Sudan
- Tunisia
- United Arab Republic
- Upper Volta

North America
- Canada
- United States

Central and South America
- Argentina
- Chile
- Mexico
- Paraguay
- Peru
- Uruguay

Asia
- Afghanistan
- Taiwan
- Hong Kong
- Iran
- Japan
- Korea, Republic of
- Nepal
- Pakistan

Middle East
- Iraq
- Israel
- Jordan
- Kuwait
- Lebanon
- Saudi Arabia
- Southern Yemen
- Syria

Europe
- Austria
- Belgium
- Cyprus
- Denmark
- Finland
- France
- Germany, Federal Republic of
- Greece
- Iceland
- Ireland
- Italy
- Luxembourg
- Malta
- Netherlands
- Norway
- Portugal
- Spain
- Sweden
- Switzerland
- Turkey
- United Kingdom

Oceania
- Australia
- New Zealand

Data Sources

Food and Agriculture Organization of the United Nations (1967). *Production Yearbook 1966*, FAO, Rome, Table 1, "Land use," pp. 3-8.

Espenshade, E.B. (ed.), *Goode's World Atlas*, 13th ed., Rand McNally, Chicago, pp. 12-13.

United Nations (1965). *Statistical Yearbook 1964*, New York, Table 187, "Illiteracy rate among population 15 years of age and over, by sex," pp. 694-697.

United Nations (1968). *Statistical Yearbook 1967*, New York, Table 185, "Estimates of total and per capita national income and gross domestic product at factor cost," pp. 576-580; Table 200, "Net official flow of external resources to individual developing countries from developed market economies and from multi-lateral agencies," pp. 692-695.

United Nations (1971). *Demographic Yearbook 1970*, New York, Table 20, "Expectation of life at specified ages for each sex: latest available year," pp. 710-728.

United Nations (1968). *Demographic Yearbook 1967*, New York, Table 2, "Population by sex, rate of population increase, area and density for each country of the world: latest census and mid-year estimates for 1963 and 1967"; Table 5, "Population by age, sex, and urban/rural residence: latest available year 1955-1967."

Correlation Matrix

	P	R	PF	I	FA	LE	AL	PL	PA
P	1.00	−0.34	−0.54	−0.31	−0.08	0.46	−0.07	−0.06	−0.23
R	−0.34	1.00	0.52	0.35	0.11	−0.25	−0.06	−0.06	0.28
PF	−0.54	0.52	1.00	0.42	0.12	−0.39	−0.05	−0.07	0.15
I	−0.31	0.35	0.42	1.00	0.23	−0.39	0.04	−0.08	0.22
FA	−0.08	0.11	0.12	0.23	1.00	−0.09	0.54	0.00	0.18
LE	0.46	−0.25	−0.39	−0.39	−0.09	1.00	0.06	0.10	−0.25
AL	−0.07	−0.06	−0.05	0.04	0.54	0.06	1.00	0.45	0.10
PL	−0.06	−0.06	−0.07	−0.08	0.00	0.10	0.45	1.00	0.23
PA	−0.23	0.28	0.15	0.22	0.18	−0.25	0.10	0.23	1.00

Index

A

Accumulation, 7
Acreage, irrigated, 234
Acreage reductions, 240
Addictive, 6
Adsorption, 86
  negative, 74
Aeration, 67
Agrarian structural reforms, 125
Agribusinesses, 121
Agricultural
  acreage reduction, 216, 232
  policies, 250
  processing, miscellaneous, 226
  production function, 216, 327, 397
  productivity,
  structures, 250
  technology, 250
Agriculture,
  irrigated, 21, 230
  miscellaneous, 226
  subsidization, 385
Alfalfa, 59, 62-63, 166, 179, 181, 192-193, 202, 211, 235
Alfalfa hay, 179, 189
Alkaline ions, 110
Alkalinity, 337
Allegheny County Authority, 287
Alternative export levels, 252
Aluminosilicates, 67
Aluminum, trivalent, 67
Ammonium, 8, 152
  salts, 8, 109
Ammonium nitrate, 8

Analysis, input-output models, 221
Anhydrite, 337
Animal wastes, 245
Anion, 67, 132 (*see also* individual anions)
  adsorption, 70
    negative, 71
  diffusion, 70
  divalent, 69
  exclusion, 70-71, 74, 88, 90
  monovalent, 69
Anthropology, 130
Approach
  interdisciplinary, 2
  voluntary, 17
Aquifers, 2, 9, 202, 393
  groundwater, 20, 146
Arable land, 385
Arid
  areas, 395
  coastal regions, 394
  hinterlands, 386-387
  uses, 387
  lands, 19, 379, 389, 399-400
  regions, 19, 161, 364, 399
  zone, 363
Arid land civilizations, 400
Arid/semiarid regions, 65
  Australia, 1
  India, 1
  Middle East, 1
  United States,
    southwestern, 1
Aridity, 365, 380, 385, 387, 400
  effects, 386
Aridity hypothesis, 385

407

Arizona, 219, 308
Arizona vs. California, 308
Arkansas River, 23
Aromatic polyamide, 346
Avodados, 37, 55

B

B, 5, 37 (*see also* Boron)
Bacteria, 22
  coliform, 132
Bacterial content, 131
Bactericides, 132
Barley, 53, 113, 168, 253
  salt-tolerant crop, 5, 168
Beans, 57
Beef, 253
Beets, 53, 57
Bell peppers, 52, 57, 61
Benefit project, 274
Benefits
  industrial, 272
  municipal, 272
  recreational, 272
Bermuda grass, 116
Bicarbonate, 34, 36, 39, 65, 152
  ions, 33
  soluble, 35
Biocides, 395
Biological attack, 347
Blackheart of celery, 52
Blaney-Criddle
  equations, 151
  method, 146
Blossom-end rot, 52
Blue Spring, Arizona, 273
Boron, 5, 37–39, 56
  hazards, 37
  tolerant crops, 38
Boulder Canyon Project Act, 307–308
Boundary condition, 82, 177
Brine, 326, 341, 345
  concentration, 350
  disposal, 353

[Brine]
  heater, 336
  pH, 337
  rejection, 331
  temperature, 330
  wastes, 139
Broccoli, 59

C

C, 27 (*see also* Carbon dioxide)
Ca, 5, 27, 31, 32, 37, 42, 56, 65, 70, 104, 113, 152 (*see also* Calcium)
$CaCO_3$, 37, 337, 350
Calcium, 5, 27, 31, 32, 37, 42, 56, 65, 70, 104, 113, 152
  bicarbonate, 337
  carbonate, 33–34, 337, 350
  concentration, 33, 36
  deficiency, 52, 57
  ions, 110
  phosphates, 8
  precipitates, soluble, 37
  salts, 109
  sulfate, 33, 337, 350
  sulfate scale, 350
California, 219
California-Arizona region, 227
Canal lining, 133, 212
Canning, preserving, and freezing, 226, 228
Capillary action, 9
Carbon dioxide, 27, 35–36
Carbonate, 27, 29, 33–34, 106, 152, 337, 350
Carrot, 52, 57, 111
Cation, 67, 132 (*see also* individual cations)
  adsorption, 88
  divalent, 67, 69, 72
  electrostatic attraction, 69
  exchange capacity, 67
  exchangeable, 108
  monovalent, 69

Index

[Cation]
  nutrition, 52
  soil-exchangeable, 108
Cationic imbalance, 57
CEC, 67
Celery, 52, 57
Cellulose acetate, 346
Center for Agricultural and Rural
    Development, 246
Chemical
  amendments, 66
  biological model, 150, 154
  biological program, 152
  delivery, 247
  quality, 149, 153-154
  soil conditioning, 96
Chemicals and fertilizers, 226
Chemistry
  equilibrium, 152
  soil modeling, 156
Chloride, 5-6, 38, 54, 65, 90, 109, 152, 341
  accumulation, 50
  movement, 90
  salinities, 49
  salts, 49, 107
Chlorine attack, 347
Citrus, 37, 39, 55
Cl, 5, 39, 54, 57, 63, (see also Chlorine attack, Chloride)
Clay, 80, 160
  crystal lattice, 67
  kaolinite, 70
  loam, 104, 209
  montmorillonite, 67, 70
  silty, 209
  soil, 108
  water system, 74
Climate, 6, 160, 379
  arid, 387
  desert, 380
  differences, 381
  humid continental, 380
  humid subtropical, 380
  hypothesis, 381
  Mediterranean-type, 97, 380

[Climate]
  steppe, 380
  tropical, 380
  variables, 384
Climatic criteria, 363
Climatic cycle, 372
Climatic requirements, 50
Closed conduits, 133
Coachella Canal, 315
Coachella Valley, 38, 216, 315
Coions, 67
Colloidal material, 132
Colorado, 308, 323
Colorado Basin, 321, 323
Colorado Basin Salinity Control
    Forum, 321
Colorado River, 165-166, 171, 173, 198-199, 215, 308, 318, 392
Colorado River Basin, 201, 210, 235, 272, 306
Colorado River Basin Salinity
    Control Act, 314
Colorado River Board, 159
Colorado River Compact, 307
Colorado River Project, 340
Columbia River, 315
Commodity demand, 246
Commodity market regions, 246
Commodity transportation, 252
Compaction, 349
Comprehensive development
    planning, 376
Computer simulation, 97
Concentrated solution, 349
Concentration polarization, 349
Condenser surfaces, 330
Condenser tubes, 339
Conduction fluid, 67
Conductivity
  electrical (see EC)
  hydraulic, 69, 78, 96, 148, 177
  hydraulic soil, 7, 70, 80-81
  swelling, 69
Conservation, 13
Consumer demand, 258

Consumer demand fulfillment, 252
Consumptive use deficit, 146
Contour, 252, 264
Convective transport, 152
  of solute, 75
Conveyance System, 133
Copper, 109
Corn, 53, 112-113, 136, 178-179, 181, 189, 192, 202, 253
Corn grain, 211
Corrosion, 352
Cost
  allocation, 320
  estimates, 325
  of production, 263
  pricing, 295
  sharing, 271, 273-274, 277, 281, 286, 298, 301
Costs
  operating, 272
  treatment, 272
Cotton, 53, 56, 112, 164-165, 168, 226, 253, 263, 265
  salt-tolerant crop, 5, 168
Crop, 12
  annual nonwoody, 56
  beans, 28
  berry, 37-38
  budjets, 219
  combination, 191
  composition, optimal, 12
  conditions, 176
  cool-season, 50
  cover, 137
  dryland, 121
  failures, 398
  farm subsystem, 127
  feed, 211
  field, 5
  forage, 5, 211
  fruit, 6, 54
  grain, 262
  grasses, 28
  growing
    conditions, 90
    period, 96
    season, 92

[Crop]
  growth, 1, 91
  irrigated production, 209
  irrigation reduction, 203
  high-value, 37
  loss, 218, 221
  loss analysis, 221
  management, 251
    systems, 252
  nurse, 189
  optimal compositions, 12
  orchard, 58
  output, 218
  prices, 218
  production, 29, 103, 127, 175, 218
  reduced-tillage cultivation, 252
  residues, 131
  response, 5, 11, 272
  response-salinity function, 97
  rotation, 14, 189
  row, 8, 58, 205
  salt sensitivity, 23, 161, 166
  salt tolerance, 28, 41, 56
  sensitive, 4, 47, 49
    vegetables, 5
  silage, 211
  speciality, 39, 202, 211
  specific, 92
  substitution, 206
  tillage method, 251
  tolerance, 7, 41
  tolerant, 4, 53, 59
  toxic effects, 5
  tree, 54
  vegetable, 5, 28, 52, 53, 57, 59, 111, 135
  wilting point, 146
  yield, 4, 91, 176, 317, 374
    reduction, 65
Cropping
  management, 252
  methods, 264
  patterns, 204, 216, 239-240
Crystal Geyser, Utah, 315
Cultural practices, 125, 127, 134

Index

D

Dairy products, 226
Darcy steady-state equation, 146
Decision processes, 375
Decision space, 363, 374
Deep injection, 273
Deforestation
  India, 389
  Pakistan, 389
Delaware River Basin, 272
Delta County, 210
Demand pattern, 294
Deposition, 132
Desalination, 13, 19, 139, 166, 191, 212, 273, 325, 343 (*see also* Distillation)
  brackish water, 18
  Hong Kong, 18
  Israel, 18
  Oklahoma, 18
  plants, 353, 359–360
  processes, 18, 325, 353
    distillation, 18, 325
    membrane, 18
    multieffect distillation, 18
    multistage flash evaporation, 18
    vapor compression, 18
  research, 325
  seawater, 18, 346, 356
  systems, 352
Desalting (*see* Desalination)
Desertification
  India, 389
  Middle East, 389
  North Africa, 389
Detergents, 132
Diffusion path, 74
Dilution, 166, 273, 368
Dilution flows, 373
Direct current, 345
Dispersion coefficient, 77, 83
Dissolved solids, 345
Dissolved solids concentrations, 131
Distillation, 333 (*see also* Desalination)
  plants, 339

[Distillation]
  processes, 326, 335, 351
  systems, energy consumption, 335
Ditch lining, 198
Ditch losses, 122
Diverting, 273
Downstream benefits, 218
Drain
  tile, 36, 139, 161
  water, 234
Drainage, 1–2, 9, 11, 40, 127, 129, 131, 161, 216
  artificial, 172
  control, 139
  ditches, 324
  outlets, 17
  practices, 160
  pump, 139
  rate, 40
  requirements, 127
  surface, 47
  tile, 139, 161
  water, 23, 36, 62, 86, 179, 206
    sampling devices, 36
Dry land, 14
Dual-purpose installations, 336

E

EC, 25, 29, 40, 59, 62, 104, 108, 112–113, 159, 164, 169, 218 (*see also* Electrical conductivity)
Econometric recursive simulation, 246
Economic (*see also* Economics)
  alternatives, 189
  analysis, 190
  cost and trade-offs, 245
  costs, 271
  credit availability, 267
  development, 247
  efficiency, 318–319
  evaluation of irrigation with saline water, 159

[Economic]
  impacts, 218
  impacts of regional saline
    irrigation, 201
  income calculations, 221
  income reduction, 178
  model, 177
  multiregional entity, 4
  parameters, 371
  physical evaluation, 173
  profit-maximizing position, 221
  profitability criterion, 219
  regional effects, 215
  regional income, 229
  socioeconomic entities, 3
  structure, 247
  water projects, 275
Economic and hydrologic impacts, 232
Economics, 130, 134
  agricultural, 42–43
  agricultural productivity, 121
  cost effectiveness, 154
  cost effectiveness of technological solutions, 154
  cost of seawater desalination, 18
  cost structure, desalination, 18
  cost-effective curves, 156
  desalination, nuclear dual-purpose plants, 18
  differential prices, 17
  domestic market, 121
  economies of large-scale farming, 172
  food shortages, 121
  fuel consumption in desalination, 18
  incentives, 17
  income distribution, 15
  income transfer payments, 15
  income transfers, 14
  land rent, 169
  marked value price (MVP), 168
  market mechanisms, 14
  national income, 229

[Economics]
  prices
    water inputs, 10
    water outputs, 10
  profitability, 179, 189
  reduced labor costs, 136
  value of irrigation water, 18
  water cost, 125, 134
  water price, 170
Economy, region, 13
Ecotypes
  grass, 52
  tidelands, 52
Effective salinity, 37
Effluent, 301
  charges, 287
  limitations, 316
  permits, 203
  standards, 203
Electric power, 397
Electric power losses, 346
Electrical conductivity, 6, 24, 43, 164, 172 (*see also* EC)
Electrical resistance, 345
Electroconductivity (*see* Electrical conductivity)
Electrodialysis, 340–341, 353
  membranes, 344
Employment, 247
Energy consumption, 335, 356
Energy crisis, 312
Energy requirements, 356
Engineering, agricultural, 42–43
Environmental contamination, 267
Environmental contaminants, 264
Environmental degradation, 271
Environmental dimensions, 246
Environmental economics, 286
Environmental equilibrium, 19
Environmental models, 250
Environmental protection, 319
Environmental Protection Agency, 315, 318, 322 (*see also* EPA)
Environmental quality, 247, 250
Environmental quality goals, 264

Index

EPA, 315, 318, 322
Equilibrium solution concentration, 71
Erosion susceptibilities, 267
Erosion-control practices, 263
ESP, 31–32, 37, 39, 42, 70, 88
ESR, 31
Evans, 1
Evaporation, 53, 83, 94, 131, 144, 184, 239, 365
  ponds, 273
  reduction, 133
  surface, 48
Evaporator
  chambers, 339
  design, 360
  internals, 339
  shells, 339
Evapotranspiration, 1, 11, 32, 40, 42, 62, 84, 122–123, 127, 131, 133, 135, 144, 151, 173, 176–177, 180, 182, 202–203
  control, 136
  losses, 205
  salinity effects on, 114
Exchange equilibria, 72
Exchangeable sodium control, 37
Exchangeable sodium percentage (*see* ESP)
Exchangeable sodium ratio (*see* ESR)
Exclusion, 86
Exogenous factors, 4
Export capacity, 246
Export levels, 258

F

Farm
  dairy products, 226
  level, 4
  machinery, 125
  management, 135
    practices, 221
    technologies, 155
  planning analyses, 11

[Farm]
  subsystem
    climate, 127
    crops, 127
    soils, 127
Farming methods, 264, 267
Feather River, 23
Federal Water Pollution Control Act, 319, 322
Federal Water Pollution Control Act Amendments, 315
Feedwater, 353
  pretreatment, 337, 350, 352
Fertility, 116
  salinity effects, 108
  salinity interaction studies, 114
  salinity relationships, 103
  salinity studies, 117
  salinity treatments, 113
Fertilization, 8, 117 (*see also* Fertilizer)
  crop response to, 103
  effects, 113
  nitrogen, 109, 374
  NPK, 104
  phosphorus, 104
  salinity, 103
Fertilizer, 131, 135, 246, 252, 398 (*see also* Fertilization)
  levels, 14
  mixture, 106
  nitrogen, 8, 109, 374, 393
  NPK, 104
  phosphorus, 104
  slow release, 135
  transformations, 8, 109
Field
  capacity, 41
  plots, salinized, 48
  leveling, 198
  tests, 26
Fish, 394
Fl, 39 (*see also* Fluoride)
Flax, 253
Flocculation, 56
Flood, 318

[Flood]
  storage, 315
  irrigation, 53
Flooding, 94
Flow
  convective, 76, 86
  diffusion, 76
Flow measurement, 134
Flow pattern, average, 73
Fluoride, 39
Foliar
  damage, 38, 39
  spray, 52
Food, demand, 20
Food and feed grains, 226
Food costs, 246
Food supply, 246
Forage crops, 226
Forages, 245
Forward linkages, 227, 229
Fruit, 226, 253
  citrus, 226
  stone, 37
Fuel consumption, 357
Fungicides, 132
Furrow-irrigated ridges, 49

G

Gapon
  -type constant, 31
  equation, 39
Geometry, 363
  demand, 367
  pollution, 364, 366
  resource, 367
Geothermal activity, 39
Germination, 49, 53, 57
Gila River, 23, 308
Glen Canyon Dam, 309
Gradient hydraulic pressure, 78
Grain, 235
  feed, 266
  small, 202, 211

Grain mill products, 226
Grain sorghum, 168
Grand Valley, Colorado, 201, 202, 203, 205, 209, 211, 234, 239–240
Grand Valley Trade Area, 210
Grand Valley Users Association, 209
Grapes, 37, 55
Gravity segregation, 86
Green Revolution, 393, 395, 398
Green River Basin, 235
Gross module plugging, 347
Groundwater, 23, 47, 127, 138, 149, 154, 194–195, 307, 366–367
  aquifers, 146, 318
  codes, 307
  degradation, 66
  deterioration, 13
  flows, 145, 156
  formation, 368–369, 371, 374
  irrigation model, 146
  levels, 154
  management, 273
  modeling, 148
  outlet, 139
  overdraft, 393
  problems, 393
  pumping, 239
  quality flow, 121
  recharge area, 24
  reservoir, 123, 131, 142
  return flows, 139, 144
  rising tables, 123
  salination, 9
  saline, 39
  salinity, 140
  system, 149
Growth stages, 53
Guayule, 57
Gulf of California, 309
Gypsum, 32–33, 42, 152, 337

H

Hay, 253, 262
  wild, 253
Health, life expectancy, 385
Heat energy, 345
Heat-transfer rates, 336
Hemihydrite, 337
Herbicides, 132, 264
High erosion regions, 266
Highline Canal, 209
Holland, 1
Homology, 375
Hong Kong, 351
Hoover Dam, 232, 315–316
Horizontal-tube, falling-film evaporator, 330
Human activity, 20
Humid region, 19, 23, 379, 386
Hungary, 1
Hydraulic
  capacity, 345
  conductivity, 69, 78, 96, 148, 177
    of soil, 7, 70, 80–81
  gradients, 146
  pressure, 340
Hydrochloric acids, 337
Hydrologic models, 235
Hydrologic system, 132, 142
Hydrolysis, 347
Hydrosalinity model, 149, 154, 156
Hysteresis, 82

I

Immobile improvements, 210
Imperial Dam, 173, 309, 316
Imperial Irrigation District, 159
Imperial Valley, California, 159, 166, 169–170, 216, 318
Income, 247
  distribution, 299
  redistribution, 318
  transfers, 299
Income-generating mechanism, 216
Indirect impact, 229
  analysis, 221
Indus Basin, 306
Indus Valley, 324
Industrial use, 132
Infiltration, 83
  capacity, 83
Inflow-outflow analysis, 156
Input-output
  models, 227
  table, 218
Insecticides, 132, 264
Insoluble carbonates, 37
Institutional modifications, 140
Integrative planning process, 367
Interaction effect, 116
Interceptor sewers, 287
International Boundary and Water Commission, 309
International trade, 253
Interregional allocation, 14
Interregional approach, 245
Ion, 5–6, 25, 27, 33, 39, 55, 70, 341 (*see also* Anion, Cation, Salt; individual elements)
  alkaline, 110
  ammonium, 8, 109
  calcium, 8, 33–34, 110, 337, 350
  carbonate, 27, 29, 33–34, 152, 337, 350
  chloride, 5–6, 38, 39, 54, 57, 63, 65, 90, 109, 110, 152, 341
  concentration, 47, 48, 77, 343
  constituents, 153
  counter ions, 67
  damage, 39
  diffusion, 75
  exchangeable, 31, 108
  ferrous, 67
  fluoride, 39
  hazards, 38
  magnesium, 8, 67, 113, 337, 350
  metallic, 67

[Ion]
  nitrate, 106, 132, 291
    mobile, 113
  phosphate, 8, 132, 267, 350
  potassium, 8, 57, 104, 110, 113
  precipitation, 32
  silicate, 37, 67
  sodium, 8, 43, 66, 106, 110, 337, 350
    adsorbed monovalent, 67
  soil interaction, 90
  sulfate, 33, 110, 337, 350
  toxic concentrations, 43
  toxicities, 42
  transfer, 345
  trivalent aluminum, 67
  uptake, 63
Ion-exchange equilibrium, 32
Ionic
  composition, 25
  concentrations, 47, 48
  distribution, 67
  proportions, 47
Iran, 390
Iron, 67, 350
Irrigated soil, 85
Irrigation, 184, 215, 305, 320, 322–324, 369, 372, 392, 399
  agriculture, 21, 23, 42
  annual, 218
  application efficiencies, 135
  basin, 128
  canal regulation, 144
  canals, 317
  changes in water salinity, 215
  cropland diversions, 144
  cultures, 20
  cycle, 58–59
  deficiencies, 63
  delivery system, 134
  development, 130
  dissolved solids, 133
  district, 4, 209
  districts, 138
  drainage systems, 317
  drip, 60, 172, 317

[Irrigation, drip]
  lines, 22
  systems, 58
  duration, 40, 95
  efficiency, 9, 17, 64, 128, 134, 136, 313, 315, 323
  excess, 9
  extensive, 2
  farm layout, 128, 129, 156
  flood, 11, 53, 92, 172, 179, 192, 195
  flow measurement, 128
  frequency, 6, 11, 40, 57, 58, 59, 218
  furrow, 57, 60, 128, 172
  gated pipe, 206
  gravity system, 180
  groundwater model, 146
  high-frequency, 60
  horizontal subsurface flow, 131
  hydrosalinity model, 142
  impacts, 121
  inadequate, 63
  laterals, 317
  layout design, 128, 129, 156
  leaching, 9
  management, 1, 4, 7, 12-13, 17, 91, 174, 178, 316
    decisions, 7
    practices, 5, 175, 198
    system, 177
  methods, 57, 64, 92, 136, 191, 218
    flooding, 11, 53, 92, 179, 192, 195
    quantity, 7
    timing, 7
  on-farm, 239–240
  overirrigation, 9
  percolation, 128, 132
  percolation losses, 129
  physical relationships, 4
  policy implications, 198
  practices, 65, 203–205, 209, 212, 216
  production, 14

[Irrigation]
  projects, 8, 309
  quotas, 17
  rate of application, 191
  regime, 166, 219
  region, 10
  regional saline, 201
  return flow, 131, 173
  saline, 1, 11, 12, 20, 53, 62, 103, 109, 117, 201, 206, 215
    return flows, 206
  salinity, 20, 201
    relationships, 66
  scheduling, 137–138, 273
  season, 11, 92
  seasonal water, 92
  seepage losses, 9, 133
  set, 205
  societies, 400
  soil surface, 91
  source, 19
  sprinkler, 11, 23, 55, 57, 92, 136, 172, 179, 180, 192, 197, 218–219, 317
    equipment, 22
    lines, 39
    nozzles, 22
  surface, 61
    runoff, 131
    system, 156
  system, 2, 4, 9, 126, 133, 156
    administration, 130
    canals, 9
    conveyance, 9
    definition, 125
    design, 126, 156
    pipes, 9
    trickling, 6
  techniques, 206
  technology, 172, 204, 273
  transmission capabilities, 22
  treatment, 164
  trickle, 7, 92
  trickle-drip, 94
  use, 132
  wage, 205

[Irrigation]
  water, 5, 7, 23, 32, 35–37, 39, 47, 49, 55, 62–63, 65, 92, 106, 109, 159, 161, 170, 219
    composition, 38
    input, 164, 169
    levels, 38
    management, 173
    quality, 56, 108, 117, 132, 179
    quality classification, 39
    requirements, 22
    return flow, 132
    reuse, 22
    users, 212
    waterlogging, 128
Isomorphous substitution, 67
Isothermal conditions, 86
Israel, 42, 91, 299, 393
  Negev, 390

J

Jenson-Haise method, 146

K

K, 8, 57, 104, 110, 113 (*see also* Potassium)

L

Land
  contours, 252, 264
  dry, 14
  erosion, 22, 263, 267
  low-productivity, 210
  optimal, 14
  optimal use, 250
  productivity, 22
  retirement, 208, 211
  rows, 252
  straight rows, 252

[Land]
   strip cropping, 252
   terraces, 252
   use, 247
   use controls, 212
Langelier index, 43
Las Vegas Wash, Nevada, 315
Leaching, 7, 29–30, 34, 37–38, 40–43, 53, 62–63, 66, 85, 92, 94, 97, 122, 128–129, 132, 135, 146, 154, 161, 164, 166, 172, 194, 203, 317
   efficiency, 96
   irrigation, 9
   nitrate, 201–202
   plant nutrient, 29
   ponded, 38
   rate, 49
   ratios, 188
   salt, 89, 195, 312
   soil, 9
Leaf
   burns, 57
   injury, 61
Legal
   irrigation quotas, 17
   restrictive regulations, 17
   water allotments, 17
Legal systems
   absolute ownership, 307
   prior appropriation, 307
   rule of correlative rights, 307
   rule of reasonableness, 307
Legumes, 110, 245
Lettuce, 52, 57, 165, 168
Li, 39, 56
Linear programming, 11, 97
   formulations, 97
   models, 12, 165–166, 170, 204, 246, 289
   model of salt outflow, 178
   study, 189
Linear regression model, 380
Lining of ditches, 234
Lithium, 39, 56
Livestock feed sources, 245

Livestock production, 252
Low-salt zone, 41
Lower Colorado River, 218, 316
Lower Colorado River Basin, 216

M

Magnesium, 5, 27, 31, 32, 35, 56, 65, 104, 152
   bivalent, 67
   hydroxide, 337, 350
   oxide, 337
   phosphates, 8
   salt, 113
Man made deserts, 390
Management
   patterns, 368
   policies, 369
   practices, 40, 42–43, 273
Mancos Shale, 202, 209
Manganese, 343, 350
Marginal benefit, 274
Marginal cost, 274, 283
Marine environment, 394
Matric potential, 6, 57
Maturation, 58
Meat and poultry processing, 226
Meat animals, 221
Mechanical dispersion coefficient, 86
MED, 328, 331, 335, 357–359
Membrane
   cellulose acetate, 348
   configurations
      hollow-fiber, 347
      spiral-wound, 347
      tubular, 347
   creep, 349
   fouling, 350
   hydrolysis, 349
   polyamide fiber, 348
   processes, 340
      electrodialysis, 325
      reverse osmosis, 325
      thickness, 349
Mesa County, 210

Index

Mexico, 308, 310, 319
Mg, 5, 31, 56, 104 (*see also* Magnesium)
Microbial activity, 8, 110, 118, 152
Microbial decomposition, 110
Middle East irrigation cultures, 20
Mineral
  composition of soil, 36, 39
  dissolution, 24
  elements, 47
  nutrients, 114
  nutrition of plants, 113
  weathering, 23, 36
Mining, 226
Mixing-cell concept, 152
Mn (*see* Manganese)
Model
  capabilities, 247
  description, 174
  nature, 250
  regression, 116
Module configurations, 346
Moisture, 5
  high status, 6
  mean stress, 162
  soil, 6, 11
  soil reservoir, 144
Molecular diffusion, 73
Monoculture, 245
Montrose County, 210
Mountain environment deterioration, 391
MSF, 326, 331, 335, 357, 358-359
Mulching, 136
Multieffect distillation, 328, 331, 335, 357-359
Multistage flash, 326, 331, 335, 357, 358-359
Municipal use, 132
Municipal discharges, 316
Muskmelon, 56

N

N, 8, 57, 104, 109, 112, 135, 152, 253, 267 (*see also* Nitrogen)

Na, 5-6, 30, 34, 38, 42, 54-55, 57, 63, 65, 70, 109, 113, 152, 341 (*see also* Sodium)
National markets, 247
Necrotic spots, 38
Nematocides, 132
Nematode control, 165
Nevada, 308
New Mexico, 308
Nitrate, 106, 132, 291 (*see also* Nitrogen)
  fertilizer, 8, 104, 109, 374, 393
  leaching, 202
  mobile ion, 113
Nitrification, 109
Nitrogen, 8, 57, 104, 109, 112, 135, 152, 253, 267 (*see also* Nitrate)
  atmospheric, 110
  balance, 252
  carriers, 8
  effects on yields, 111
  fertilization, 8, 104, 109, 374, 393
  fixation, 8, 110
  salinity, 110
  gaseous, 110
  organic, 110
  response, 117
  urea, 152
No-tillage studies, 136
Nomads
  Israel, 396
  Navaho, 396
Noneconomic vegetation, 136
Noninteracting solute, 86
Nonsaline conditions, 116
Norfolk sand, 108
Nuclear dual-purpose plant, 358
Nuclear plant, 397
Numerical dispersion, 86
Nutrient
  absorption, 8, 118
  concentrations, 47, 48
  deficiency, 113
  dissolved, 22
  leaching, 29, 85

[Nutrient]
  losses, 134
  macronutrients, 8, 104
  micronutrients, 8, 104
  mineral, 114
  mobility, 112
  plants, 4
  status, 108
Nuts, 226, 253

O

Oats, 179, 189, 253
Ohio River, 272
Oil solution, 37
On-farm water management, 156
Onions, 57
Operation practices, 273
Optimal
  land, 14
  water use, 14
Optimization, 10
  social, 15
Orchard, 202, 211
Organic
  compounds, 132
  substances, 132
  wastes, 22
Organisms, pathogenic, 132
Orthodox planning, 366
Osmotic, 37
  effect, 110, 118
  gradients, 78, 88
  potential, 6, 8, 27, 47, 49, 57, 91, 103, 106, 114, 176–177
  pressure, 27, 82, 108, 340, 349, 350
  pressure gradients, 78
  properties, 27, 43
  shock, 49
Overirrigation, 138, 144

P

P, 57, 104, 109, 112, 113, 132 (*see also* Phosphorus)

Painted Rock Dam, 315
Pakistan, 139, 306
Paradox Valley, 315
Parker Dam, 316
Pasture, 235, 253
  permanent, 202, 211
Pathogenic organisms, 132
Peak-load pricing, 292
Pecos River, 23
Penman method, 146
Per capita income, 380–381, 384–385, 387, 389, 396
Percolation, 122, 202–204, 209, 318
  loss, 203–204, 206
Perennials, 6
Permanent pasture, 202, 211
Permeability, 30
Permissible soil loss, 14
Pest control, 127
Pesticides, 131–132, 246, 252, 264, 267, 395, 398
Pests, 22
Petroleum, 226
pH, 337
  brine, 337
  soil, 104
Phase space, 363–364, 370
Phosphate, 132, 267, 350
  calcium, 8
  magnesium, 8
  polyphosphate, 337
Phosphorus, 57, 104, 109, 112, 113, 132 (*see also* Phosphate)
  effects on yields, 111
  fertilizers, 8
Phreatophyte, 144
  eradication, 133
Physiochemical aspects, 66
Planning space, 363, 370, 377
Plant
  growth, 37, 47, 111, 122
    inhibition, 28
  hormones, 132
  injury, 106
  mineral composition, 113
  nutrition, 52, 104

Index

[Plant]
  pathogens, 22
  physiology, 42–43
  property, 177
  relative growth, 162
  root, 108
  salt tolerance, 108
  soil processes, 108
  soil-water management techniques, 125
  soil-water regime, 132
  toxic effects, 56
  woody, 54
    leaf injuries, 54
    tip burns, 54
Planting beds modified, 53
Plots, two-dimensional, 114
Plow layer, 53
Poiseuille's law, 80
Polarization current, 345
Political science, 130
Pollutants, 123, 132, 134, 138, 372
  treatment, 291
Pollution, 19, 301, 365, 370
  abatement, 321
  air, 287
  biological, 19, 376
  coastal water, 394
  control, 203, 305, 321–322, 367
  damages, 288
  downstream waters, 9
  "externality," 374
  geometry, 364, 365, 366
  municipal, 305
  nonpoint, 247, 258
  nonpoint control, 14
  oil, 394
  physicochemical, 19, 376
  polluters, 15, 288, 292, 319
  rights, 286
  scare, 373
  soil, 85, 368
  stream, 367–369
  stream geometries, 369
  thermal, 272
  water, 272, 283
Polyelectrolytes, 337

Polyphosphates, 337
Population
  age, 384
  change, 384
  growth, 20
  level of education, 384
Potassium, 8, 57, 104, 110, 113
  effects on yield, 111
Potatoes, 50
Poultry, 253
Poultry and eggs, 226
Power production, 336
Precipitation, 133
Pressure head, 83
Prices, 247
Pricing, 271, 277, 286
Pricing policy, 273, 294, 298
Process thermodynamics, 334
Processes
  drainage, 86
  evaporation, 86
  infiltration, 86
  redistribution, 86
Production
  efficiency, 164
  patterns, 247
  technologies, 372
Profit-maximization model, 218
Profitability, 229
Programming (*see also* Linear programming, Quadratic programming)
  models, multiregional, 14
Project
  benefit, 274
  cost, 274
  efficiency, 134, 274
  evaluation, 274
  irrigation
    delivery phase, 8
    farm irrigation phase, 8
    water-removal phase, 8
  scale, 275, 277, 282
  types, 277
  water, 299
  water-quality improvement, 15
Psychic costs, 271

Public projects, 287
Public resource development, 274
Pumping energy, 345

Q

Quadratic programming, 246
Quality restraints, 258

R

Rain, 91
Rainfall, 1, 371
　low, 1-2
　patterns, 252
　winter, 42
　winter inputs, 42
Rainwater, 24, 42, 65
Reclamation, 56, 92, 96
　potential, 34
Regional
　agency, 313
　economy, 313
　infrastructure, 10
　management, 323
　planning, 376
Regression modeling, 116
Regression surface, 116
Regulative storage, 371
Reservoirs, 394
Residual chemicals, 245
Residual sodium carbonate, 37
Resource
　allocation, 174, 295
　constraints, 179
　demand, 365
　land, 127
　models, 250
　pricing sector, 246
　use, 247
　use patterns, 250
　system, 363
Response surfaces, 116

Return flow, 9, 234, 267, 313, 315, 318, 369
Reverse osmosis, 340-341, 349-350, 353
Reverse osmosis membranes, 346
Rice, 53, 253, 398
　flooded, 56
Richards equation, 150
Rio Grande, 29
Riparian, 215, 307
River, 199
　Arkansas, 23
　basin, 4
　Colorado, 23, 33, 122, 159, 165-166, 173
　Columbia, 23
　Delaware, 272
　Dolores, 239
　Feather, 23
　flow impact on irrigation, 121
　Gila, 23, 308
　low, 294
　Ohio, 272
　Pecos, 23
　Rio Grande, 29
　Sacramento, 23
　Salt, 23
　Snake, 23
　systems, 391
　upper Colorado, 201
River Basin
　Colorado, 165-166, 171, 173, 198-199, 215, 308, 318, 392
　Delaware, 272
　Green, 235
　San Juan, 235
　Upper Main Stem, 235
Root, 8
　alfalfa, 192
　distribution, 151
　　function, 177
　growth, 114
　interface, 113
　medium, 47
　nodules, 110

Index

[Root]
  plant extraction, 180
  profile, 179
  soil interface, 82
  system growth, 177
  uptake, 176
  water potential, 177
  zone, 7, 9, 32, 34-35, 41, 48-49, 53, 62-63, 66, 96, 122, 125, 127, 129, 134-135, 139, 144, 149, 154, 156, 162, 183
    flows, 144
    plant, 40
    salinity levels, 41
    wetted, 94

Rootstocks, 54
  tolerant, 39
Rotation, crop, 14, 189
RSC, 37
  ranges, 37
Ruhr water system, 287
Runoff, 133
  surface, 22, 122, 125, 201
  water, 245, 247, 261, 267

S

S, 29 (*see also* Sulfate)
Sacramento River, 23
Saline (*see also* Salinity, Salinization, Salt)
  aquifers, 202
  fields, 94
  high water tables, 48
  irrigation, 109, 206
    water, 9, 103, 109, 117
  return flows, 211
  soil, 7, 56, 58, 66, 92, 209, 318, 392
    water, 47
  water, 9, 94, 309, 316, 334, 344
Salinity, 2, 6, 13, 21, 39-41, 52-53, 112-113, 116, 131-132, 134, 140, 171, 173, 202, 212, 216,

[Salinity]
218, 227, 283, 309, 319, 320, 341, 346, 356, 391-392 (*see also* Saline, Salinization, Salt)
  accumulation, 41, 272
  appraisal, 25, 27, 37
  average, 62
  concentration, 149, 215, 234, 315
  concentrations of elements, 41
  conditions, 48
  control, 23, 32, 37, 41-42, 92, 96, 138, 139, 154, 204, 234, 307, 312, 316, 318
    measures, 149
    program, 321
    projects, 315
    reclamation, 7
  criteria, 316
  damage, 12
  detriments, 57, 156
  discharge, 203
  dissolved solids, 201
  effects, 41, 109
    on evapotranspiration, 114
    on fertility, 108
    on nitrogen fixation, 110
  fertility
    effects, 108
    interaction studies, 114
    relationships, 103
    studies, 114, 117
    treatments, 113
  fertilization, 103
  field tests, 26
  groundwater, 9, 48, 140
  hazard, 22, 29
  high levels, 172
  increased, 134, 229
  increments, 62
  index, 96
  industrial, 317
  initial, 94
  intervals, 240
  irrigation, 1, 11, 12, 20, 62, 103, 109, 117, 201, 206

[Salinity, irrigation]
    relationships, 66
    water, 53, 206, 215
legal, 15
level, 5, 7, 41, 60, 105, 112, 166, 189, 312, 322
loadings, 234
management, 17, 22, 319, 321–323 (*see also* Salinity, control)
    institutional approaches, 305
    legal approaches, 305
maximal zone, 62
mean integrated, 59
minimum, 60
osmotic effects, 57
parameters, 142
physical relationships, 4
pickup, 318
problems, 42, 123, 324
    Iran, 1
    South Africa, 1
    Syria, 1
    Uzbekistan, 1
profiles, 64
quality standards, 315
reduction, 10, 235, 320
regime, 96
return flows, 211
soil, 5–6, 48, 56, 60, 62, 66, 77, 85, 91–92, 97, 104, 109–110, 117, 164, 318
soil water, 47
tolerance, 49, 62, 63
total, 54
toxic effects, 57
values, 314
water, 2, 4, 94, 309, 316, 334, 344, 392
Salinization, 1, 122, 161
    groundwater, 9
    soil, 122
Salt (*see also* Ion, Saline, Salinity, individual ions)
    absorption, 23
        foliar, 57
    accumulation, 5, 11, 20, 29, 36,

[Salt, accumulation]
    42, 48, 49, 53, 62, 65–66, 85, 89, 92, 195
adverse effects, 106
affected soils, 30, 103
ammonium, 109
atmospheric, 23
balance, 8, 29, 34, 36, 90, 127–128, 133, 139, 141–142, 156
budget, 148
calcium, 109
chloride, 49, 107
concentration, 5, 22, 24, 30, 34, 38, 47, 49, 66, 69, 74, 77–78, 81, 86, 88, 91, 98, 103, 122, 131, 144, 173, 175, 179, 180, 181, 190, 234
concentration profile, 185
condition, 191
conservation, 59, 63
content, 5
damage, 118
dilute solutions, 24
discharge, 206, 212
dissolved, 1, 8, 23–25, 65
distribution, 7, 11, 85, 94
domes, 239
excess, 4
excess soluble, 65
exclusion, 80
flow, 149
flow model, 175
fluctuation, 86
heavy metals, 132
index, 8, 106, 108
inorganic, 4, 47, 132, 152
leaching, 5, 85, 89, 195, 312
leaching models, 7
level, 23, 112, 199
load, 21, 131–132, 232, 318
    reduction, 154, 216, 235
loading, 12, 215–216, 239, 240, 312, 392
loading effect, 316
natural, 139
outflow, 11, 179, 187, 189, 191,

Index

[Salt, outflow]
  198
  model, 178
  output, 198
  pickup, 12, 131, 133, 138, 141, 149, 153-154, 202, 204, 206, 209, 234, 316
  precipitation, 49, 63, 85
  profile, 91-92
  reduction, 198, 209
  rejection, 349, 351
  removal, 198, 239
  removal from seawater, 350
  residual, 133
  return, 316
  savings, 240
  sensitive species, 57
  sensitivity, 5, 117, 161
  separation, 341
  soil, 29, 54, 62, 108
  soil solution, 5
  soluble, 33, 41, 66, 73, 337
  solution, 27, 333, 340
  springs, 201, 392
  status, 109
  sulfate, 29
  surface crust, 30
  system, 149
  tolerance, 5-6, 49, 50, 52, 57, 62, 103, 115, 135, 161
    of crop plants, 5, 7, 28, 54, 56, 57, 168
    of fruit crops, 54
  tolerant crops, 56, 57
    barley, 5, 168
    cotton, 5, 168
    sugar beets, 5, 168
  total dissolved, 104
  transport, 53, 149, 152
  uptake, 90
  water conversion, 325
Salt River valley, 23
Sand cultures, 48
San Juan River Basin, 235
SAR, 30-32, 36, 40, 42-43

Saturation
  extract data, 41
  extracts, 49
  index, 34
Scale formation, 352
Se, 39, 56
Seasonal variation, 10
Seawater, 325, 328, 330, 337, 341, 353
  desalting, 346, 350, 356, 397
  intrusion, 393
  makeup, 331
Sectorial planning, 376
Sediment, 22, 132
  delivery, 246-247, 268
  loads, 268
  transport, 261
Sedimentation, 247, 258, 261
Seeding practices, 127
Seepage, 202, 204, 315
  losses, 122, 206, 209
Selenium, 39, 56
Semiarid zones, 47
Semiarid regions, 215
Service pricing, 301
Sesbania macrocarpa, 53
Sesquioxides, 70
Sewage treatment, 287
Sewer surcharge, 287
Shadow prices, 198
Shadow pricing, 250
Shale, 318
  formation, 12
Sheep, 253
Silage, 189
Silicate, 37
Silicon, tetravalent, 67
Silt, 245
  permeability, 32
Siltation, 394
Simulation model, 97
Social
  evaluation, 10
  psychology, 130
Socioeconomic variables, 384
Sociology, 130

Sodic soils, 30, 32, 56, 65–66, 81
Sodium, 5–6, 30, 34, 38, 42, 54–55,
   57, 63, 65, 70, 109, 113, 152,
   341
  adsorbed, 65
  adsorbed monovalent ions, 67
  adsorption, 30
  adsorption ratio, 30–32, 36, 40,
   42–43
  carbonate, 43
   residual, 37
  exchangeable, 56, 66
  hazards, 39
  hexametaphosphate, 350
  hydroxide, 337
  ions, 66
  nitrate, 8, 106
  toxicity, 56
Sodium-calcium system, 70
Soil, 41
  aeration, 137
  alkalinity, 85
  amorphous minerals, 38
  characteristics, 128, 137
  chemistry, 42–43, 66
  chemistry modeling, 156
  clay, 108, 160
  clay fraction, 67
  coarse textured, 32
  colloidal material, 70
  conditioning chemicals, 96
  conditions, 176
  conservation methods, 250
  cultures, 57
  dispersion, 32
  drainage waters, 36
  erosion, 14
  exchangeable cations, 108
  extract, 26, 218
  farm subsystem, 127
  fertility, 6, 103
  fine sands, 160
  fine-textured, 23
  Gilat loam, 81
  hydraulic conductivity, 7, 80–81
   saturated, 70

[Soil, hydraulic conductivity]
   unsaturated, 70
  infiltration characteristics, 127
  infiltration rates, 203
  ion interaction, 90
  irrigated, 32, 85
  Israeli, 72
  leaching, 9
  loss, 246, 250, 252–253, 258,
   263, 266
  loss reduction, 251
  low infiltration, 134
  low permeability, 134
  macroscopic properties, 66
  matrix, 86, 88
  matrix suction, 96–97
  mineral solubilization, 62
  minerals, 39, 132
   weathering, 47
  moisture, 6, 11, 136, 146, 194
   measurements, 137
   movements, 150
   reservoir, 144
   status, 103
   storage, 146
   stress, 164
   tension, 172
  movement, 175
  nonsaline, 49
  particles, 268
  permeability, 1, 42, 80, 161
  pH, 31, 104
  physics, 66
  plant moisture-salinity relationships, 171–172
  plant processes, 108
  plant-water management techniques, 125
  plant-water regime, 132
  pollution, 85, 368
  pressure head, 177
  productivity, 210, 245
  profile, 11, 38, 42, 73, 88, 131,
   134, 138, 141–142, 150, 153
   176, 177, 191, 317
  properties, 118, 177

Index

[Soil]
  reclamation, 7, 94
  root interface, 82
  saline, 7, 56, 58, 66, 92, 209, 318, 392
  salt-affected, 41, 56, 103
    black alkali, 30
    profile, 38
    slick spot
    sodic
    solonetz
  salinity, 5–6, 48, 56, 60, 62, 66, 77, 85, 91–92, 97, 104, 109–110, 117, 164, 318
    control, 92
    index, 97
    salinization, 122
    salts, 29, 54, 108
    sandy, 108
    loam, 108
  saturation extract, 55, 164
  silts, 160
  sodic, 30, 32, 56, 65–66, 81
  solute interaction, 88
  solute transport, 73
  solution, 4, 6, 24, 29, 33, 35, 38–39, 58, 65–66, 69, 72, 82, 98, 106, 108, 114, 175
    concentration, 97, 177
  stratification, 135
  structure, 37, 56, 135
  surface, 131
  swelling, 80
  texture, 135
  tillage practices, 135
  transport of solutes, 73
  unproductive, 209
  unsaturated, 74
  water, 47, 53, 58, 62
    composition, 38
    content, 83
    flow, 175, 180
    potential, 114
    regime, 97
    status, 50
    systems, 34

[Soil]
  weathered, 39
Solid-solution interface, 66
Solids, total dissolved (*see* TDS)
Solubility relationships, 8, 118
Solution
  composition, 69
  concentration, 69
Sorghum, 253
Soybeans, 50, 110, 245, 253, 263, 265–266
Sprays, foliar, 52
Sprinkler, 94, 234
  equipment, 22
  irrigation, 11, 23, 55, 57, 92, 136, 172, 179, 180, 192, 197, 218–219, 317
  lines, 39
  nozzles, 22
Stability, aggregate, 67
Statistical analysis, 380
Steady-state safe yields, 367
Stone fruits, 55
Storage system, 133
Strawberry, 55
Streamflows, 307
Strip cropping, 262, 264
Substitution relationships, 11–12
Subsystems, 2
  farm, 127
  goals and objectives, 4
  water delivery, 126, 128, 132, 133, 156
  water removal, 126, 129, 132, 138
Sugar beet, 53, 165, 168, 202, 211, 253
  salt-tolerant crop, 5, 168
Sugar cane, 56
Sulfate, 33, 65, 152
  calcium, 33, 337, 350
  calcium scale, 350
  concentration, 33
  salinities, 49
  salts, 29
Sulfur, 29 (*see also* Sulfate)

Surface
 flows, 366
 runoff, 22, 122, 125, 201
  fish growth, 22
  industrial cooling, 22
  recreation, 22
Sweden, 1
System
 analysis, 324
 conveyance, 12
 description, 3
 drainage, 10
 dynamic, 3
 stationary conditions, 3

T

Tactoids, 70
Tailwater, 205, 234
TDS, 4, 22-23, 25, 218, 232, 241, 341, 346, 350, 356
 Beth She'an, Israel, 4
 Dead Sea, Israel, 5
 N'eot Ha'kikar, Israel, 4
 Pecos River, Oila, Texas, 4
 southwestern United States, 4
Technological changes, 140
Temperature-heat transfer, 333
Tensiometer, 58, 60
Terracing, 262, 264
Thermal energy, 356
Threshold values, 97
Tile drain effluents, 34
Tillage, 14, 127, 136, 205, 251, 258
 reduced, 262, 264
Tolerance, 62, 63
 levels, 368
 limits, 62
 salt, 5-6, 49, 50, 52, 57, 62, 10 103, 115, 135, 161
 of crops, 5, 7, 28, 54, 56, 57, 168
Tomato, 52, 57
Toxic
 concentrations, 38, 40, 41, 57
 effects, 5

[Toxic]
 element concentrations, 41
 Na, 56
Trade and transportation, 226
Transpiration, 11, 85, 114, 131, 176-177, 182, 184
Transport diffusion, 73
Transport processes, 73
Transportation network, 246
Treatment costs, 291, 294
Treatment plant, 273, 282, 291, 294, 299
Tree fruits, 38-39
Tree nuts, 226, 253
Trewartha's classification, 380
Trickle
 discharge, 94
 irrigation, 7, 92
Trickle-drip irrigation, 94
Tube bundles, 339

U

U.N. Conference on Environment, 395
U.S. Salinity Laboratory, 5, 48, 164
U.S.S.R., 1
Uncompaghre Valley, 209, 239
Upper Colorado River, 232
Upper Main Stem River Basin, 235
Upstream costs, 218
Urea, 109
 nitrogen, 152
Utah, 308
Utilities, 226

V

Valley
 Grand, Colorado, 201
 Imperial, California, 159, 166
 Paradox, 239
 San Joaquin, 139
 Yuma, 139

Vapor, 333
  compression, 330, 335
  pressure, 84
Variables
  decision, 5
  soil state, 5
  target, 5
Vegetable, 211, 226, 253 (*see also* individual vegetables)
Vegetative (stover), 53
Vernal, Utah, 179
Vertical tube evaporator, 330
Vines, 38

W

Waste, 271
  accumulation, 271
  control device, 287
  disposal, 367
  treatment, 294
    plant, 301
Water
  allocation, 250, 267, 365
  allotments, 17, 42
  analyses, 37
  application, 9, 41
    flooding, 189
    sprinkling, 189
  application rate, 187
  aquifers, 2, 9, 202, 393
  availability, 8
  brackish, 56, 61, 325, 341, 343, 346, 349, 353
    sprinkling, 61
    technology, 356
  budget, 148
  canal seepage, 129
  changes in salinity, 215
  charges, 210
  clay system, 74
  composition, 25, 31
  conservation methods, 250
  constrained supplies, 206
  consumption, 127, 250

[Water]
  content, 177
  control, 135
  control deliveries, 130
  control of flows, 17
  conveyance system, 206, 317
  corrosiveness, 22
  cost, 125, 134, 198, 352, 357, 359, 372, 379
  cultures, 48
  deep percolation, 144
  degraded return flow, 123
  deliveries, 206
  delivery, 136
  delivery and farm subsystems, 132
  delivery subsystem, 126, 128, 132, 133, 156
  demand, 19, 371
    agricultural, 3
    domestic, 3
    industrial, 3
    recreational, 3
  demineralized, 105
  depletion, 58, 60
  desalination, 18 (*see also* Desalination)
    plant, 281
  desalted, 19, 352
    costs, 397
  deterioration of quality, 19, 122
  development, 325
  differential pricing, 17
  dilute, 349
  discharge, 203
  discharge of wastes, 374
  distillate, 330
  distilled, 33
  distribution, 63
  distribution uniformity, 11
  downstream pollution, 9
  drainage, 9, 23, 24, 29, 36, 38, 40-41, 62, 86, 206
  drainage compositions, 34, 36
  economic projects, 275
  economy, 360
  effective management, 125

[Water]
   effluent, 283
   evaporation (*see also* Evaporation)
      changes in, 76
      losses, 136
   excess, 3
   extraction period, 84
   farm-water management, 206
   flow, 77, 86, 149, 368
   flow equations, 81
   flow system, 149
   fresh flows, 368
   good quality, 2
   groundwater (*see* Groundwater)
   improved, 281
   improving costs, 300
   increases, competition for, 17
   infiltration, 41, 76, 176, 205
   infiltration rates, 30, 146
   inflow-outflow analysis, 140
   intake, 283
   integrated management, 29
   interregional transfers, 267
   irrigation, 5, 7, 23, 24, 26, 32, 33,
      35-37, 39, 47, 49, 55, 62-63,
      65, 92, 106, 109, 159, 161,
      170, 215, 219
      composition, 38
      input, 164, 169
      levels, 38
      management, 173
      quality, 56, 108, 117, 132, 179
      quality classification, 39
      requirements, 22
      return flow, 132
      reuse, 22
      saline, 53, 62, 109, 117, 206,
      215 (*see also* Irrigation, saline;
      Salinity, irrigation)
      users, 212
   law systems, 307
   laws, 126, 312, 376
   losses, 318
   low-quality national sources, 15
   management, 121, 125, 137, 140,
      142, 173, 198, 201, 212,

[Water, management]
   215-216, 306
   objectives, 130
   on-farm, 134
   practices, 41, 128, 206
   matric potential content, 177
   mineralized springs, 142
   mixing, 10
   movement, 175
   natural, 25, 132
   natural source improvement, 274
   nonsaline, 4
   optimal use, 250
   partly polluted, 373
   percolation losses, 135-138, 155-156
   planning
      demographic, 19
      economic expansion, 19
   pollution, 272, 283
   potable, 351
   potential, 57-58, 175
   precipitation, 127
   price, 170
   pricing, 17, 282
   pricing procedures, 203
   profile content, 88
   program, 376
   projects, 299, 301, 318
   quality, 2, 4, 19, 21, 42-43, 108,
      129, 133, 161, 166, 169, 174,
      212, 215, 218, 232, 245, 267,
      271-272, 282, 293, 299, 313,
      368, 373, 377
      classification for irrigation, 39
      control legislation, 209
      degradation, 19, 42, 122, 126,
      135, 139, 156
      downstream, 154
      externality, 286
      improvement projects, 15
      level, 164, 170
      management, 134, 202
      monitoring programs, 27
      of irrigation, 56, 108, 117,
      132, 179

[Water, quality]
  of source, 376
  policies, 301
  problems, 131
  recommendations, 43
  standards, 288, 315, 373
quantity, 2, 7, 134, 313
quantity allocation, 307
recirculation, 139
reclaimed waste, 373
redistribution changes, 76
regime, 97
removal subsystem, 126, 129, 132, 138
removal system, 134
requirements, 22, 91
resources, 2, 364, 370, 372, 392, 398
  allocation, 366
  development, 372, 375
  geometry, 366, 368
  management, 19-20, 132, 363, 365, 376
  planning, 19, 377
  planning space, 19, 373
  sector, 376
  subspace, 370
  system, 367, 369, 374
  use, 42
return flows, 9, 129, 132
reuse, 22, 132, 173
rights, 165, 205, 318
rights purchase, 17
rising tables, 134
river, 5, 23
rotation system, 137
runoff, 245, 247, 261, 267
saline, 2, 4, 94, 309, 316, 334, 344, 392
  irrigation, 62, 103, 117, 177
  soil, 4, 47
salt concentrations, 41
salt content, 10
savings, 17, 61, 317
scarcity, 19, 364, 372
seasonal irrigation, 92

[Water]
seepage, 138, 144, 146
  loss, 133, 136, 154, 156
  loss reduction, 154
soil, 47, 53, 58, 62 (*see also* Soil, water)
  content, 83
  potential, 114
  status, 50
soil-plant management techniques, 125
soil-plant regime, 132
steam gauging stations, 140
storage, 135, 184
stress, 58
subsurface return flow, 144, 146, 148, 149, 155
  estimate, 148
supply, 3, 18, 121, 127, 171, 206, 364
  constraints, 204
  regions, 246
surface drainage interception, 146
surface flow system, 149
surface flux, 83
surface return flows, 138
surface runoff, 22, 83, 127, 132, 138
table, 131, 139, 177, 181, 184
  elevation data, 146
  high, 48, 144
  high-saline, 48
  levels, 172
tailwater runoff, 144
technology, 247
tensions, 59
transfers, 14, 252, 397
transmission, 22
transpiration, changes, 76
transportation, 125, 313, 322
treatment, 216, 273, 292, 340, 343
treatment plants, 282, 288
tube wells, 139
underground sources, 195, 372
uptake, 81-82, 90, 176

[Water]
  use, 239, 258, 365, 368, 374
    controls, 212
    efficiency, 132, 134, 136, 212
    management, 323
    regulations, 323
    users, 126, 201
    vapor transfer, 136
    vaporization, 84
    viscosity, 74
    volume saturation, 59
    waste, 287, 301
Waterlogging, 9, 40, 122–123, 128, 132–134, 138, 140, 144, 154, 156, 306, 324
Watershed, 2, 131, 133
  industrial community, 292
Waterways, 267
Welton-Mohawk
  desalination plant, 319
  drain extensions, 319
Wellton-Mohawk Project, 309, 315, 323
Wetting
  depth, 92, 97
  front, 92, 94
  rate, 92

Wheat, 53, 353, 266
Wilamette Basin, 299
Wilting, 58
Winnipeg Sanitary District, 287
Wyoming, 308

Y

Yields, 92, 103
  crop, 91
  declination curves, 219
  depressions, 59
  fiber, 53
  grain, 53
  losses, 59, 61
  moderate stress, 61
  nitrogen effects, 111
  phosphorous effects, 111
  potassium effects, 111
  reduction, 28, 41
    estimated, 5
  response, 7, 115, 216

Z

Zero pollutant discharge, 209